Detection and Feature Extraction in Acoustic Sensor Signals

Detection and Feature Extraction in Acoustic Sensor Signals

Editors

Yuxing Li
Luca Fredianelli

MDPI

Basel • Beijing • Wuhan • Barcelona • Belgrade • Novi Sad • Cluj • Manchester

Editors

Yuxing Li
School of Automation and Information Engineering
Xi'an University of Technology
Xi'an
China

Luca Fredianelli
Institute of Chemical and Physical Processes
Italian National Research Council
Pisa
Italy

Editorial Office
MDPI
St. Alban-Anlage 66
4052 Basel, Switzerland

This is a reprint of articles from the Special Issue published online in the open access journal *Sensors* (ISSN 1424-8220) (available at: www.mdpi.com/journal/sensors/special_issues/AcousticSensorSignals).

For citation purposes, cite each article independently as indicated on the article page online and as indicated below:

Lastname, A.A.; Lastname, B.B. Article Title. *Journal Name* **Year**, *Volume Number*, Page Range.

ISBN 978-3-0365-9191-9 (Hbk)
ISBN 978-3-0365-9190-2 (PDF)
doi.org/10.3390/books978-3-0365-9190-2

Contents

Preface

Acoustic sensors have an extremely wide range of applications in many fields, including underwater acoustics, architectural acoustics, engineering acoustics, physical acoustics, environmental acoustics, psychoacoustics, and so on. The signals collected by high-sensitivity acoustic sensors contain a large amount of valid information that facilitates further processing of the collected acoustic signals. In particular, detection and feature extraction, as two important measures of acoustic sensor signal processing, can capture more information regarding the target and extract features with separability.

Yuxing Li and Luca Fredianelli
Editors

sensors

Editorial

Detection and Feature Extraction in Acoustic Sensor Signals

Yuxing Li [1,2,*] and Luca Fredianelli [3]

1 School of Automation and Information Engineering, Xi'an University of Technology, Xi'an 710048, China
2 Shaanxi Key Laboratory of Complex System Control and Intelligent Information Processing, Xi'an University of Technology, Xi'an 710048, China
3 Institute of Chemical and Physical Processes of National Research Council, Via G. Moruzzi 1, 56124 Pisa, Italy; luca.fredianelli@cnr.it
* Correspondence: liyuxing@xaut.edu.cn

Our advances in detection and feature extraction in the processing of acoustic signals allow us to capture more information about a target and extract features with separability. Various trends suggest that detection and feature extraction play increasingly important roles in the processing of acoustic sensor signals. The eleven papers in this Special Issue and an Editorial signed by the book's Editors cover a number of important topics and innovative approaches towards acoustic transducer signal processing, providing valuable techniques and ideas for related research.

The first work [1] in this volume focuses on seafloor Scholte waves. The authors proposed a method to detect these waves in an acoustic pressure field and found it to be effective even with sediment layers present. This study is important for understanding and detecting seafloor Scholte waves, and it provides valuable techniques for researching acoustic wave propagation in marine environments.

In reference [2], the authors proposed a deep learning hydroacoustic recognition method using the channel attention mechanism to address the issue of Doppler frequency shift in underwater targets caused by their motion speed and trajectory. The experimental results reveal that this method has significant advantages over other methods.

For conduit ranging and temperature measurements, the authors proposed an acoustic time-of-flight (TOF) estimation method based on digital lock-in filtering (DLF). The experimental results show the effectiveness of the proposed DLF method under different tube length and temperature conditions. The method has higher accuracy and robustness compared to conventional methods [3].

The fourth study proposes a delay estimation optimisation algorithm based on SVD and improved GCC-PHAT-$\rho\gamma$ for efficient delay estimation under low signal-to-noise ratio conditions. The algorithm improves the signal quality via noise reduction and weighting methods and determines the delay difference using peak detection. The experiments confirmed that the algorithm significantly improved the delay estimation accuracy under a low signal-to-noise ratio with excellent performance [4].

Ascari, et al. [5] described methods to assess the effectiveness of low-noise pavements: the Close Proximity Index (CPX) methods allows measure noise near the wheels through sensors, while the Sound Pass-By (SPB) solves the problem through real-time events and roadside data feature extraction. The authors proposed a methodology improvement with the U-SPB, allowing to evaluate low-noise pavements in urban areas through unattended measurements and laboratory procedures. It considers long-term noise levels and traffic parameters.

Reference [6] describes the application of a partially updated adaptive algorithm PU to an extremely demanding structural active noise control ANC system to achieve global noise reduction and save computational power. The study discusses an improvement in the ANC system using the PU algorithm and verifies its high performance in the laboratory.

Citation: Li, Y.; Fredianelli, L. Detection and Feature Extraction in Acoustic Sensor Signals. *Sensors* **2023**, 23, 8030. https://doi.org/10.3390/s23198030

Received: 17 September 2023
Accepted: 21 September 2023
Published: 22 September 2023

The seventh paper [7] proposes an FrFB-based inverse convolution beamforming method to improve the spatial resolution of DOA estimation by converting the signal to fractional-order Fourier domain via fractional-order Fourier transform, followed by delay and beamforming. The experimental results show that the method has higher resolution at low signal-to-noise ratios.

This study presents a new rolling bearing fault diagnosis method using dual-optimisation of WSO-HSlopEn and WSO-SVM [8]. By introducing HSlopEn as a new feature and dual-optimisation HSlopEn and SVM using WSO, the method demonstrates a high fault recognition rate in both single and multi-feature cases. The highest recognition rate is up to 100% [8].

For composite fault diagnosis of metro traction motor bearings, the authors used multi-signal fusion, MTF and optimised ResNet to improve both accuracy and effectiveness under complex working conditions. The method is able to extract the composite fault features under complex working conditions and improve the diagnostic accuracy and efficiency [9].

In the tenth [10] study presented in this paper, the authors proposed an adaptive parametric bearing fault detection method. By improving the grey wolf optimisation algorithm and optimising the structural parameters of the multi-stable stochastic resonance, the effective detection of bearing fault signals was achieved.

In order to improve the estimation accuracy problem of the cost reference particle filter (CRPF), the authors proposed an intelligent cost reference particle filtering algorithm based on multiple swarm co-operation. The simulation results show that the method has lower RMSE and MAE, reduces sensitivity to the initial values of the particles and improves the diversity of particles during resampling [11].

Conflicts of Interest: The authors declare no conflict of interest.

References

1. Liang, M.; Wang, L.; Yu, G.; Ren, Y.; Peng, L. Study on a Detection Technique for Scholte Waves at the Seafloor. *Sensors* **2022**, *22*, 5344. [CrossRef] [PubMed]
2. Xue, L.; Zeng, X.; Jin, A. A Novel Deep-Learning Method with Channel Attention Mechanism for Underwater Target Recognition. *Sensors* **2022**, *22*, 5492. [CrossRef]
3. Liu, Q.; Zhou, B.; Zhang, J.; Cheng, R.; Zhao, X.; Zhao, R.; Dai, M.; Wang, B.; Wang, Y.A. Time-of-Flight Estimation Method for Acoustic Ranging and Thermometry Based on Digital Lock-In Filtering. *Sensors* **2022**, *22*, 5519. [CrossRef]
4. Wang, S.; Li, Z.; Wang, P.; Chen, H. Optimization Algorithm for Delay Estimation Based on Singular Value Decomposition and Improved GCC-PHAT Weighting. *Sensors* **2022**, *22*, 7254. [CrossRef] [PubMed]
5. Ascari, E.; Cerchiai, M.; Fredianelli, L.; Licitra, G. Statistical Pass-By for Unattended Road Traffic Noise Measurement in an Urban Environment. *Sensors* **2022**, *22*, 8767. [CrossRef] [PubMed]
6. Bismor, D. Leaky Partial Update LMS Algorithms in Application to Structural Active Noise Control. *Sensors* **2023**, *23*, 1169. [CrossRef] [PubMed]
7. Liu, Z.; Tao, Q.; Sun, W.; Fu, X. Deconvolved Fractional Fourier Domain Beamforming for Linear Frequency Modulation Signals. *Sensors* **2023**, *23*, 3511. [CrossRef] [PubMed]
8. Li, Y.; Tang, B.; Huang, B.; Xue, X. A Dual-Optimization Fault Diagnosis Method for Rolling Bearings Based on Hierarchical Slope Entropy and SVM Synergized with Shark Optimization Algorithm. *Sensors* **2023**, *23*, 5630. [CrossRef] [PubMed]
9. He, K.; Xu, Y.; Wang, Y.; Wang, J.; Xie, T. Intelligent Diagnosis of Rolling Bearings Fault Based on Multisignal Fusion and MTF-ResNet. *Sensors* **2023**, *23*, 6281. [CrossRef] [PubMed]
10. Huang, W.; Zhang, G. Bearing Fault-Detection Method Based on Improved Grey Wolf Algorithm to Optimize Parameters of Multistable Stochastic Resonance. *Sensors* **2023**, *23*, 6529. [CrossRef] [PubMed]
11. Zhang, X.; Ren, M.; Duan, J.; Yi, Y.; Lei, B.; Wu, S. An Intelligent Cost-Reference Particle Filter with Resampling of Multi-Population Cooperation. *Sensors* **2023**, *23*, 6603. [CrossRef] [PubMed]

sensors

Article

An Intelligent Cost-Reference Particle Filter with Resampling of Multi-Population Cooperation

Xinyu Zhang [1,2,*], Mengjiao Ren [1,2], Jiemin Duan [1,2], Yingmin Yi [1,2], Biyu Lei [1,2] and Shuyue Wu [1,2]

1 Shaanxi Key Laboratory of Complex System Control and Intelligent Information Processing, Xi'an University of Technology, Xi'an 710048, China; 2220320137@stu.xaut.edu.cn (M.R.); 2210321191@stu.xaut.edu.cn (J.D.); yiym@xaut.edu.cn (Y.Y.); lby_zj1998@163.com (B.L.); wsy152292@163.com (S.W.)
2 Faculty of Automation and Information Engineering, Xi'an University of Technology, Xi'an 710048, China
* Correspondence: xhyzzxy@126.com

Abstract: Although the cost-reference particle filter (CRPF) has a good advantage in solving the state estimation problem with unknown noise statistical characteristics, its estimation accuracy is still affected by the lack of particle diversity and sensitivity to the particles' initial value. In order to solve these problems of the CRPF, this paper proposed an intelligent cost-reference particle filter algorithm based on multi-population cooperation. A multi-population cooperative resampling strategy based on ring structure was designed. The particles were divided into multiple independent populations upon initialization, and each population generated particles with a different initial distribution. The particles in each population were divided into three different particle sets with high, medium and low weights by the golden section ratio according to the weight. The particle sets with high and medium weights were retained. Then, a cooperative strategy based on Gaussian mutation was designed to resample the low-weight particle set of each population. The high-weight particles of the previous population in the ring structure were randomly selected for Gaussian mutation to replace the low-weight particles in the current population. The low-weight particles of all populations were resampled in turn. The simulation results show that the intelligent CRPF based on multi-population cooperation proposed in this paper can reduce the sensitivity of the CRPF to the particles' initial value and improve the particle diversity in resampling. Compared with the general CRPF and intelligent CRPF with adaptive MH resampling (MH-CRPF), the RMSE and MAE of the proposed method are lower.

Keywords: state estimation; unknown statistical characteristics of noise; cost-reference particle filter; multi-population cooperation; intelligent resample; Gaussian mutation

Citation: Zhang, X.; Ren, M.; Duan, J.; Yi, Y.; Lei, B.; Wu, S. An Intelligent Cost-Reference Particle Filter with Resampling of Multi-Population Cooperation. *Sensors* **2023**, *23*, 6603. https://doi.org/10.3390/s23146603

Academic Editor: Antonio Moschitta

Received: 16 June 2023
Revised: 9 July 2023
Accepted: 17 July 2023
Published: 22 July 2023

1. Introduction

State estimation has interested scholars around the world for a long time. It is an important problem in the fields of parameter detection, automatic control, fault diagnosis and navigation guidance [1–3]. A particle filter (PF) is a recursive Bayesian estimation method based on the Monte Carlo idea. This method approximates the posterior probability distribution of the system by sampling a large number of particles (random samples). The weighted sum of these particles is used instead of the high-dimensional integral for system state estimation. Due to its good solving ability for nonlinear non-Gaussian systems, the application scope of filtering technology in state estimation is greatly expanded [4–6]. However, the particle filter and its improved algorithms can only obtain good filtering results when the statistical characteristics of the process noise and measurement noise are known [7]. In practical applications, such as the growth of single-crystal silicon by the Czochralski method and the prediction of lithium–ion's remaining lifetime, the system state measured by the sensor contains noise. Due to the interference of some variable factors or the external environment, the noise statistical characteristics are difficult to identify.

The standard particle filter may reduce the accuracy of state estimation, which affects the analysis and control of the system [8]. In view of the fact that the statistical characteristics of the noise are unknown, the cost-reference particle filter (CRPF) was proposed for the first time [9]. This method does not need to know the statistical characteristics of process noise and measurement noise [10]. It calculates the weight of each particle by a self-defined cost function and risk function, and then resamples according to the weight. Finally, the system state is approximated by the set of resampled particles. Since the cost-reference particle filter does not need to know the statistical characteristics of the noise, it has been widely used since it was proposed. It has gradually become an effective method to solve the problem of filtering nonlinear systems with unknown noise statistical characteristics. However, the CRPF still uses traditional polynomial resampling. The particles are selected according to the weight: the particles with high weights are copied in high numbers and the particles with low weights are eliminated. Therefore, the particles are concentrated in the same area, and the number of the same particles gradually increases, resulting in the loss of offspring diversity. The particles cannot completely cover the posterior probability distribution, that is, there is a lack of particle diversity, so that the filtering results have a large deviation. In addition, the CRPF is sensitive to the initial value of the particle distribution. Filtering results of the CRPF will further deteriorate when the initial value of the particle is unknown. Therefore, how to develop an efficient and reliable resampling method under the framework of the CRPF, to further improve the diversity of particles when the initial distribution of particles is unknown, and then improve the accuracy of the CRPF, becomes an important issue in the research of the CRPF.

At present, many scholars have conducted a significant amount of research and proposed some improved resampling algorithms to solve the problem of lack of particle diversity. The paper [11] proposed a deterministic resampling strategy. Instead of blindly copying particles with high weights and discarding particles with low weights as in traditional resampling methods, this strategy replicates particles selectively. It divides particles based on their state values and weights. Therefore, the number of effective particles in this method was improved. The paper [12] proposed a particle filter algorithm based on error ellipse resampling. In this method, the particle set is divided hierarchically by defining an error ellipse, and then the particles are copied according to the level of division. These resampling methods suppress the problem of lack of particle diversity to a certain extent. However, they still adopt the traditional resampling framework and replace the abandoned particles by copying the existing particles, which cannot fundamentally solve the problem. In recent years, the development of swarm intelligence optimization algorithms has provided new ideas for improving the particle diversity. In the literature [13–15], the artificial fish swarm algorithm, the firefly algorithm and the self-controlled bat algorithm are, respectively, used in the particle filter, attempting to optimize all the particles in each resampling by the swarm intelligence optimization method. Although these methods have a certain effect on increasing the diversity of the particles, they will increase the complexity of the algorithm and seriously reduce its real-time performance. In addition, these methods run the risk of introducing the problem of falling into a local optimum that exists in most swarm intelligence optimization methods, which affects the estimation accuracy of the particle filter. In 1997, the paper [16] proposed that the particle filter and evolutionary algorithm have a similar structure, and introduced the crossover operation of the genetic algorithm into the particle filter to increase the diversity of the particles. The paper [17] proposed an intelligent particle filter algorithm based on a real-coded crossover and mutation strategy. This method uses a real-coded crossover and mutation strategy to increase the diversity of the particles. Based on this idea, the paper [18] proposed a sequential evolutionary filtering algorithm without traditional resampling, which improved the running speed of the algorithm. Subsequently, the paper [19] introduced selection, crossover and mutation operators into resampling, which enriched the resampling strategy based on genetic operation and further improved the particle diversity. Although the resampling method based on the crossover and mutation strategy can improve the diversity of the particles, it is difficult to obtain an ideal particle distribution due to the dynamic change in particle distribution. Therefore, the estimation accuracy of the particle filter needs to be further improved. The paper [20] proposed an adaptive resampling particle filter based on Student's t distribution. The two subsets divided according to the weight are adaptively crossed. After that, some particles are mutated randomly to improve the diversity of the particles. The paper [21,22] also proposed an adaptive genetic particle filter and enhanced mutation particle filter, respectively. They improved the efficiency of particle resampling by adaptively adjusting the mutation rate. The above methods introduce the crossover and mutation operations of the genetic algorithm to increase the diversity of the particles. Although this is of great significance for the development of resampling algorithms, it seriously affects the computational complexity and realizability of the algorithm. As the resampling strategy improves, the number of parameters to set increases. The paper [23] proposed an adaptive Metropolis–Hastings (M–H) resampling algorithm, which introduced the accept–reject mechanism of M–H into resampling. It adaptively selected Gaussian mutation or crossover of the high weight and low weight to resample particles according to the particle distribution, effectively improving the diversity of the particles. The above methods have made good progress in improving resampling, but they are all based on the traditional particle filter and have not been extended to the CRPF. Therefore, they still assume that the particles' initial value and noise distribution are known. However, in practice, the particles' initial value and noise distribution are often not accurately obtained, which affects the estimation accuracy. Since the particle filter and swarm intelligence evolutionary algorithm have the same structure, the latest research results in swarm intelligence methods can provide a new development direction for the research of resampling strategy [24], such as the multi-swarm cooperation mechanism. The multi-population cooperative mechanism means that the population is divided into several populations in the evolution process, each population evolves independently and then the information is shared among the populations. Thus, the convergence speed of the algorithm is improved and the algorithm is kept from falling into local optimum. The multi-population cooperation mechanism provides a new idea for further improving the diversity of particles and solving the problem of low estimation accuracy, when the initial values of the particles and the statistical characteristics of the noise are unknown.

Therefore, this paper proposes an intelligent resampling method based on multi-population cooperation, which improves the previous resampling method and further improves the particle diversity. The new resampling method is applied in the cost-reference particle filter (CRPF) to solve the problem of lack of particle diversity and sensitivity to the particles' initial value in the CRPF, which is good at estimating the system state when the statistical characteristics of noise are unknown. Firstly, the particles are divided into several independent populations, and each population performs importance sampling using the distribution of different initial values. Then, the particles of each population are sorted according to the weight from large to small, and the golden section ratio is used to divide the particle set into three parts: high, medium and low weight; while retaining the set of particles with high and medium weights, Gaussian mutation is performed on the particles with high weights with a certain probability. By using the cooperative strategy, the low weight particles in the current population are replaced by the high weight particles after mutation in the previous population, so as to realize the resampling of the low weight particle set in turn. In this paper, a new method is proposed to realize information sharing among populations through the ring coordination mechanism, which makes the particles in the population closer to the posterior probability density of the state. The superimposed Gaussian mutation effectively increases the diversity of the particles in the resampling process. The final state estimate is obtained according to all particles and their corresponding weights. Finally, the effectiveness of the proposed method was verified by common one-dimensional and multidimensional models.

2. Materials and Methods

Suppose the nonlinear discrete system model is described as follows:

$$x_k = f(x_{k-1}) + w_k \tag{1}$$

$$y_k = h(x_k) + v_k \tag{2}$$

where, $x \in R^{n_x}$ and $y \in R^{n_y}$ are system state vector and observation vector at time k. $w \in R^{n_x}$ is process noise, $v \in R^{n_y}$ is observation noise, $f(x_{k-1})$ and $h(x_k)$ are bounded nonlinear mappings, respectively.

Unlike the traditional particle filter, the CRPF does not need to know the statistical characteristics of noise when estimating the state. Instead, it proposes a cost function and a risk function, and finally calculates the weight of the particle by them. The cost function of the i-th particle x_k^i at time k is defined as follows:

$$C_k^i = \lambda C_{k-1}^i + \Delta C_k^i \tag{3}$$

$$\Delta C_k^i = \Delta C\left(x_k^i \mid y_k\right) = \left\| y_k - h\left(x_k^i\right) \right\|^q \tag{4}$$

In the formula, $0 \leq \lambda < 1$ is the forgetting factor, $q \geq 1$. The risk function of the i-th particle at time k is defined as follows:

$$\Re_k^i = \lambda C_{k-1}^i + \left\| y_k - h\left(f\left(x_{k-1}^i\right)\right) \right\|^q \tag{5}$$

According to Equations (3)–(5), when the particle is closer to the true state value, the cost due to the estimation error is smaller. This indicates that the risk to be taken when selecting the particle is smaller, so the probability that the particle is selected to be retained is larger. That is, its weight is higher. The algorithm steps for the standard CRPF are as Algorithm 1.

According to the above steps of the CRPF algorithm, the process noise and measurement noise of the system are not used in the whole process of system state estimation. Instead, it is achieved through the cost function and risk function defined by the CRPF algorithm. Therefore, compared with the conventional particle filter, the CRPF has a significant advantage in solving the state estimation problem with unknown noise statistical properties.

Although the CRPF has a good filtering effect when the statistical characteristics of noise are unknown, it still uses the traditional resampling method. It copies high-weight particles and discards low-weight particles, so the problem of lack of particle diversity also exists in the CRPF. When the particles with high weight are copied and most particles with small weight are discarded, this leads to the result of state estimation deviating from the real state, due to the entire set of particles being unable to completely cover the posterior distribution, as shown in Figure 1. At the same time, the initial value of the particle cannot be accurately obtained in practical problems, which makes the problem more prominent in the estimation process of the CRPF. Therefore, how to improve the resampling algorithm of the CRPF, so that it can effectively suppress the lack of particle diversity and improve the estimation accuracy when the particles' initial value is not accurate, has become a key issue in the study of the CRPF algorithm.

Algorithm 1 The algorithm steps for the standard CRPF.

(1) Initialize $k = 0$

(1) N particles are extracted from the initial distribution $p(x_0)$, the initial cost is set to 0;

(2) The initial particle set is $\{x_0^i, C_0^i\}_{i=1}^N$, where $p(x_0)$ is uniformly distributed;

(2) Resample

(1) The risk function value of the particle is calculated by Formula (5);

(2) The weight of the particle is calculated from the risk function value according to the following formula:

$$w_k^i = \frac{(\Re_k^i)^{-\beta}}{\sum_{i=1}^N (\Re_k^i)^{-\beta}} \tag{6}$$

where $\beta > 1$;

(3) Polynomial resampling is performed on the particle according to the weight value, and the particle set after resampling is $\{\bar{x}_k^i, \bar{C}_k^i\}_{i=1}^N$;

(3) Update

(1) Carry out particle update according to the following formula:

$$x_k^i \sim N\left(f\left(\bar{x}_k^i\right), \Sigma\right) \tag{7}$$

(2) Update particle cost by Formula (3) and (4);

(3) Calculate the updated particle weight according to the following formula:

$$w_k^i = \frac{(C_k^i)^{-\beta}}{\sum_{i=1}^M (C_k^i)^{-\beta}} \tag{8}$$

(4) State estimation

The estimated system status is

$$\hat{x}_k = \sum_{i=1}^N x_k^i w_k^i \tag{9}$$

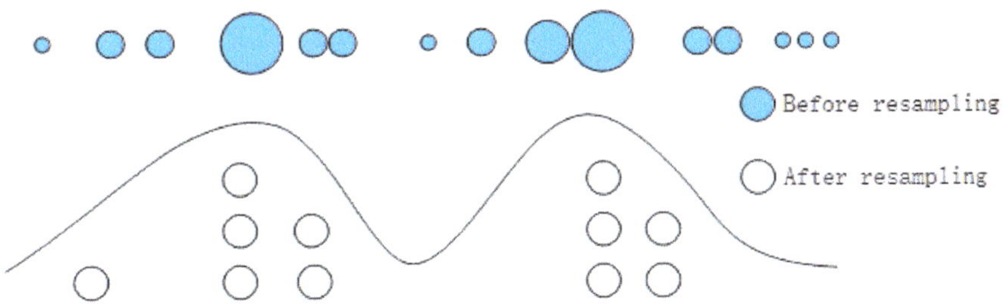

Figure 1. Schematic diagram of traditional resampling particle distribution.

3. Intelligent CRPF Based on Multi-Population Cooperation

3.1. Multi-Population Cooperative Intelligent Resampling Mechanism Based on Ring Structure

In order to solve the problem of low estimation accuracy due to the lack of particle diversity in the CRPF when the initial value is not accurate, an intelligent cooperative resampling mechanism based on ring structure is proposed in this paper. The structure of the intelligent cooperative resampling mechanism is shown in Figure 2. The particles are extracted from the importance distribution of M different initial values, and the extracted particles form a population with N particles which are mutually independent. The particles in each population were sorted according to their weights and divided into three parts according to the golden proportion. The golden proportion is strictly proportional, artistic and harmonious, and is considered to be the most ideal proportion in architecture and art. The first 0.382N particle set is X_H^s with a larger weight, the 0.382N particle set is X_L^s with the smallest weight and the remaining particle set is X_M^s with a middle weight, where $s = 1, 2, \ldots M$. The high weight particle set X_H^s and medium weight particle set X_M^s with a larger weight in each population are directly retained. However, since the particles in X_L^s have little influence on the estimated results, all the particles in it are resampled using the cooperative strategy. In order to ensure the particle resampling order and particle diversity, the cooperation strategy between various swarms is completed based on the ring structure. When the low-weight particles of the s-th population are resampled, the high-weight particles of the s−1 population are selected for cooperation. When s = 1, then the particles are drawn from the set of high-weight particles of the M-th population for cooperation, and the details of the cooperation strategy will be described in Section 3.2. After resampling all the particles with low weights in M populations, the weights of the particles are recalculated. The weighted sum of all particles in each population is used to obtain the state estimate value, and then the final state estimate value of the whole particle set is obtained.

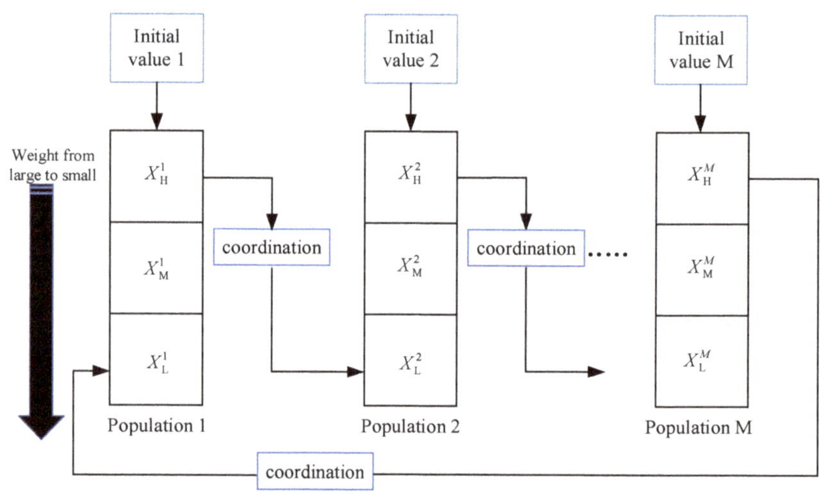

Figure 2. Intelligent resampling mechanism based on ring structure.

Different from the traditional resampling methods used in the CRPF, new resampling mechanisms are proposed in this paper. Instead of simply copying the high-weight particles to replace the low-weight ones, this method adopts an intelligent cooperative strategy to improve the diversity of the particles. Compared with the previous intelligent resampling methods, the intelligent cooperative resampling mechanism adopts a multi-population ring cooperative structure. Each population uses the importance distribution of different initial values to generate particles, which reduces the sensitivity of the particle filter to the initial values. In addition, the collaborative strategy was used to exchange high-weight particle information between different populations, which improved the information sharing ability of the particles and effectively improved the diversity of the particles. At the same time, the traditional resampling step is canceled to reduce the complexity of the algorithm.

3.2. Cooperative Strategy Based on Gaussian Mutation

Since the particles in the high weight and medium weight particle sets have large weights and are all effective particles, these sets must be retained when resampling the particles. Only the low weight particles in each population are resampled by the collaborative strategy. This can not only improve the efficiency of resampling, but also increase the diversity of the particles. However, the traditional resampling method can no longer be used for cooperative operation. Copying particles with high weight and directly replacing particles with low weight will cause the same particles to be copied in large numbers, resulting in the serious loss of particle diversity and the reduction in particle filtering accuracy.

In this paper, a collaborative strategy based on Gaussian mutation is designed to resample low-weight particles in each population, avoiding blindly copying high-weight particles to improve the diversity of the particles. This strategy is based on the ring structure of Section 3.1. Particles are randomly selected from the high-weight particle set of the previous population, and Gaussian mutation is performed on them. Then, the low-weight particles of the local population are replaced with the mutated particles.

Gaussian mutation is calculated as follows:

$$x_{kS}^{ls} = x_{kH}^{ns-1} + q^{*}\lambda \tag{10}$$

In Formula (10), x_{ks}^{ls} is the particle after mutation at time k, $l = 1, 2, \ldots, N_{kL}^{s}$ and N_{kL}^{s} are the particle numbers of the low-weight particle set in the s-th population at time k, x_{kH}^{is-1} is the particle number of the high-weight particle set randomly selected from the (s−1)-th population at time k, $n \in \left\{1, 2, \ldots, N_{kH}^{s-1}\right\}$ and N_{kH}^{s-1} are the particle numbers of the high-weight particle set in the (s−1)-th population at time k, while $\lambda \sim N(0,1)$ and q are the mutation rate.

According to the above equation, when q is set appropriately, the particle with Gaussian mutation must be located near the particle with high weight. Different from the existing high-weight particles, this method can effectively improve the diversity of the particles and ensure the quality of the whole set of particles. At the same time, q should not be too large; otherwise, the newly generated particles will be far away from the high probability region, and then reduce the accuracy of filtering. The organic integration of cooperative strategy and ring structure ensures that the information of high-weight particles generated by different importance distributions of each population is shared. After several iterations, the particles generated by the importance distribution far away from the true posterior distribution are gradually abandoned, and more and more particles close to the posterior distribution are retained. This strategy effectively improves the diversity of the particles, thereby reducing the impact of inaccurate initial value settings on the state estimation results.

3.3. Steps of Intelligent CRPF Algorithm Based on Multi-Population Cooperation

The multi-population cooperative intelligent resampling proposed above is introduced into the CRPF algorithm to obtain the intelligent CRPF based on multi-population cooperation. The algorithm steps are as Algorithm 2.

Algorithm 2 Steps of Intelligent CRPF Algorithm Based on Multi-Population Cooperation.

(1) Initialize

(1) Set the population number M, the particle number N of each population and the uniform distribution parameter $[a_s, b_s]$ of each population, $s = 1, 2, \ldots, M$;

(2) Randomly extract the initial particle of each population from the uniform distribution x_0^{is} , $i = 1, 2, \ldots, N$; and set the initial value corresponding to each particle to 0, that is $C_0^{is} = 0$, to obtain the particle-cost set $\{x_0^{is}, C_0^{is}\}_{i=1}^{N}$;

(2) Resample

(1) The risk function $\Re_k^{is}$ of each particle is calculated according to Formula (5), and the particle weight w_k^{is} is calculated according to Formula (6);

(2) Arrange the particles in each population according to the weight from large to small, and retain X_{kH}^s and X_{kM}^s in each population;

(3) Resampling the particle x_{kL}^{is} in the low-weight particle set X_{kL}^s of each population according to Formula (10) and Section 3.2 to obtain the particle set X_{kS}^s. If $s = 1$, the high-weight particle is extracted from X_{kH}^M for resampling according to Formula (10);

(4) The high-weight particle set X_{kH}^s and medium-weight particle set X_{kM}^s of each population are combined with the resampling particle set X_{kS}^s to obtain the new set of each population, while each population remains independent;

(3) Update

(1) The resampled particles are updated according to Formula (7), the updated particle cost C_k^{is} is calculated according to Formulas (3) and (4) and the updated particle weight ω_k^{is} is calculated according to Formula (8);

(2) Calculate the estimated value $\hat{x}_k^s$ of each population according to Formula (9);

(3) Finally, calculate the final state estimate of the system according to the estimated value $\hat{x}_k^s$ of each population according to the formula below:

$$\hat{x}_k = \frac{1}{M}\sum_{s=1}^{M}\hat{x}_k^s \tag{11}$$

4. Results and Discussion

In order to verify the effectiveness of the proposed method, three simulations were designed. The standard CRPF and the adaptive MH resampling-CRPF (MH-CRPF, Intelligent CRPF) are used as comparison algorithms, and each experiment was independently repeated 100 times. The three simulation models are the mathematical model of one-dimensional non-stationary economic growth [17,18], the one-dimensional nonlinear univariate time series model [3,21] and the lithium–ion battery remaining life prediction model [1], respectively. Due to their strong nonlinear characteristics, these models are widely used to verify the effectiveness of the particle filter. Meanwhile, root mean square error (RMSE) and mean absolute error (MAE) are selected to evaluate the performance of the algorithm [25]. RMSE and MAE are, respectively, calculated according to the following formula:

$$RMSE = \sqrt{\frac{1}{KL}\sum_{i=1}^{K}\sum_{k=1}^{L}(\hat{x}_{k,i} - x_{k,i})^2} \tag{12}$$

$$MAE = \frac{1}{KL}\sum_{i=1}^{K}\sum_{i=1}^{L}|\hat{x}_{k,i} - x_{k,i}| \tag{13}$$

where K represents the number of repeated experiments, L is the length of time series, $\hat{x}_{k,i}$ and $x_{k,i}$ are, respectively, the estimated value and the real value of the system state at the time k in the i-th simulation.

4.1. Mathematical Model of One-Dimensional Non-Stationary Economic Growth

The one-dimensional non-stationary economic growth mathematical model has the characteristics of being strong nonlinear and bimodal, and it is difficult to estimate its system state, so the model is widely used to verify the effectiveness of the particle filter. Its model is described as follows:

$$x_k = \frac{1}{2}x_{k-1} + \frac{25x_{k-1}}{1+x_{k-1}^2} + 8\cos[1.2(k-1)] + w_k \tag{14}$$

$$y_k = \frac{1}{20}x_k^2 + v_k \tag{15}$$

where $w_k \sim N(0,\sigma_w^2)$ and $v_k \sim N(0,\sigma_v^2)$ are process noise and observation noise, respectively.

In this simulation, $\sigma_v^2 = 1$, σ_w^2 is the random number subject to uniform distribution $U(0,9)$, and the initial value of x_k, x_0 is the random number subject to uniform distribution $U(0,12)$. Each independent repeated experiment is randomly generated. The population number of this method is M=3, and the particle number of each population is N = 80. The initial state values of the three populations are $x_0^1 \sim U(0,4)$, $x_0^2 \sim U(4,8)$ and $x_0^3 \sim U(8,12)$, respectively, and the variation rate of Gaussian mutation is 2. The particle numbers of the standard CRPF and MH-CRPF are both 240, and the initial particle values are random numbers that follow uniform distribution $U(0,12)$.

Figure 3 shows the comparison of the average value of the 100-time state estimation results of the mathematical model of one-dimensional non-stationary economic growth by the three methods, and Figure 4 shows the RMSE comparison of the 100-time estimation results of the three methods. It can be seen from Figures 3 and 4 that the three methods can track the real state, but the RMSE of the CRPF and MH-CRPF (Intelligent CRPF) is larger than that of the proposed method. Table 1 shows RMSE and MAE of 100-time estimation results of the mathematical model of one-dimensional non-stationary economic growth by the three methods. According to Table 1, compared with the other two comparison methods, the RMSE and MAE of the method in this paper are both minimal. According to these experimental results, the proposed method in this paper adopts a resampling strategy based on multi-population cooperation. It strengthens the interaction of high-weight particle information between different populations and reduces the dependence of the algorithm on the initial state value. This strategy can improve the diversity of the particles and the accuracy of the CRPF. Therefore, compared with the standard CRPF and MH-CRPF (Intelligent CRPF), the proposed method has higher estimation accuracy when the initial distribution of the particles is unknown.

Table 1. State estimation results of the three methods of model Section 4.1.

Performance Index	Multi-CRPF (Proposed)	MH-CRPF (IntelligentCRPF)	CRPF
RMSE	2.6220	3.2517	4.2156
MAE	1.9961	2.5345	3.5527

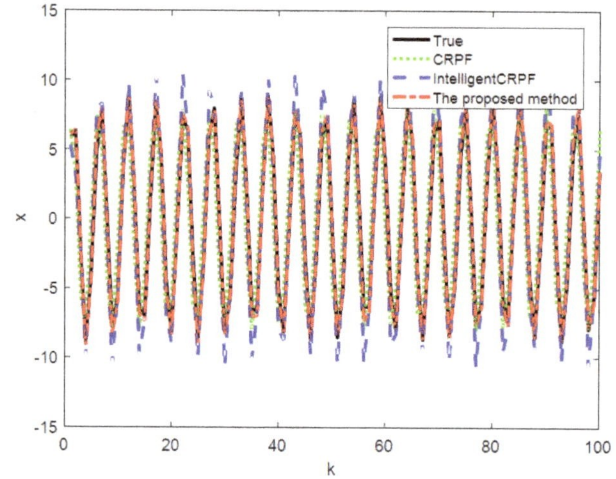

Figure 3. Comparison of three methods for state estimation of model Section 4.1.

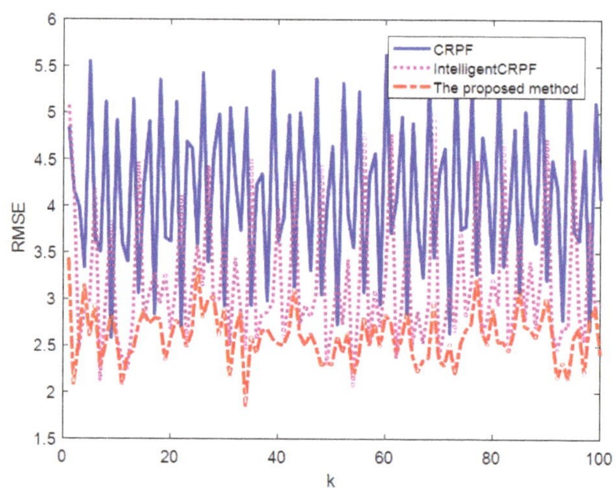

Figure 4. RMSE of state estimation results of three methods of model Section 4.1.

4.2. One-Dimensional Nonlinear Univariate Time Series Model

The one-dimensional nonlinear univariate time series model is also widely used to verify the effectiveness of the particle filter due to its strong nonlinearity. The model is described as follows:

$$x_k = 1 + \sin(0.04\pi t) + 0.5x_{k-1} + w_k \tag{16}$$

$$y_k = \begin{cases} 0.2x_k^2 + v_k & k \leq 30 \\ 0.5x_k - 2 + v_k & k > 30 \end{cases} \tag{17}$$

where $w_k \sim N(0, \sigma_w^2)$ and $v_k \sim N(0, \sigma_v^2)$ are process noise and observation noise, respectively.

In this simulation, σ_ω^2, σ_v^2 and the initial values x_0 of x_k as well as other parameter settings are consistent with those in Section 4.1.

Figure 5 shows the comparison of the average values of the 100-time state estimation results of the one-dimensional nonlinear univariate time series model by the three methods. Figure 6 shows the RMSE comparison of the 100-time state estimation results of the three methods. It can be seen from Figures 5 and 6 that the proposed method can track the real state well, but MH–CRPF (Intelligent CRPF) has a certain error in the estimation process, while the CRPF has a large error. Table 2 shows the RMSE and MAE of 100 independent repeated experiments estimated by the three methods for the one-dimensional nonlinear univariate time series model. As can be seen from Table 2, the RMSE and MAE of the proposed method are always the smallest, followed by MH-CRPF (Intelligent CRPF) and the largest CRPF. Therefore, when the initial state value is unknown, the state estimation accuracy of the proposed method for the one-dimensional nonlinear univariate time series model is higher than that of the other two methods, due to the multi-population intelligent cooperative resampling mechanism.

Table 2. State estimation results of the three methods of model Section 4.2.

Performance Index	Multi-CRPF (Proposed)	MH-CRPF (IntelligentCRPF)	CRPF
RMSE	2.4747	2.6284	2.8030
MAE	1.8861	2.0164	2.1852

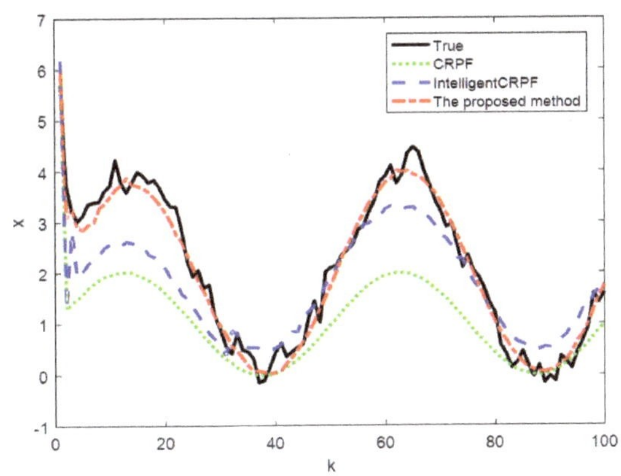

Figure 5. Comparison of three methods for state estimation of model Section 4.2.

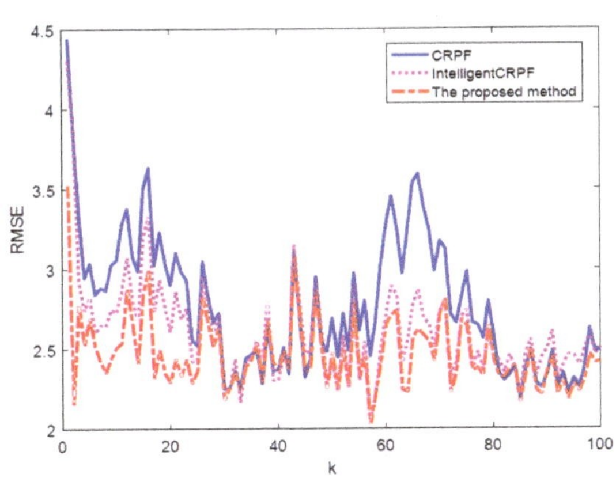

Figure 6. RMSE of state estimation results of three methods of model Section 4.2.

4.3. Lithium–Ion Battery Remaining Useful Life Prediction Model

Due to the high-dimensional, nonlinear and non-Gaussian characteristics of the remaining useful life prediction model of a lithium–ion battery, the nonlinear filtering algorithm is usually used to predict the remaining life of a lithium–ion battery. Therefore, this model can also be used to verify the effectiveness of the filtering algorithm. The model is described below.

$$x_k = x_{k-1} + w_k \tag{18}$$

$$Q_k = a_k \exp(b_k * k) + c_k \exp(d_k * k) + v_k \tag{19}$$

$$x_{k-1} = \begin{bmatrix} a_{k-1} \\ b_{k-1} \\ c_{k-1} \\ d_{k-1} \end{bmatrix} \tag{20}$$

where k is the number of cycles of a lithium–ion battery, Q_k is the capacity of the lithium–ion battery at the k-th cycle, $w_k \sim N(0, \sigma_w^2)$ and $v_k \sim N(0, \sigma_v^2)$ are process noise and observation noise, respectively.

This simulation selects the data of the B0007 lithium–ion battery, which is widely used in the verification experiment of lithium battery remaining life prediction [26]. These data from NASA laboratories are common to international simulation experiments. In this simulation, the population number and particle number are the same as in Section 4.1, the rate of Gaussian mutation is $\sqrt{0.05}$, the particle number of the other two methods is 240 and the process variances are all $U(0, 1.5)$ random numbers.

Figure 7 shows the comparison curve of prediction results of 100 independent repeated experiments on the remaining life of the B0007 lithium battery by the three methods. It can be seen from Figure 7 that the three methods can track the actual curve well. However, with the increase in the service cycles, the CRPF and MH-CRPF gradually deviate from the actual curve. The proposed method in this paper can always track the actual curve well. Figure 8 shows the comparison of the MAE of the remaining life prediction results of the B0007 lithium battery by the three methods. As can be seen from Figure 8, the deviation between the CRPF and MH-CRPF keeps increasing with the increase in time. The deviation of the proposed method (Multi-CRPF) is smaller than that of the other two methods. The proposed method shows better prediction accuracy. Table 3 shows the RMSE and MAE of 100 independent repeated experiments predicted by the three methods for the remaining life of the B0007 lithium battery. As can be seen from Table 3, the RMSE and MAE of the predicted results of the proposed method are both smaller than those of the other two methods. It can be concluded that, compared with the standard CRPF and MH-CRPF, the proposed method has better performance in predicting the remaining useful life of lithium batteries, especially when the initial value of the system state is unknown.

According to the above three simulation results, each population uses different initial state value to generate particles, and the multi-population cooperative intelligent resampling mechanism based on ring structure and Gaussian mutation strategy is proposed in this paper. It can effectively reduce the dependence of state estimation results on the initial state value. In addition, the high-weight particle information of each population is shared to other populations based on the ring structure. The high-weight and medium-weight particles of each population are retained, and the low-weight particles are resampled, which can effectively improve the diversity of the particles, so as to improve the estimation accuracy of the CRPF.

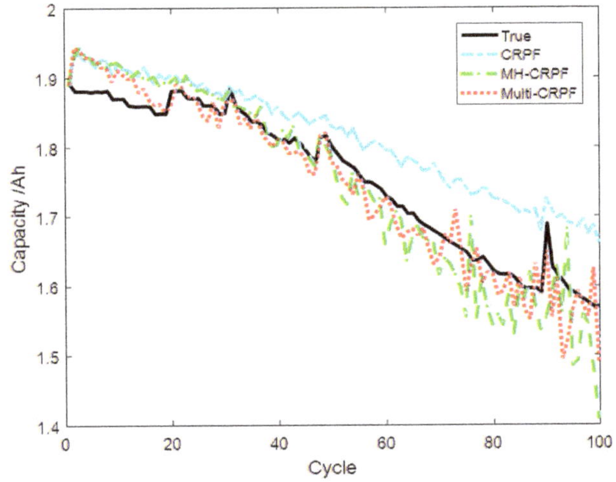

Figure 7. Prediction results of model Section 4.3 by three methods.

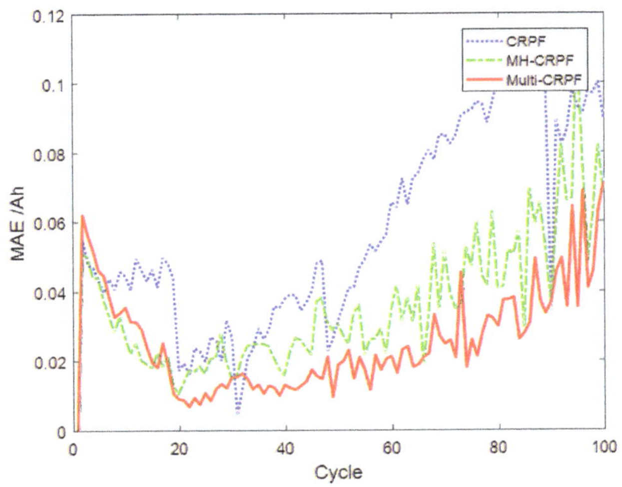

Figure 8. Comparison of state estimation MAE by three methods of model Section 4.3.

Table 3. Prediction results of remaining useful life of lithium battery by three methods.

Performance Index	Multi-CRPF (Proposed)	MH-CRPF (IntelligentCRPF)	CRPF
RMSE	0.0363	0.0503	0.0646
MAE	0.0258	0.0360	0.0574

5. Conclusions

When the statistical characteristics of noise are unknown, the CRPF shows good filtering performance. An intelligent CRPF method based on multi-population cooperation is proposed to solve the problem of low estimation accuracy caused by the lack of particle diversity in the CRPF when the initial value of the particles is not set accurately.

The main contribution of this method is that the multi-population cooperative intelligent resampling method based on ring structure and the cooperative strategy based on Gaussian mutation are introduced into the CRPF. The particles are divided into several independent populations with different initial state values, and after the importance sampling is completed, the resampling process of low-weight particles is improved based on the ring structure. This method promotes the cooperation of particles among various populations and enhances the interaction ability of particle information. At the same time, Gaussian mutation is introduced to improve the particle diversity. The results of three simulation experiments show that, compared with the standard CRPF and MH-CRPF, the proposed method can effectively improve the estimation accuracy of the CRPF when the initial value of the state is not set accurately.

Although the proposed method has good estimation results, how to better set the particle initial value of each population and the mutation rate of Gaussian mutation, and how to enrich the cooperative strategy between multiple populations, so as to achieve more accurate estimation results, are the future research directions and research goals.

Author Contributions: X.Z. designed the proposed framework and performed the simulations, M.R. helped improve the optimization method, J.D. helped revise the manuscript in terms of both the structure and grammar, Y.Y. helped check and analyze the experimental data, B.L. helped conduct the simulations, and S.W. helped enhance the design of the analytical method. X.Z. is the corresponding author of this paper. All authors have read and agreed to the published version of the manuscript.

Funding: This work is supported by National Natural Science Foundation of China (No.62003261, 61702410, 62073258, 61871318), Natural Science Basic Research Plan in Shaanxi Province of China (No.2020JQ-645, 2020JQ-650, 2021JQ-487) and the Science and Technology Innovation Team of Shaanxi Province (No.2023-CX-TD-01).

Institutional Review Board Statement: Not applicable.

Informed Consent Statement: Not applicable.

Data Availability Statement: Not applicable.

Acknowledgments: The authors thank the editor and anonymous reviewers for providing helpful suggestions for improving the quality of this manuscript.

Conflicts of Interest: The authors declare no conflict of interest.

References

1. Jiao, R.; Peng, K.; Jie, D. Remaining useful life prediction of lithium-ion batteries based on conditional vibrational auto encoders-particle filter. *IEEE Trans. Instrum. Meas.* **2020**, *69*, 8831–8843. [CrossRef]
2. Qiao, D.; Zhang, X.; Wang, Y. Fault and state estimation for discrete linear variable parameter systems with integral measurement and delay. *Control Theory Appl.* **2021**, *38*, 587–594.
3. Niu, F.; Liu, J.; Xiong, J.; Li, J.; Shen, L. Research on ground multi-target guidance method of UAV group cooperative tracking. *Sci. China (Tech. Sci.)* **2020**, *50*, 403–422.
4. Doucet, A.; Godsill, S.J.; Andrieu, C. On sequential monte Carlo sampling methods for bayesian filtering. *Stat. Comput.* **2000**, *10*, 197–208. [CrossRef]
5. Dai, J.; Xu, P.; Li, X. Second order central difference particle filter Fast SLAM algorithm. *Control Theory Appl.* **2018**, *35*, 1382–1390.
6. Stephen, G.P.; Takemasa, M. A local particle filter for high-dimensional geophysical systems. *Nonlinear Process. Geophys.* **2015**, *2*, 1631–1658.
7. Wang, X.; Cao, J.; Li, W. Intelligent optimization CRPF algorithm. *Syst. Eng. Electron. Technol.* **2017**, *39*, 2857–2862.
8. Zhang, X.; Liu, D.; Jiang, H.; Liang, J. Particle Filter with Unknown Statistics to Estimate Liquid Level in the Silicon Single-Crystal Growth. *IEEE Trans. Instrum. Meas.* **2020**, *69*, 2759–2770. [CrossRef]
9. Míguez, J.; Bugallo, M.F.; Djurić, P.M. A new class of particle filters for random dynamic systems with unknown statistics. *EURASIP J. Adv. Signal Process.* **2004**, *15*, 2287–2294. [CrossRef]
10. Zhong, L.; Li, Y.; Cheng, W.; Zheng, Y. Cost-Reference Particle Filter for Cognitive Radar Tracking Systems with Unknown Statistics. *Sensors* **2020**, *20*, 3669. [CrossRef]
11. Li, T.C.; Sattar, T.P.; Sun, S.D. Deterministic resampling: Unbiased sampling to avoid sample impoverishment in particle filters. *Signal Process.* **2012**, *92*, 1637–1645. [CrossRef]
12. Xu, C.; Wang, X.; Duan, S.; Wan, J. Particle filter tracking algorithm based on error ellipse resampling. *J. Instrum.* **2020**, *41*, 76–84.

13. Li, Z.; Liu, T. Improved particle filter based soft sensing of room cooling load. *J. Instrum.* **2017**, *142*, 56–61.
14. Zhou, W.; Lu, L.; Hou, J. Firefly algorithm-based particle filter for nonlinear systems. *Circuits Syst. Signal Process.* **2019**, *38*, 1583–1595. [CrossRef]
15. Chen, Z.; Wu, P.; Bo, Y.; Tian, M.; Yue, C.; Gu, F. Maneuvering target tracking based on self-control bat algorithm intelligent optimization particle filter. *Acta Electron. Sin.* **2018**, *46*, 886–894.
16. Higuchi, T. Monte Carlo filter using the genetic algorithm operators. *J. Stat. Comput. Simul.* **1997**, *59*, 1–23. [CrossRef]
17. Yin, S.; Zhu, X. Intelligent particle filter and its application to fault detection of nonlinear system. *IEEE Trans. Ind. Electron.* **2015**, *62*, 3852–3861. [CrossRef]
18. Yin, S.; Zhu, X.; Qiu, J.; Gao H. State estimation in nonlinear system using sequential evolutionary filter. *IEEE Trans. Ind. Electron.* **2016**, *63*, 3786–3794. [CrossRef]
19. Zhou, N.; Lau, L.; Bai, R.; Terry, M. A genetic optimization resampling based particle filtering algorithm for indoor target tracking. *Remote Sens.* **2021**, *13*, 132. [CrossRef]
20. Teng, F.; Xue, L.; Li, X. Adaptive resampling particle filter based on Student's t distribution. *Control Decis.* **2018**, *33*, 361–365.
21. Qiu, Z.; Qian, H. Adaptive genetic particle filter and its application to attitude estimation system. *Digit. Signal Process.* **2018**, *81*, 163–172. [CrossRef]
22. Lin, M.; Yang, C.; Li, D. An improved transformed unscented FastSLAM with adaptive genetic resampling. *IEEE Trans. Ind. Electron.* **2019**, *66*, 3583–3594. [CrossRef]
23. Zhang, X.; Liu, D.; Yang, Y.; Liang, J. An intelligent particle filter with adaptive M-H resampling for liquid-level estimation during silicon crystal growth. *IEEE Trans. Instrum. Meas.* **2021**, *70*, 1–12. [CrossRef]
24. Wang, H.; Yang, F.; Yao, H. Evolutionary particle filter inference algorithm for discrete dynamic Bayesian networks. *J. Comput. Res. Dev.* **2008**, *45*, 295–299.
25. Lee, S.; Kwon, J.; Park, D. Optimized Replication of ADC-Based Particle Counting Algorithm with Reconfigurable Multi-Variables in Pseudo-Supervised Digital Twining of Reference Dust Sensor Systems. *Sensors* **2023**, *23*, 5557. [CrossRef]
26. Jafari, S.; Byun, Y.-C. XGBoost-Based Remaining Useful Life Estimation Model with Extended Kalman Particle Filter for Lithium-Ion Batteries. *Sensors* **2022**, *22*, 9522. [CrossRef]

sensors MDPI

Article

Bearing Fault-Detection Method Based on Improved Grey Wolf Algorithm to Optimize Parameters of Multistable Stochastic Resonance

Weichao Huang [1,2,*] and Ganggang Zhang [2]

1 Shannxi Key Laboratory of Complex System Control and Intelligent Information Processing, Xi'an University of Technology, Xi'an 710048, China
2 School of Automation and Information Engineering, Xi'an University of Technology, Xi'an 710048, China; zgg@stu.xaut.edu.cn
* Correspondence: huangwc@xaut.edu.cn

Abstract: In an effort to overcome the problem that the traditional stochastic resonance system cannot adjust the structural parameters adaptively in bearing fault-signal detection, this article proposes an adaptive-parameter bearing fault-detection method. First of all, the four strategies of Sobol sequence initialization, exponential convergence factor, adaptive position update, and Cauchy–Gaussian hybrid variation are used to improve the basic grey wolf optimization algorithm, which effectively improves the optimization performance of the algorithm. Then, based on the multistable stochastic resonance model, the structure parameters of the multistable stochastic resonance are optimized through improving the grey wolf algorithm, so as to enhance the fault signal and realize the effective detection of the bearing fault signal. Finally, the proposed bearing fault-detection method is used to analyze and diagnose two open-source bearing data sets, and comparative experiments are conducted with the optimization results of other improved algorithms. Meanwhile, the method proposed in this paper is used to diagnose the fault of the bearing in the lifting device of a single-crystal furnace. The experimental results show that the fault frequency of the inner ring of the first bearing data set diagnosed using the proposed method was 158 Hz, and the fault frequency of the outer ring of the second bearing data set diagnosed using the proposed method was 162 Hz. The fault-diagnosis results of the two bearings were equal to the results derived from the theory. Compared with the optimization results of other improved algorithms, the proposed method has a faster convergence speed and a higher output signal-to-noise ratio. At the same time, the fault frequency of the bearing of the lifting device of the single-crystal furnace was effectively diagnosed as 35 Hz, and the bearing fault signal was effectively detected.

Keywords: multistable stochastic resonance; adaptive parameter; improved grey wolf algorithm; bearing fault detection

Citation: Huang, W.; Zhang, G. Bearing Fault-Detection Method Based on Improved Grey Wolf Algorithm to Optimize Parameters of Multistable Stochastic Resonance. *Sensors* **2023**, *23*, 6529. https://doi.org/10.3390/s23146529

Academic Editor: Luca De Marchi

Received: 21 June 2023
Revised: 14 July 2023
Accepted: 17 July 2023
Published: 19 July 2023

1. Introduction

The failure rate of rolling bearings accounts for about 30% of all rotating machinery failures, which is the main reason affecting the operating efficiency, productivity, and life of mechanical equipment. Almost all rolling bearing fault signals are in a very noisy environment, resulting in early weak faults that are difficult to find. Therefore, how to enhance the signal-to-noise ratio of fault signals under extreme conditions has become a key issue in the direction of fault diagnosis. At the same time, monitoring the status of rolling bearings, promptly identifying faults, and conducting equipment maintenance are of great practical significance for ensuring the smooth working of rotating machinery systems [1]. Nowadays, the main methods used for rolling bearing fault detection are: wavelet decomposition [2], empirical mode decomposition [3], variational mode decomposition [4], principal component analysis [5], stochastic resonance [6], etc. The stochastic resonance algorithm overturns the view that noise is harmful for a long time. It uses the resonance principle to transfer noise energy to the fault signal, thus improving the detection and diagnosis of the fault signal, and opening up a new method and idea for weak bearing fault-signal detection submerged in strong noise.

Benzi raised the concept of stochastic resonance (SR) in 1981 when studying the changes of the Earth's ice ages [7]. After 40 years of development, SR theory has been widely used in fault diagnosis [8], optics [9], medicine [10], image denoising [11], and other fields, and has achieved many remarkable results. The SR algorithm makes use of the synergy generated by the joint excitation of nonlinear systems, input signals, and noise to make Brownian particles oscillate, improve the output signal-to-noise ratio, and effectively detect the measured signal, which is a typical method to enhance the measured signal. Therefore, it is widely concerned with the domain of signal detection [12]. Classical bistable and monostable SR models have been extensively used in the study of signal detection [13]. However, for the signal to be measured with ultra-low amplitude, due to the potential function structure constraints, particles are often unable to effectively jump between potential wells, and SR-detection methods for bistable and monostable models are also powerless. When studying multistable stochastic resonance systems, Li et al. found that the multistable model can better enhance the output signal-to-noise ratio and improve the noise utilization ratio than the bistable and monostable models [14]. Therefore, more and more scholars have carried out relevant research on multistable SR [15]. For example, Zhang et al. proposed a piecewise unsaturated multistable SR (PUMSR) method which overcomes the weakness of tri-stable SR output saturation and enhances the ability of weak signal detection [16].

However, whether it is a monostable, bistable, or multistable SR algorithm, it is inevitably difficult to select model parameters in practical applications. Mitaim et al. [17] put forward the adaptive SR theory to enhance useful signals by automatically adjusting the structural parameters of nonlinear systems. But, the adaptive SR method, which takes a single parameter of the system as the optimization object, often ignores the interaction between the parameters of the system. With the rise of the swarm intelligence optimization algorithm, finding the global optimal solution through the swarm intelligence algorithm can solve the limitations of traditional adaptive SR systems, and this concept has been extensively used in the domain of bearing fault detection [18]. However, in the existing research results, the adaptive selection of SR model parameters still depends on the performance of intelligent optimization algorithms, so there are generally issues such as a low solving accuracy and being prone to falling into local optima [19]. Therefore, the feasible method to effectively enhance the parameter performance of adaptive selection of SR systems is to improve the defects of the intelligent optimization algorithm, so that it can more quickly and accurately optimize the parameters of the SR system. The grey wolf optimization algorithm can find the optimal solution by simulating the tracking, encircling, pursuit, and attack stages of the group predation behavior of the grey wolf. With few parameters and a simple structure, it is easy to integrate with other algorithms for improvement, but there are also the problems that it is easy to fall into local optimal solutions and low computational efficiency [20]. Therefore, it is of great research value to improve the basic grey wolf algorithm and improve its optimization performance [21]. Vasudha et al. proposed a multi-layer grey wolf optimization algorithm to further achieve an appropriate equivalence between exploration and development, thereby improving the efficiency of the algorithm [22]. Rajput et al. proposed an FH model based on the sparsity grey wolf optimization algorithm, which helps to minimize the computational overhead and improve the computational accuracy of the algorithm [23].

This article takes bearing fault-signal detection as the research object. Aiming at the problem of difficult parameter selection of multistable SR systems, a bearing fault-detection method based on an improved grey wolf algorithm to optimize multistable SR parameters is raised. This method improves the basic grey wolf optimization algorithm. Firstly, considering the quality of the initial solution, a Sobol-sequence initialization population strategy is proposed to make the distribution of the initial grey wolf population more uniform. Secondly, a convergence-factor adjustment strategy based on exponential rules is proposed to coordinate the global exploration and local development stages of the algorithm. Meanwhile, an adaptive position-update strategy is proposed to improve the accuracy of the algorithm, and Cauchy–Gaussian mixture mutation is used to enhance the algorithm's ability to escape from local optima. Experimental verification is conducted on the performance of the improved grey wolf algorithm using fifteen benchmark test functions from the CEC23 group of commonly used test functions. The verification results display that the multi-strategy improved grey wolf optimization algorithm (MSGWO) has a faster convergence speed and a higher convergence accuracy. Then, on the basis of the model of the multistable SR system, the parameters of the multistable SR system are optimized through the MSGWO, so as to enhance the fault signal and realize the effective detection of the bearing fault signal. Finally, the bearing fault-detection method raised in this article is used to analyze and diagnose a bearing data set from Case Western Reserve University (CWRU) and a bearing data set from the Mechanical Fault Prevention Technology Association (MFPT), and is compared with the optimization results of other improved algorithms. Meanwhile, the method raised in this article is used to diagnose the fault of the bearing of the lifting device of a single-crystal furnace. The test results display that the bearing fault-detection method raised in this article has a fast convergence speed and a large output signal-to-noise ratio, and can detect bearing fault signals accurately and efficiently.

The rest of this article is arranged as below: The Section 2 introduces the specific cases of bearing failure in rotating machinery in different industries. The Section 3 introduces the basic principle of multistable SR. The Section 4 introduces the principle of the basic grey wolf optimization algorithm and the MSGWO, and compares it with some basic optimization algorithms and improved optimization algorithms, respectively. At the same time, the population diversity and the exploration and development stage of the MSGWO are analyzed. The Section 5 introduces the bearing fault-diagnosis method based on the MSGWO to optimize the multistable SR parameters, and uses the proposed method to analyze and diagnose the bearing data sets from CWRU and the MFPT. Meanwhile, the raised method is used to diagnose the bearing fault of the monocrystal furnace lifting device. The Section 6 is the summary.

2. Specific Cases of Bearing Failure

Due to the diverse working environments of bearings during the operation of rotating machinery, they are easily affected by wear, corrosion, and other factors, making it easy for various faults to occur. For example, in June 1992, during the overspeed test of a 600 MW supercritical active generator set at the Kansai Electric Power Company Hainan Power Plant in Japan, the bearing failure of the unit and the critical speed drop caused strong vibration of the unit, resulting in a crash accident and economic losses of up to JPY 5 billion. From September 2003 to October 2004, the China Railway Beijing–Shanghai Line, Shitai Line, and Hang-gan Line had a total of five traffic incidents. According to relevant statistics, four of these accidents were caused by train bearing-fatigue fracture, with a total economic loss of up to CNY 2 billion. In April 2015, China Dalian West Pacific Petrochemical Co., LTD., due to the serious distortion and fracture of the inner ring of the driving end bearing and the serious wear and deformation of the bearing ball, the seal of the bottom pump of the stripping tower of a hydrocracking unit quickly failed, and the medium leaked, which caused a fire. The accident caused three pumps, the frame above the pump, and a small number of meters and power cables to set fire; a local pipeline to crack; and direct economic losses of CNY 166,000. In 2018, the US Navy's "Ford" aircraft carrier had to return to the shipyard for maintenance due to a thrust bearing failure during a mission. In August 2019, when a drone was spraying pesticides at a farm in Hebei, China, its motor rolling bearing failed, causing the drone to lose control, and a large amount of pesticides were spilled into the river, causing serious pollution. In December 2021, there were two recessive cracks in the bearing of unit #33 of a wind farm in Liaoning, China. Due to the limited installation position, the appearance inspection could not find them. As a result, the shaft cracks were promoted by the wind wheel's alternating load during operation, resulting in a spindle fracture and the impeller's overall fall. Therefore, the research on fault-diagnosis technology of rolling bearings is very necessary and has great practical significance.

3. Basic Principles of Multistable SR

3.1. The Basic Theory of Multistable SR

The principle of SR is that weak characteristic signals can be enhanced and detected by noise transfer mechanism under the action of nonlinear system. In general, when interpreting the SR model, we should first consider Langevin's dynamic equation [24], which is as follows:

$$\frac{d^2x}{dt^2} + \frac{dx}{dt} = -U'(x) + s(t) + n(t) \tag{1}$$

where x is the system response of SR, $U(x)$ is a class of nonlinear multistable potential function, $s(t)$ is the external incentive, $n(t)$ is the noise excitation, m is the mass of the particle, and k is the drag coefficient.

The definition formula of the nonlinear multistable potential function is:

$$U(x) = \frac{a}{2}x^2 - \frac{1+a}{4b}x^4 + \frac{c}{6}x^6 \tag{2}$$

In the formula, a, b, and c are parameters of the nonlinear multistable model, and they are all greater than 0. The potential function model image of the multistable system is displayed in Figure 1.

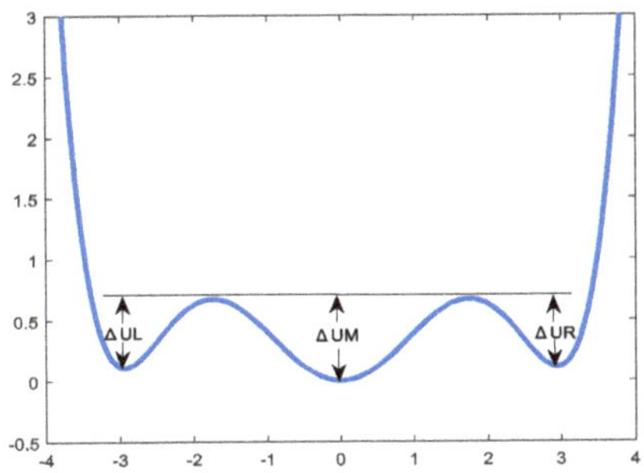

Figure 1. Potential function curve of multistable system.

Substitute the potential function of the multistable model into Formula (1), add noise with intensity D in the system, and then obtain the Langevin equation of the nonlinear multistable system as follows:

$$\frac{dx}{dt} = -ax + \frac{1+a}{b}x^3 - cx^5 + s(t) + \sqrt{2D}n(t) \tag{3}$$

When periodic signal and noise signal are used as excitation simultaneously, the inclination of potential well in the multistable system will increase. In addition, the periodic signal will also make the potential well depth of the three potential wells of the multistable potential function change periodically, and can guide the noise signal to switch synchronously. When the signal, noise, and multistable SR system reach a certain matching relationship, particles can make periodic transitions between potential wells, so that the components of the system output with the same frequency as the input signal are strengthened.

3.2. System Parameters' Range

The fourth order Runge–Kutta formula was used to solve the multistable SR model. The specific calculation formula is:

$$\begin{cases} k_1 = h(-ax(n) + \frac{1+a}{b}x^3(n) - cx^5(n) + s(n)) \\ k_2 = h(-a(x(n) + \frac{k_1}{2}) + \frac{1+a}{b}(x(n) + \frac{k_1}{2})^3 - c(x(n) + \frac{k_1}{2})^5 + s(n)) \\ k_3 = h(-a(x(n) + \frac{k_2}{2}) + \frac{1+a}{b}(x(n) + \frac{k_2}{2})^3 - c(x(n) + \frac{k_2}{2})^5 + s(n)) \\ k_4 = h(-a(x(n) + k_3) + \frac{1+a}{b}(x(n) + k_3)^3 - c(x(n) + k_3)^5 + s(n)) \\ x(n+1) = x(n) + \frac{1}{6}(k_1 + 2k_2 + 2k_3 + k_4) \end{cases} \quad (4)$$

where $x(n)$ is the nth sampling value of the system output, $s(n)$ is the nth sampling value of the noise-added input signal, h is the sampling step, and $k_i(i = 1, 2, 3, 4)$ is the slope of the output response at the relevant integration point.

Normally, due to noise, particles jump over higher barrier heights by accumulating energy, so b, c, and h take the real numbers of [0, 10]. As the target signal is relatively weak, the interval in [25] is quoted; the range of a is set to [0, 0.5].

4. Multi-Strategy Improved Grey Wolf Optimization Algorithm

4.1. The Primary Theory of Grey Wolf Optimization Algorithm

Grey Wolf Optimizer (GWO) is a new intelligent swarm optimization algorithm proposed by Mirjalili et al. [26], whose main ideas are the leadership hierarchy and group hunting mode of grey wolf groups. The grey wolf population has a strict hierarchy. The head of the population is α, which represents the most coordinated individual in the wolf pack, and is mainly responsible for the decision-making affairs of the group's predation behavior. The β wolf is second only to α in the population, and its role is to serve the α wolf to make decisions and deal with the behavior of the population. The third rank in the population is the δ wolf, which obeys the instructions issued by the α and β, but has command over other bottom individuals. The lowest individual in the group, known as ω, is submissive to the instructions of other higher-ranking wolves and is primarily responsible for balancing the relationships within the group. GWO defines the three solutions with the best fitness as α, β, and δ, while the remaining solutions are defined as ω. The hunting process (optimization process) is guided by α, β, and δ to track and hunt the prey (position update), and finally complete the hunting process, that is, obtain the optimal solution. Grey wolf groups gradually approach and surround their prey through several formulas:

$$D = |C \cdot X_p(t) - X(t)| \quad (5)$$

$$X(t+1) = X_p(t) - A \cdot D \quad (6)$$

where t represents the number of iterations, $X(t)$ and $X_p(t)$ represent the position vector between the wolf and its prey, A and C represent the cooperation coefficient vector, and D is the distance between the individual wolf pack and the target. The formula for calculating coefficient vectors A and C is:

$$A = 2f \cdot r_1 - f \quad (7)$$

$$C = 2 \cdot r_2 \quad (8)$$

where, as the number of iterations increases, f decays linearly from 2 to 0. To enable some agents to reach an optimal position, r_1 and r_2 take values between [0, 1].

When hunting, GWO thinks that α, β, and δ are better at predicting the location of prey. Therefore, individual grey wolves will judge the distance D_α, D_β, and D_δ between themselves and α, β, and δ; calculate their moving distances X_1, X_2, and X_3 toward the three, respectively; and finally move within the circle of the three. The moving formula is shown in Equation (9).

$$\begin{cases} D_\alpha = |C_1 \cdot X_\alpha - X(t)| \\ D_\beta = |C_1 \cdot X_\beta - X(t)| \\ D_\delta = |C_1 \cdot X_\delta - X(t)| \end{cases} \tag{9}$$

$$\begin{cases} X_1 = X_\alpha - A \cdot D_\alpha \\ X_2 = X_\beta - A \cdot D_\beta \\ X_3 = X_\delta - A \cdot D_\delta \end{cases} \tag{10}$$

$$X(t+1) = (X_1 + X_2 + X_3)/3 \tag{11}$$

4.2. *Multi-Strategy Improved Grey Wolf Optimization Algorithm*

4.2.1. Sobol-Sequence Initialization Population Strategy

In the swarm intelligence algorithm, whether the initial population distribution is uniform will have a great impact on the optimization performance of the algorithm. GWO initializes the population randomly, resulting in the distribution of the initial population being extremely scattered, which will have a great impact on the algorithm's solving speed and optimization accuracy. Therefore, this paper initializes population individuals through the Sobol sequence. The Sobol sequence is a kind of low difference sequence [27], which is based on the smallest prime number, two. To produce a random sequence $X \in [0, 1]$, an irreducible polynomial of the highest order k in base two is first required to produce a set of predetermined directional numbers $V = [V_1, V_2, \cdots, V_k]$, and then the index value of the binary sequence $i = (\cdots i_3 i_2 i_1)_2$ is required; then, the nth random number generated by the Sobol sequence is:

$$X_i = i_1 V_1 \oplus i_2 V_2 \oplus \cdots \quad i = (\cdots i_3 i_2 i_1)_2 \tag{12}$$

The distribution of individuals with the same population size in the same dimensional space is shown in Figure 2. From Figure 2, it can be seen that the distribution of the population initialized using the Sobol sequence is more uniform than that generated randomly, which enables the population to traverse the entire search space better.

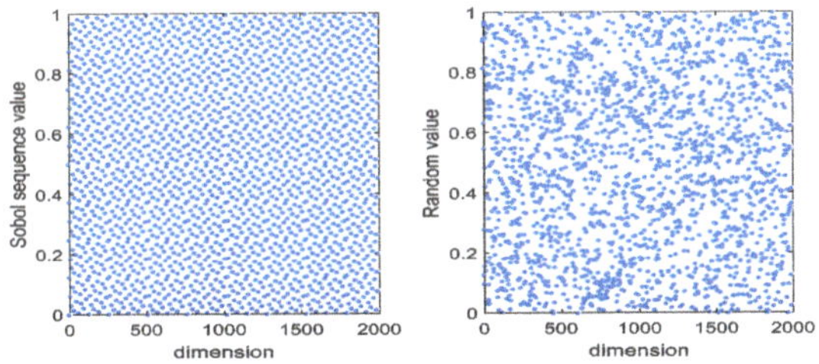

Figure 2. The Sobol sequence and random method to generate individual distribution maps.

4.2.2. Exponential Rule Convergence-Factor Adjustment Strategy

The parameter A is an important parameter regulating global exploration and local development in GWO, which is mainly affected by convergence factor f. In GWO, when $|A| > 1$, the grey wolf population searches the entire search domain for potential prey, and when $|A| \leq 1$, the grey wolf population will gradually surround and capture prey.

In GWO, the value of convergence factor f decreases linearly from 2 to 0 with the increase in the number of iterations, which cannot accurately reflect the complex random search process in the actual optimization process. In addition, in the process of algorithm iteration, the same method was used to calculate the enveloping step length for grey wolf individuals with different fitness, which did not reflect the differences among individual grey wolves. Therefore, this paper introduces an updated mode of convergence factor based on exponential rule changes, whose equation is as follows:

$$f' = 2e^{-t/T} \tag{13}$$

The curves of the linear convergence factor and exponential regular convergence factor proposed in this paper with the number of iterations are shown in Figure 3. As can be seen from Figure 3, the convergence factor f in GWO decreases linearly with the increase in iterations, resulting in incomplete prey searches in the early stage and slow convergence in the later hunting process. The convergence factor f', which varies exponentially, can thoroughly search for prey in the early stages of the algorithm, thereby enhancing its global optimization performance.

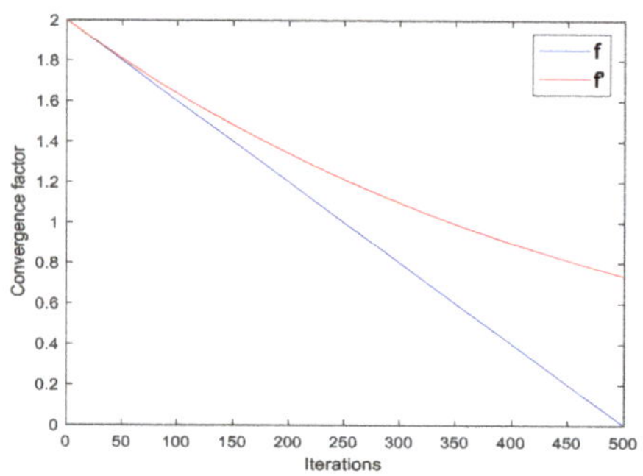

Figure 3. The convergence factor comparison curve.

4.2.3. Adaptive Location-Update Strategy

In GWO, the initializing α, β, and δ solutions are recorded and retained until they are replaced by a better-fitting individual in the iterative process. In other words, if there is no better α, β, and δ solution in the population than that recorded in the t generation, the new population will still update its position toward wolves α, β, and δ. But when these three are in the local optimal area, then the whole population cannot obtain the optimal solution. Moreover, the average value of X_1, X_2, and X_3 in GWO cannot show the importance of α, β, and δ. Therefore, a new adaptive location-update strategy is proposed, which is expressed as follows:

$$\begin{cases} W_1 = \frac{|X_1|}{|X_1|+|X_2|+|X_3|+\varepsilon} \\ W_2 = \frac{|X_2|}{|X_1|+|X_2|+|X_3|+\varepsilon} \\ W_3 = \frac{|X_3|}{|X_1|+|X_2|+|X_3|+\varepsilon} \end{cases} \tag{14}$$

$$g = \frac{T-t}{T}(g_{initial} - g_{final}) + g_{final} \tag{15}$$

where g is the inertia weight. The mathematical expression of grey wolf position update is shown in Equation (16).

$$X(t+1) = \frac{W_1X_1 + W_2X_2 + W_3X_3}{3}g + X_1\frac{t}{T} \tag{16}$$

4.2.4. Cauchy–Gaussian Hybrid Mutation Strategy

In order to avoid the local optimization of the basic GWO algorithm, this paper introduces the Cauchy–Gaussian hybrid mutation strategy combining Cauchy and Gaussian distribution, and gives the best individuals the Cauchy–Gaussian perturbation. The

Cauchy–Gaussian operator can generate a large step length to avoid the algorithm falling into local optimality, and its expression is as follows:

$$X^*_{new}(t) = X^*(t) \cdot (1 + \lambda_1 cauchy(0,1) + \lambda_2 Gauss(0,1)) \quad (17)$$

$$\lambda_1 = 1 - \frac{t^2}{T^2_{\max}} \quad (18)$$

$$\lambda_2 = \frac{t^2}{T^2_{\max}} \quad (19)$$

where $X^*_{new}(t)$ is the value obtained using Cauchy–Gaussian perturbation, $cauchy(0,1)$ is the Cauchy operator, and $Gauss(0,1)$ is the Gaussian operator.

The pseudocode of MSGWO is shown in Figure 4.

```
Sobol sequence is used to generate the initial grey wolf population Xi (i = 1, 2, ..., n)
Initialize f, A, C, count = 0, t = 0, limit=10
while (t < Max number of iterations)
        Calculate the fitness of each agent
        Xα = the best search agent
        Xβ = the second search agent
        Xδ = the third search agent
        if  (count > limit)
             Individuals underwent Cauchy Gaussian mixed mutation by equation (17)
             Repeat the operation in the else
        else
             Update the exponential rule convergence factor f by equation (13)
        end if
        for each search agent
             Update A by equation (7)
             Update the position of the current search agent by equation (16)
        end for
        Last best fitness = best fitness
        if  best fitness == Last best fitness
             count = count + 1
        else
             count = 0
        end if
        t = t + 1
end while
return Xα
```

Figure 4. The pseudocode of MSGWO.

4.3. Improved Performance Test of Grey Wolf Optimization Algorithm

CEC23 sets of commonly used test functions are important examples of testing algorithm performance [28]. In an effort to test the performance of the MSGWO raised in this article, fifteen test functions in the CEC23 group of commonly used test functions were selected for verification, in which F_1 to F_7 were single-peak benchmark functions, F_8 to F_{13} were multi-peak benchmark functions, and F_{14} to F_{15} were fixed-dimensional multi-peak test functions. The computing platform performance was based on Intell CITM) i5-6500 CPU, 3.20 GHz main frequency, and 8 GB memory. The details of the test function are shown in Table 1.

Table 1. Benchmark functions.

Function	Dim	Range	Optima
$F_1(x) = \sum_{i=1}^{n} x_i^2$	30	$[-100, 100]$	0
$F_2(x) = \sum_{i=1}^{n} \lvert x_i \rvert + \prod_{i=1}^{n} \lvert x_i \rvert$	30	$[-10, 10]$	0
$F_3(x) = \sum_{i=1}^{n} \left(\sum_{j-1}^{i} x_j \right)^2$	30	$[-100, 100]$	0
$F_4(x) = \max_i \{ \lvert x_i \rvert, 1 \le i \le n \}$	30	$[-100, 100]$	0
$F_5(x) = \sum_{i=1}^{n-1} \left[100(x_{i+1} - x_i^2)^2 + (x_i - 1)^2 \right]$	30	$[-30, 30]$	0
$F_6(x) = \sum_{i=1}^{n} ([x_i + 0.5])^2$	30	$[-100, 100]$	0
$F_7(x) = \sum_{i=1}^{n} i x_i^4 + random[0, 1)$	30	$[-1.28, 1.28]$	0
$F_8(x) = \sum_{i=1}^{n} -x_i \sin(\sqrt{\lvert x_i \rvert})$	30	$[-500, 500]$	$-418.98 \times \mathrm{Dim}^n$
$F_9(x) = \sum_{i=1}^{n} [x_i^2 - 10\cos(2\pi x_i) + 10]$	30	$[-5.12, 5.12]$	0
$F_{10}(x) = -20\exp\left(-0.2\sqrt{\frac{1}{n}\sum_{i=1}^{n} x_i^2}\right) - \exp\left(\frac{1}{n}\sum_{i=1}^{n}\cos(2\pi x_i)\right) + 20 + e$	30	$[-32, 32]$	0
$F_{11}(x) = \frac{1}{4000}\sum_{i=1}^{n} x_i^2 - \prod_{i=1}^{n}\cos\left(\frac{x_i}{\sqrt{i}}\right) + 1$	30	$[-600, 600]$	0
$F_{12}(x) = \frac{\pi}{n}\left\{10\sin(\pi y_1) + \sum_{i=1}^{n-1}(y_i - 1)^2[1 + 10\sin^2(\pi y_{i+1})] + (y_n - 1)^2\right\}$ $+ \sum_{i=1}^{n} u(x_i, 10, 100, 4)$ $y_i = 1 + \frac{x_i + 1}{4}$ $u(x_i, a, k, m) = \begin{cases} k(x_i - a)^m & x_i > a \\ 0 & -a < x_i < a \\ k(-x_i - a)^m & x_i < -a \end{cases}$	30	$[-50, 50]$	0
$F_{13}(x) = 0.1\left\{\sin^2(3\pi x_1) + \sum_{i=1}^{n}(x_i - 1)^2[1 + \sin^2(3\pi x_i + 1)] + (x_n - 1)^2[1 + \sin^2(2\pi x_n)]\right\}$ $+ \sum_{i=1}^{n} u(x_i, 5, 100, 4)$	30	$[-50, 50]$	0
$F_{14}(x) = \left(\frac{1}{500} + \sum_{j=1}^{25}\frac{1}{j + \sum_{i=1}^{2}(x_i - a_{ij})^6}\right)^{-1}$	2	$[-65, 65]$	1
$F_{15}(x) = \sum_{i=1}^{11}\left[a_i - \frac{x_1(b_i^2 + b_1 x_2)}{b_i^2 + b_1 x_3 + x_4}\right]^2$	4	$[-5, 5]$	0.1484

4.3.1. Comparison Experiment between MSGWO and Standard Optimization Algorithm

In an effort to objectively verify the performance of MSGWO, the population size was set to 30 times, the maximum number of iterations was set to 500 times, and each algorithm was run independently 30 times. Algorithms to be compared in the experiment included the bat optimization algorithm (BOA) [29], whale optimization algorithm (WOA) [30], grey wolf optimization algorithm (GWO), gravity search algorithm (GSA) [31], particle swarm optimization algorithm (PSO) [32], and artificial bee colony algorithm (ABC) [33]. The parameters of all the comparison algorithms in the experiment were the same as those recommended in the original literature. The mean value and standard deviation of the optimal value of the simulation results were taken as the evaluation indexes of the algorithm performance, and the results are shown in Table 2. The test results shown in bold black in Table 2 are the best for comparison.

It can be seen from the data in Table 2 that MSGWO obtained the optimal mean and variance in functions F_1–F_4, F_7, F_9–F_{13}, and F_{15}. In the function F_5, MSGWO obtained the best average value, but its stability was worse than BOA. In the function F_6, MSGWO obtained the best average value, but its stability was worse than WOA and GWO. In the function F_8, MSGWO achieved the best average, but its stability was the worst. In the function F_{14}, MSGWO obtained the best average value, but its stability was worse than that of the ABC algorithm. It can be seen that MSGWO obtained the optimal average value in all the selected test functions. Although the stability of the algorithm was worse in some individual functions than that of some comparison algorithms, MSGWO still had better optimization performance on the whole.

Table 2. The compared results of MSGWO and standard optimization algorithms.

F	Index	WOA	GWO	BOA	GSA	PSO	ABC	MSGWO
F_1	mean	4.97×10^{-74}	1.04×10^{-27}	4.06×10^{-4}	98.91	11.65	3.54	**0**
	std	2.49×10^{-73}	1.37×10^{-27}	8.97×10^{-5}	106.42	5.28	1.26	**0**
F_2	mean	2.46×10^{-52}	9.51×10^{-17}	4.54×10^{-9}	4.46	11.69	0.16	**0**
	std	5.61×10^{-52}	7.47×10^{-17}	1.26×10^{-9}	4.47	3.64	0.05	**0**
F_3	mean	3.87×10^{4}	3.15×10^{-5}	1.25×10^{-11}	1.31×10^{3}	7.25×10^{2}	3.37×10^{4}	**0**
	std	1.48×10^{4}	9.68×10^{-5}	8.97×10^{-13}	4.14×10^{2}	5.27×10^{2}	5.45×10^{3}	**0**
F_4	mean	59.16	7.78×10^{-7}	6.15×10^{-9}	10.07	6.73	51.08	**0**
	std	23.48	8.85×10^{-7}	4.28×10^{-10}	1.71	1.26	5.48	**0**
F_5	mean	27.90	28.44	28.94	3.26×10^{2}	1.87×10^{3}	1.40×10^{5}	**27.08**
	std	0.48	0.82	**0.03**	2.51×10^{2}	1.15×10^{3}	6.78×10^{4}	0.42
F_6	mean	0.42	0.90	5.75	52.25	9.89	3.96	**0.35**
	std	0.48	**0.38**	0.72	60.45	3.57	0.98	0.54
F_7	mean	2.54×10^{3}	2.07×10^{3}	1.39×10^{3}	1.36	0.68	0.25	$\mathbf{6.58 \times 10^{-5}}$
	std	2.30×10^{-3}	7.10×10^{-4}	7.65×10^{-4}	2.63	0.33	0.08	$\mathbf{6.62 \times 10^{-5}}$
F_8	mean	-1.04×10^{4}	-5.70×10^{3}	-3.77×10^{4}	-2.48×10^{3}	-2.22×10^{3}	-4.98×10^{3}	$\mathbf{-5.47 \times 10^{58}}$
	std	1.73×10^{3}	1.18×10^{3}	3.80×10^{2}	5.29×10^{2}	5.89×10^{2}	$\mathbf{3.55 \times 10^{2}}$	1.81×10^{59}
F_9	mean	0.15	3.63	6.72	38.94	92.15	2.33×10^{2}	**0**
	std	0.83	4.07	36.10	10.12	16.83	15.05	**0**
F_{10}	mean	5.51×10^{-15}	1.03×10^{-13}	5.81×10^{-9}	0.55	5.43	1.89	$\mathbf{8.88 \times 10^{-16}}$
	std	2.77×10^{-15}	2.23×10^{-14}	7.12×10^{-10}	0.61	1.18	0.57	**0**
F_{11}	mean	0.03	3.02×10^{-3}	5.22×10^{-12}	1.01×10^{3}	0.45	1.02	**0**
	std	0.09	5.70×10^{-3}	2.40×10^{-12}	11.85	0.12	0.03	**0**
F_{12}	mean	0.05	0.07	0.66	3.12	4.40	17.54	**0.05**
	std	0.13	0.27	0.16	1.10	1.98	8.64	**0.10**
F_{13}	mean	0.51	0.71	2.91	27.43	22.29	1.49×10^{4}	**0.43**
	std	0.29	0.24	0.18	10.75	16.15	2.36×10^{4}	**0.14**
F_{14}	mean	2.90	4.53	1.68	6.66	2.05	1.69	**1.55**
	std	3.20	4.03	0.94	4.61	1.63	**0**	0.70
F_{15}	mean	6.13×10^{-4}	2.47×10^{-3}	4.39×10^{-4}	1.17×10^{-2}	6.15×10^{-4}	7.04×10^{-4}	$\mathbf{3.46 \times 10^{-4}}$
	std	3.04×10^{-4}	6.00×10^{-3}	1.73×10^{-4}	6.30×10^{-3}	4.65×10^{-4}	5.80×10^{-4}	$\mathbf{1.69 \times 10^{-4}}$

The simulation results show that MSGWO had better optimization performance under different benchmark test functions. This shows that compared with GWO, MSGWO enhances the local search ability, thus increasing the solution accuracy, and for multi-modal test functions, MSGWO has a strong local optimal avoidance ability, and can better find the optimal solution. When other algorithms have low optimization accuracy or even cannot converge, MSGWO still has high solving accuracy.

In order to explore the influence of improvement strategies on the algorithm convergence speed, the convergence curves of each algorithm under 15 benchmark test functions are shown in Figure 5. As can be seen from Figure 5, MSGWO has high precision and the fastest convergence rate of the optimal solution in the comparison algorithm, which effectively saves the optimization time.

4.3.2. Comparison Experiment between MSGWO and Improved Optimization Algorithm

In an effort to further test the performance of the MSGWO, the population size was set to 30 times, the maximum number of iterations was set to 500 times, and each algorithm was independently run 30 times. Comparative experimental analysis was conducted between MSGWO and GWO, MEGWO [34], mGWO [35], IGWO [36], and MPSO [37]. The mean value and standard deviation of the optimal value of the simulation results were taken as the evaluation indexes of the algorithm performance, and the results are shown in Table 3. The test results shown in bold black in Table 3 are the best for comparison.

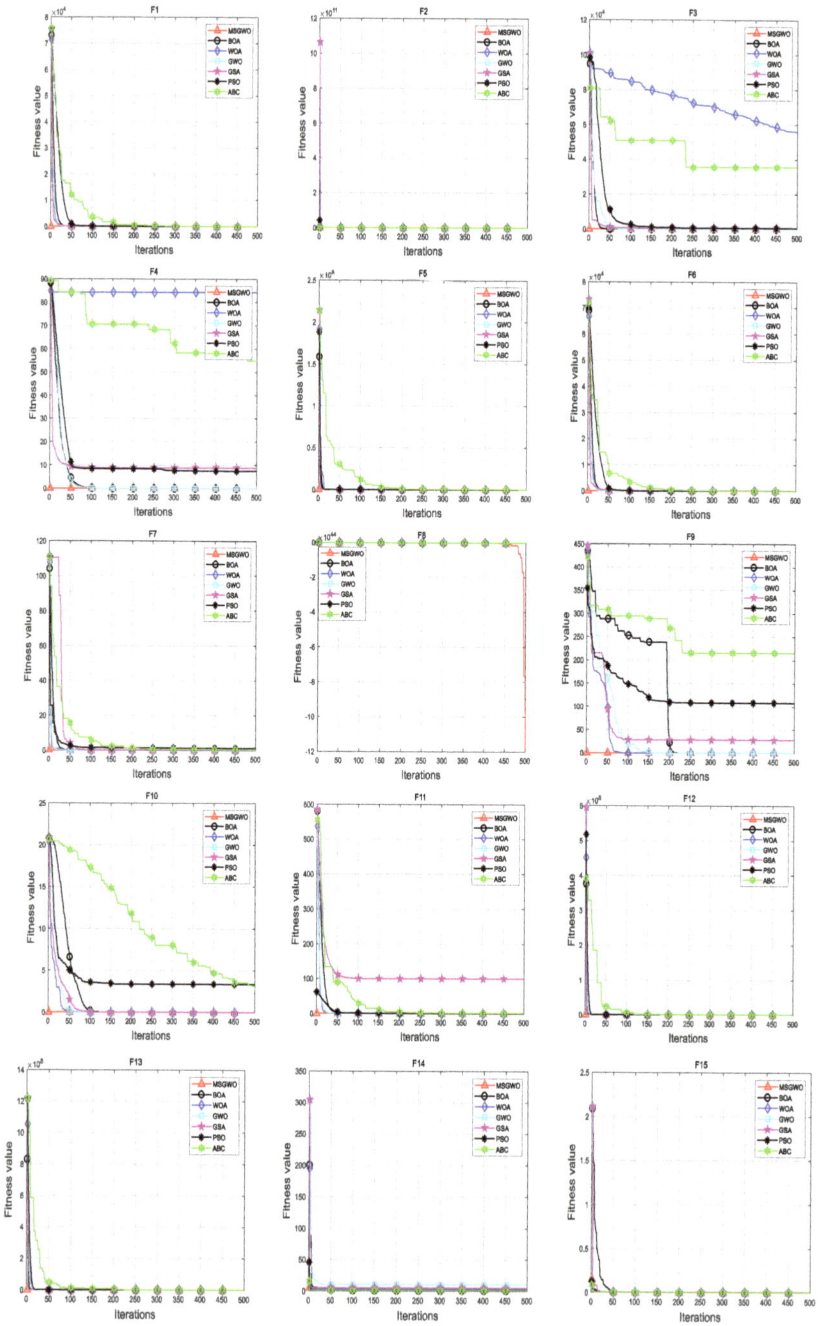

Figure 5. The convergence curve of MSGWO is compared with that of standard algorithm.

Table 3. Comparison of experimental results between MSGWO and improved algorithms.

F	Index	GWO	MEGWO	mGWO	IGWO	MPSO	MSGWO
F_1	mean	1.04×10^{-27}	4.30×10^{-64}	1.04×10^{-18}	1.33×10^{-209}	2.61×10^{-26}	**0**
	std	1.37×10^{-27}	2.09×10^{-63}	2.97×10^{-18}	0	1.12×10^{-25}	**0**
F_2	mean	9.50×10^{-17}	1.70×10^{-43}	2.65×10^{-12}	6.12×10^{-21}	1.40×10^{-16}	**0**
	std	6.40×10^{-17}	5.77×10^{-43}	1.99×10^{-12}	6.67×10^{-21}	2.86×10^{-16}	**0**
F_3	mean	3.15×10^{-5}	0.23	0.68	2.73×10^{-5}	9.63×10^{2}	**0**
	std	9.68×10^{-5}	0.48	0.81	9.57×10^{-5}	4.81×10^{2}	**0**
F_4	mean	7.78×10^{-7}	2.06×10^{-5}	0.68	2.93×10^{-7}	2.05×10^{-10}	**0**
	std	8.85×10^{-7}	5.68×10^{-5}	0.85	1.78×10^{-7}	4.81×10^{-10}	**0**
F_5	mean	28.44	27.94	27.92	27.64	88.91	**27.08**
	std	0.82	9.97	0.58	**0.32**	1.89×10^{2}	0.42
F_6	mean	0.90	0.49	0.41	0.43	0.41	**0.36**
	std	0.38	1.14	0.25	**0.19**	0.22	0.54
F_7	mean	2.07×10^{-3}	1.01×10^{-3}	4.68×10^{-3}	2.80×10^{-3}	1.68×10^{-3}	$\mathbf{6.58 \times 10^{-5}}$
	std	7.10×10^{-4}	9.10×10^{-4}	1.90×10^{-3}	1.10×10^{-3}	8.87×10^{-4}	$\mathbf{6.62 \times 10^{-5}}$
F_8	mean	-5.70×10^{3}	-1.26×10^{4}	-5.33×10^{3}	-8.28×10^{3}	-8.12×10^{3}	$\mathbf{-5.47 \times 10^{58}}$
	std	1.18×10^{3}	$\mathbf{2.15 \times 10^{-12}}$	1.11×10^{3}	1.69×10^{3}	1.12×10^{3}	1.81×10^{59}
F_9	mean	3.63	0	37.94	27.09	23.92	**0**
	std	4.07	0	30.01	22.81	22.64	**0**
F_{10}	mean	1.03×10^{-13}	5.27×10^{-15}	1.26×10^{-10}	6.25×10^{-14}	6.22×10^{-15}	$\mathbf{8.88 \times 10^{-16}}$
	std	2.23×10^{-14}	1.50×10^{-15}	9.69×10^{-11}	8.96×10^{-15}	7.38×10^{-15}	**0**
F_{11}	mean	3.02×10^{-3}	0	3.83×10^{-3}	3.37×10^{-3}	0	**0**
	std	5.70×10^{-3}	0	9.40×10^{-3}	6.00×10^{-3}	0	**0**
F_{12}	mean	0.07	0.05	0.05	6.58×10^{-2}	0.42	**0.05**
	std	0.27	0.56	0.04	$\mathbf{2.00 \times 10^{-3}}$	0.73	0.10
F_{13}	mean	0.71	0.46	0.63	0.66	0.45	**0.43**
	std	0.24	0.15	0.22	0.16	0.25	**0.13**
F_{14}	mean	4.53	1.78	2.00	1.70	1.99	**1.55**
	std	4.03	2.91	2.76	0.76	**0.36**	0.71
F_{15}	mean	2.47×10^{-3}	3.07×10^{-4}	1.04×10^{-3}	8.62×10^{-4}	5.68×10^{-4}	$\mathbf{3.46 \times 10^{-4}}$
	std	6.00×10^{-3}	$\mathbf{3.42 \times 10^{-15}}$	3.60×10^{-3}	3.00×10^{-3}	3.36×10^{-4}	1.69×10^{-4}

It can be seen from the data in Table 3 that for the optimization accuracy of the algorithm, MSGWO obtained the optimal average value in the function F_1–F_{15}. In terms of algorithm stability, the stability of the MSGWO was worse than that of the IGWO algorithm in F_5; worse than those of the GWO, mGWO, IGWO, and MPSO algorithms in F_6; the worst in F_8; worse than those of the mGWO and IGWO algorithms in F_{12}; worse than that of the MPSO algorithm in F_{14}; and worse than that of MEGWO in F_{15}. However, in the other nine test functions, its stability was better than the comparison algorithm, so the overall stability was still the best.

The convergence curves of the MSGWO algorithm and improved algorithms under 15 benchmark functions are shown in Figure 6. It can be seen from the convergence curves of each test function in Figure 6 that MSGWO has better local extreme value escape ability, overall optimization coordination, and convergence performance than the comparison algorithm.

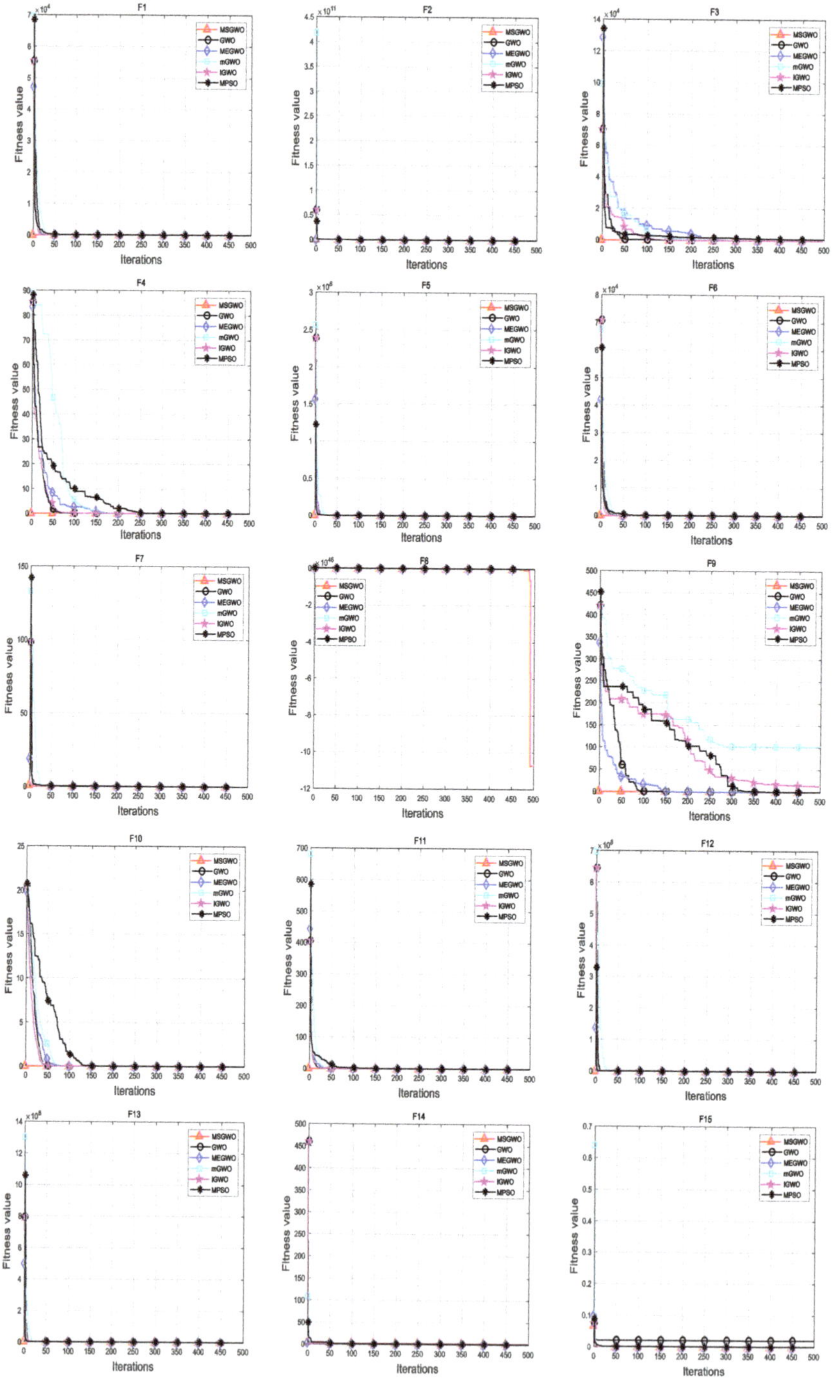

Figure 6. The convergence curves are compared between MSGWO and the improved algorithm.

4.3.3. Wilcoxon Rank Sum Test

In order to verify whether there were significant differences between MSGWO and other comparison algorithms, the Wilcoxon rank sum test was used for statistical analysis of the experimental data. For each test function, the results of 30 independent optimizations of MSGWO were compared with the 30 independent optimizations of the standard optimization algorithms (WOA, GWO, BOA, GSA, PSO, ABC) and improved optimization algorithms (MEGWO, mGWO, IGWO, MPSO) using the Wilcoxon rank sum test at a significance level of 5%. The population size of all algorithms was set to 30, with 500 iterations. The p value of the test result was less than 0.05, indicating that there were significant differences between the comparison algorithms. The symbols "+", "−", and "=" of R indicate that the performance of MSGWO was better than, worse than, and equivalent to the comparison algorithm, respectively, and N/A indicates that a significance judgment could not be made. The test results are shown in Tables 4 and 5, respectively.

Table 4. Wilcoxon rank sum test results for MSGWO and standard algorithms.

F	Index	MSGWO–WOA	MSGWO–GWO	MSGWO–BOA	MSGWO–GSA	MSGWO–PSO	MSGWO–ABC
F_1	P	1.73×10^{-6}	1.73×10^{-6}	1.73×10^{-6}	1.73×10^{-6}	1.73×10^{-6}	1.73×10^{-6}
	R	+	+	+	+	+	+
F_2	P	1.73×10^{-6}	1.73×10^{-6}	1.73×10^{-6}	1.73×10^{-6}	1.73×10^{-6}	1.73×10^{-6}
	R	+	+	+	+	+	+
F_3	P	1.73×10^{-6}	1.73×10^{-6}	1.73×10^{-6}	1.73×10^{-6}	1.73×10^{-6}	1.73×10^{-6}
	R	+	+	+	+	+	+
F_4	P	1.73×10^{-6}	1.73×10^{-6}	1.73×10^{-6}	1.73×10^{-6}	1.73×10^{-6}	1.73×10^{-6}
	R	+	+	+	+	+	+
F_5	P	1.06×10^{-4}	2.88×10^{-6}	1.73×10^{-6}	1.92×10^{-6}	1.73×10^{-6}	1.73×10^{-6}
	R	+	+	+	+	+	+
F_6	P	1.73×10^{-6}	1.73×10^{-6}	1.73×10^{-6}	7.69×10^{-6}	1.73×10^{-6}	9.37×10^{-3}
	R	+	+	+	+	+	+
F_7	P	2.13×10^{-6}	1.73×10^{-6}	1.73×10^{-6}	1.73×10^{-6}	1.73×10^{-6}	1.73×10^{-6}
	R	+	+	+	+	+	+
F_8	P	1.73×10^{-6}	1.73×10^{-6}	1.73×10^{-6}	1.73×10^{-6}	1.73×10^{-6}	1.73×10^{-6}
	R	+	+	+	+	+	+
F_9	P	1.73×10^{-6}	2.53×10^{-6}	1.82×10^{-5}	1.73×10^{-6}	1.73×10^{-6}	1.73×10^{-6}
	R	+	+	+	+	+	+
F_{10}	P	2.57×10^{-6}	1.61×10^{-6}	1.73×10^{-6}	1.73×10^{-6}	1.73×10^{-6}	1.73×10^{-6}
	R	+	+	+	+	+	+
F_{11}	P	2.57×10^{-6}	1.73×10^{-6}	1.73×10^{-6}	1.73×10^{-6}	1.73×10^{-6}	1.73×10^{-6}
	R	+	+	+	+	+	+
F_{12}	P	1.73×10^{-6}	1.73×10^{-6}	1.97×10^{-5}	1.73×10^{-6}	1.73×10^{-6}	1.73×10^{-6}
	R	+	+	+	+	+	+
F_{13}	P	1.73×10^{-6}	1.73×10^{-6}	0.0021	1.73×10^{-6}	1.92×10^{-6}	0.0047
	R	+	+	+	+	+	+
F_{14}	P	1.73×10^{-6}	1.73×10^{-6}	4.45×10^{-5}	4.86×10^{-5}	1.92×10^{-6}	0.0023
	R	+	+	+	+	+	+
F_{15}	P	1.73×10^{-6}	1.73×10^{-6}	1.73×10^{-6}	1.73×10^{-6}	1.73×10^{-6}	1.73×10^{-6}
	R	+	+	+	+	+	+

Table 5. Wilcoxon rank sum test results for MSGWO and improved algorithms.

F	Index	MSGWO–GWO	MSGWO–MEGWO	MSGWO–mGWO	MSGWO–IGWO	MSGWO–MPSO
F_1	P	1.73×10^{-6}	1.73×10^{-6}	1.73×10^{-6}	1.73×10^{-6}	1.73×10^{-6}
	R	+	+	+	+	+

Table 5. *Cont.*

F	Index	MSGWO–GWO	MSGWO–MEGWO	MSGWO–mGWO	MSGWO–IGWO	MSGWO–MPSO
F_2	P R	1.73×10^{-6} +	1.73×10^{-6} +	1.73×10^{-6} +	1.73×10^{-6} +	1.73×10^{-6} +
F_3	P R	1.73×10^{-6} +	1.73×10^{-6} +	1.73×10^{-6} +	1.73×10^{-6} +	1.73×10^{-6} +
F_4	P R	1.73×10^{-6} +	1.73×10^{-6} +	1.73×10^{-6} +	1.73×10^{-6} +	1.73×10^{-6} +
F_5	P R	2.88×10^{-6} +	1.73×10^{-6} +	1.73×10^{-6} +	1.73×10^{-6} +	1.92×10^{-3} +
F_6	P R	1.73×10^{-6} +	1.73×10^{-6} +	1.73×10^{-6} +	9.37×10^{-3} +	1.73×10^{-6} +
F_7	P R	1.73×10^{-6} +	1.73×10^{-6} +	1.73×10^{-6} +	1.73×10^{-6} +	1.73×10^{-6} +
F_8	P R	1.73×10^{-6} +	1.73×10^{-6} +	1.73×10^{-6} +	1.73×10^{-6} +	1.73×10^{-6} +
F_9	P R	2.53×10^{-6} +	0.012 =	1.73×10^{-6} +	1.73×10^{-6} +	1.73×10^{-6} +
F_{10}	P R	1.61×10^{-6} +	3.99×10^{-7} +	1.73×10^{-6} +	1.47×10^{-6} +	1.01×10^{-7} +
F_{11}	P R	1.73×10^{-6} +	0.012 =	1.22×10^{-4} +	7.8×10^{-3} +	0.012 =
F_{12}	P R	1.73×10^{-6} +	0.012 =	1.22×10^{-4} =	1.73×10^{-6} +	2.9×10^{-3} +
F_{13}	P R	1.73×10^{-6} +	1.73×10^{-6} +	1.73×10^{-6} +	1.73×10^{-6} +	1.73×10^{-6} +
F_{14}	P R	1.73×10^{-6} +	3.59×10^{-4} +	1.73×10^{-6} +	1.73×10^{-6} +	1.73×10^{-6} +
F_{15}	P R	1.73×10^{-6} +	1.73×10^{-6} +	1.7×10^{-3} +	3.11×10^{-5} +	1.73×10^{-6} +

As can be seen from Table 4, comparing the optimization results of MSGWO with those of WOA, GWO, BOA, GSA, PSO, and ABC on 15 test functions, the p values of the test results are all less than 0.05, and the R values are all +, indicating that the optimization results of MSGWO are significantly different from those of other six algorithms. Additionally, MSGWO is significantly better, which shows the superiority of the MSGWO algorithm statistically.

As can be seen from Table 5, compared with the optimization results of the five improved algorithms on 15 test functions, the p values of the test results of MSGWO are all less than 0.05, and R is +/=, which indicates that the optimization results of MSGWO are significantly different from the optimization results of the five improved algorithms, and MSGWO is significantly better. This result shows the superiority of the MSGWO algorithm statistically.

4.3.4. Population Diversity Analysis of MSGWO

In an effort to further illustrate the effectiveness of the proposed algorithm, the diversity of population particles during evolution was analyzed. Population diversity measurements can accurately evaluate whether a population is being explored or exploited [38], and the specific calculation formula is as follows:

$$I_C(t) = \sqrt{\sum_{i=1}^{N} \sum_{d=1}^{D} (x_{id}(t) - c_d(t))^2} \tag{20}$$

$$c_d(t) = \frac{1}{D} \sum_{i=1}^{N} x_{id}(t) \tag{21}$$

where I_C represents the dispersion between the population and the center of mass c_d in each iteration, and x_{id} represents the value of the d dimension of the ith individual at the time of iteration t.

A small population diversity measure indicates that particles converge near the population center, that is, develop in a small space. A large population diversity measure indicates that the particles are far from the center of the population, that is, they explore in a larger space. Unimodal function F_1 and multi-modal function F_{15} of the commonly used test functions of CEC23 were selected as representatives to analyze the population diversity measurements of MSGWO and GWO, respectively. The experimental results are shown in Figure 7a,b.

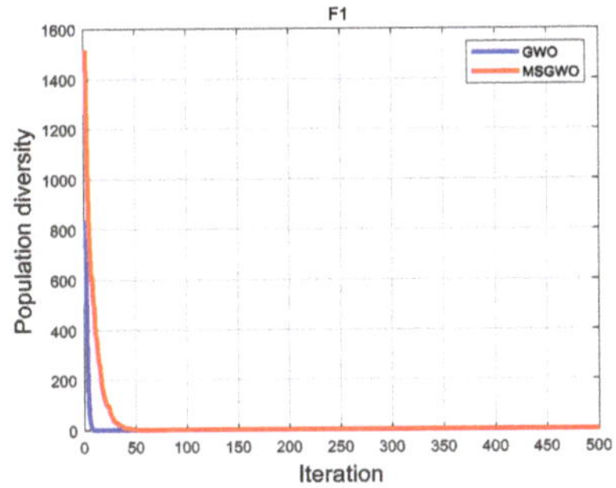

(**a**) F_1 diversity measure

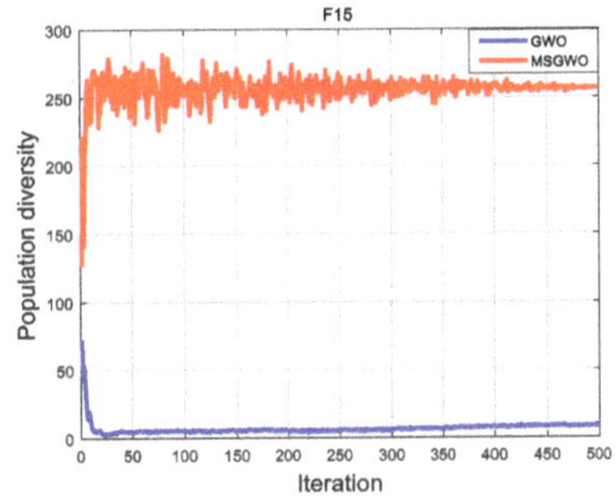

(**b**) F_{15} diversity measure

Figure 7. Population diversity measurement analysis.

As can be seen from Figure 7, the population diversity measure of the GWO algorithm decreased at the fastest speed in F_1 and F_{15}, which is not conducive to sufficient space exploration in the early stage and is easy to fall into local optimization. In F_1, the MSGWO algorithm maintained a high level of population diversity in the early stage of evolution, fully satisfying the exploration of particles in the whole space, while the population diversity decreased rapidly in the middle and late stages of evolution, indicating that the algorithm has a good development ability. In F_{15}, MSGWO population diversity fluctuated greatly and remained at a high level, indicating that the algorithm has a good global exploration ability.

5. Bearing Fault Detection

5.1. Parameter Adaptive Multistable Stochastic Resonance Strategy

In SR performance measurement indicators, signal-to-noise ratio (SNR) is commonly used and plays an important role. In this paper, the SNR is used as the target of optimization, that is, the fitness function. The formula for calculating the SNR is as follows [39]:

$$\text{SNR} = 10 \log_{10} \frac{A_t}{\sum_{n=0}^{N/2} A_n} \tag{22}$$

where A_t is the amplitude of the target frequency, A_n is the amplitude of frequencies other than the target frequency in the input signal, and N is the number of samples.

Based on the above analysis, the flow chart of the bearing fault-detection method proposed in this paper is shown in Figure 8, and its specific steps are as follows:

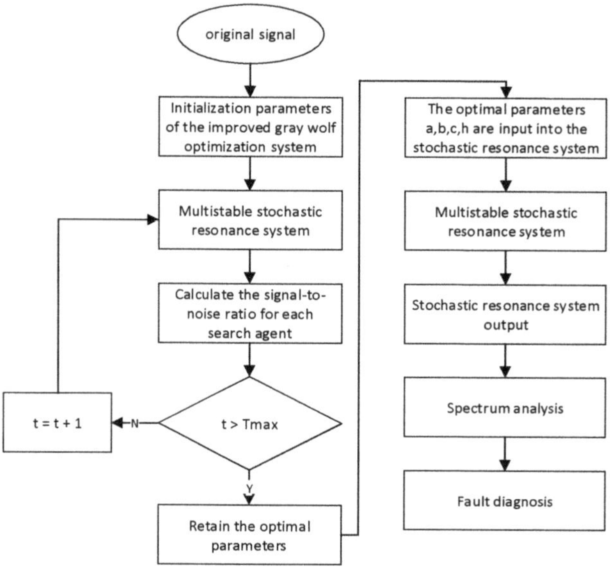

Figure 8. The flow diagram of the proposed algorithm.

Step 1: Input noisy signals and initialize MSGWO parameters. The range of a is [0, 0.5]; the range of b, c, and h are [0, 10]. The maximum number of iterations is 200 and the number of grey wolf populations is 30.

Step 2: Run the MSGWO, calculate the SNR according to Equation (22), then update the individual position, iterate to the maximum number of iterations, and finally terminate the iteration.

Step 3: Substitute the optimal solutions of a, b, c, and h into the SR system for operation, and subject the output of the SR system to fast Fourier transform to obtain the frequency domain. Then, analyze the output of the SR according to the frequency domain, and capture the fault frequency.

5.2. CWRU Bearing Data Set

In an effort to verify the applicability of the raised method in actual fault-signal detection, the open bearing-fault data set of CWRU was selected for the experiment [40], and the driving end bearing model 6205-2RS was used. Since the rotating speed of the bearing was 1750 rpm, the fault characteristic frequency of the inner ring was calculated to be 158 Hz. In the experiment, the sampling frequency was set to 12 kHz, and the data length of the signal was 12,000. The time domain and frequency domain waveforms of the input signal are shown in Figure 9, and the output signal-to-noise ratio was SNR $= -37.77$. As can be seen from Figure 9, the fault frequency of the original signal was difficult to capture in its frequency domain due to the influence of environmental noise. In order to ensure the accuracy of the experimental results, the average method of 30 experiments was adopted. The optimal parameters optimized by MSGWO were as follows: $a = 0.033$, $b = 0.567$, $c = 0.082$, and $h = 0.086$. We substituted the four parameters a, b, c, and h into the SR system to obtain the frequency domain waveform of its output, as shown in Figure 10. The output signal-to-noise ratio was SNR $= -26.92$, which was 10.85 dB higher than that of the input. According to the frequency domain waveform diagram in Figure 10, it can be observed that there was a clear spike at the target frequency, and the amplitude of the peak frequency was much larger than the amplitude of other surrounding frequencies. It can be seen that the method in this paper can effectively detect the bearing fault signal.

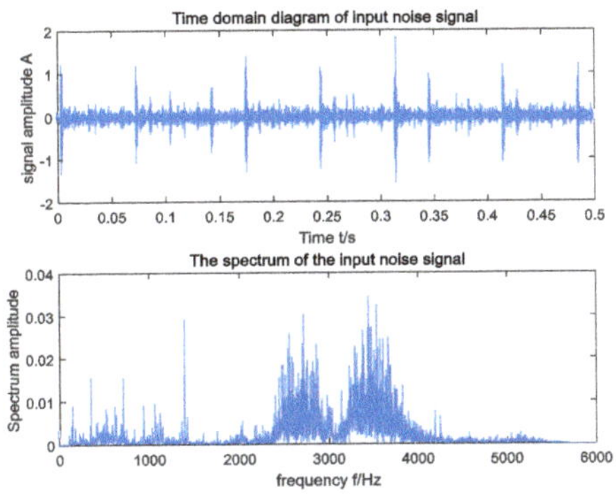

Figure 9. Time domain waveform and FFT spectrum of CWRU input signal.

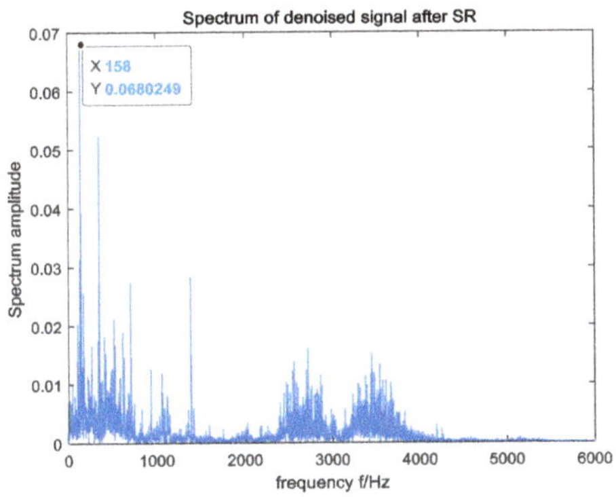

Figure 10. The FFT spectrum of the output signal processed by the raised algorithm.

In the case of the same parameters, the raised method was compared with five bearing fault-detection methods based on the improved algorithms to optimize the SR parameters. In an effort to ensure the accuracy of the experimental results, the method of averaging 30 experiments was adopted. The comparison experiment results are shown in Table 6. The test results shown in black bold in Table 6 are the best results for comparison.

Table 6. Comparison of experimental parameter results based on CWRU dataset.

	GWO	IGWO	MEGWO	mGWO	MPSO	MSGWO
a	0.077	0.080	0.065	0.101	0.055	0.033
b	4.197	6.581	6.305	6.571	8.418	0.567
c	7.206	2.830	6.028	7.417	6.160	0.082
h	0.755	0.888	0.792	0.757	0.763	0.086
Time	15.37	14.24	15.25	15.92	**10.58**	14.72
SNR	−28.35	−28.51	−28.27	−28.37	−28.32	**−26.92**

According to the data in Table 6, compared with five bearing fault-detection methods based on improved algorithms to optimize SR parameters, the raised method had the highest SNR, but the convergence speed was slower than that of bearing fault-detection methods based on IGWO and MPSO. Since the SNR was taken as the evaluation index in bearing fault detection, the proposed method had some advantages over the five bearing fault-detection methods based on the improved algorithm to optimize the SR parameters.

5.3. MFPT Bearing Data Set

In an effort to further verify the applicability of the raised method in actual fault-signal detection, the bearing data set of the MFPT in the United States was selected as the research object [41] to detect the outer-ring signal of the faulty bearing. The input shaft speed of the selected outer ring fault signal was 25 Hz, the load was 25, and the fault characteristic frequency was calculated to be 162 Hz. The time domain and frequency domain waveform of the input signal are shown in Figure 11. According to Figure 11, due to the influence of ambient noise, the fault frequency of the original signal was submerged in the noise and was difficult to be captured in its frequency domain. In an effort to ensure the accuracy of the experimental results, the average method of 30 experiments was adopted. The optimal parameters optimized by MSGWO were as follows: $a = 0.500$, $b = 9.571$, $c = 0.019$, and $h = 0.409$. We substituted the four parameters a, b, c, and h into the SR system to obtain the frequency domain waveform of its output, as shown in Figure 12. According to the frequency domain waveform diagram in Figure 12, it can be observed that the amplitude of the target frequency was the largest in its frequency domain and was much larger than the amplitude of other surrounding frequencies. This further proves that the raised method can detect the bearing fault signal effectively.

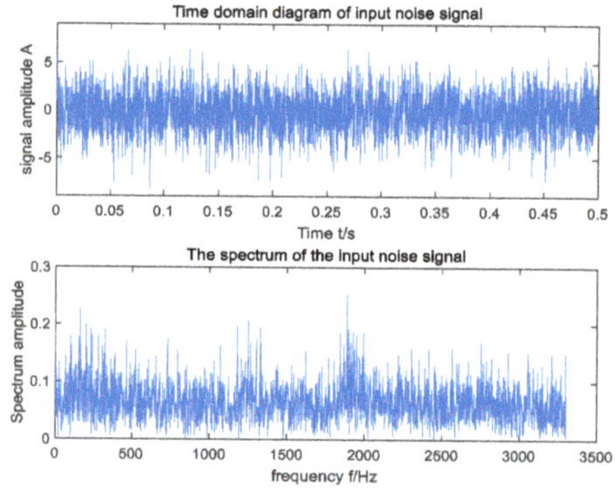

Figure 11. Time domain waveform and FFT spectrum of MFPT input signal.

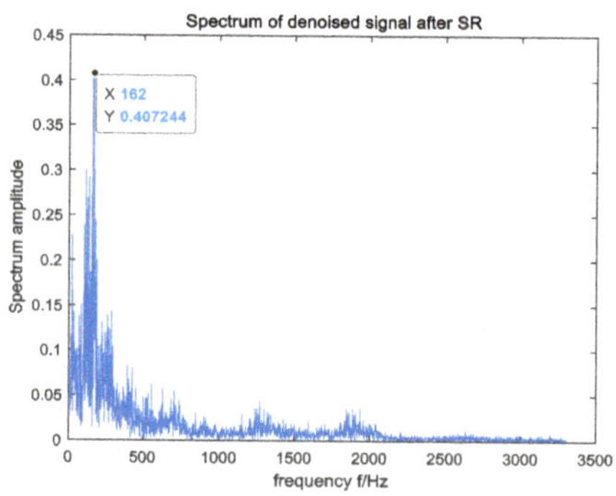

Figure 12. The FFT spectrum of the output signal processed by the proposed algorithm.

In the case of the same parameters, the raised method was compared with five bearing fault-detection methods based on the improved algorithms to optimize the SR parameters. In an effort to ensure the accuracy of the experimental results, the method of averaging

30 experiments was adopted. The comparison experiment results are shown in Table 7. The test results shown in black bold in Table 7 are the best results for comparison.

Table 7. Comparison of experimental parameter results based on MFPT dataset.

	GWO	IGWO	MEGWO	mGWO	MPSO	MSGWO
a	0.500	0.495	0.500	0.472	0.052	0.500
b	10.00	2.173	8.554	8.247	8.968	9.571
c	0.025	0.488	0.054	3.728	1.287	0.019
h	0.328	0.185	0.257	0.069	0.122	0.409
Time	21.51	25.35	35.52	34.79	22.91	**19.95**
SNR	−26.56	−27.75	−26.82	−27.21	−27.62	**−26.42**

According to the data in Table 7, compared with five bearing fault-detection methods based on the improved algorithms to optimize SR parameters, the method raised in this article had a larger SNR and better time performance. Therefore, the method proposed in this article has certain advantages over the comparative method.

5.4. Bearing-Fault Diagnosis of Crystal Growing Furnace

In this paper, the crystal lifting and rotating mechanism of a crystal growing furnace was taken as the actual test object, as shown in Figure 13. The crystal growing furnace is the major equipment for producing wafers. The mechanism is composed of two Mitsubishi HG-KR73 servo motors, the crystal lift motor is used to lift the crystal upward, and the crystal rotating motor is used to drive the crystal to spin during the growth process. Because the stability of crystal rotating is an important factor to determine the crystal formation and crystal quality, it is necessary to accurately monitor the fault of the crystal rotating motor. The experiment object was the motor of a certain type of electronic-grade silicon single-crystal growing furnace. The purpose was to detect the failure frequency of the crystal rotating motor. A certain type of three-dimensional vibration sensor was used in the experiment, and its connection with the motor is shown in Figure 14. As shown in Figure 14, the vibration sensor was adsorbed on the motor, and information such as vibration displacement, vibration speed, and vibration frequency can be collected. The deceleration ratio of the crystal rotating system was 100:1, that is, when the crystal rotating speed was 10 rad/min, the speed of the crystal rotating motor was 1000 rad/min.

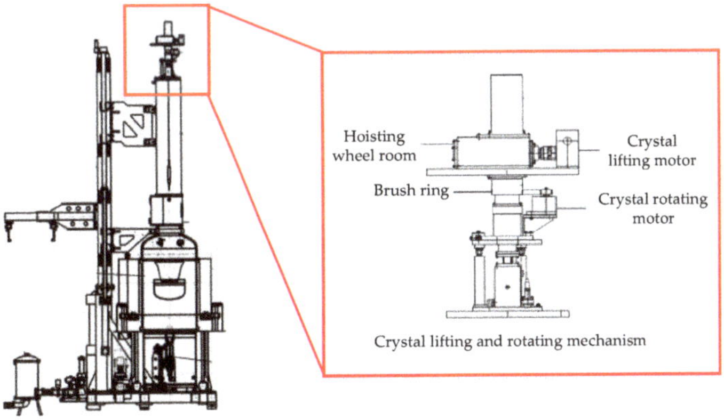

Figure 13. Crystal growing furnace and crystal lifting and rotating mechanism.

Figure 14. Vibration sensor installation position.

The vibration signal of the motor collected by the vibration sensor is shown in Figure 15. As can be seen from Figure 15, the time domain signal of the actual motor fault collected by the vibration sensor is very weak, completely submerged in the noise, and the frequency domain signal cannot distinguish the fault frequency. The method proposed in this paper was used to detect the fault frequency of the crystal rotating motor, and the test results are shown in Figure 16. It can be seen from Figure 16 that the algorithm increased the frequency domain amplitude of the fault signal and effectively detected that the fault frequency of the crystal motor was 35 Hz.

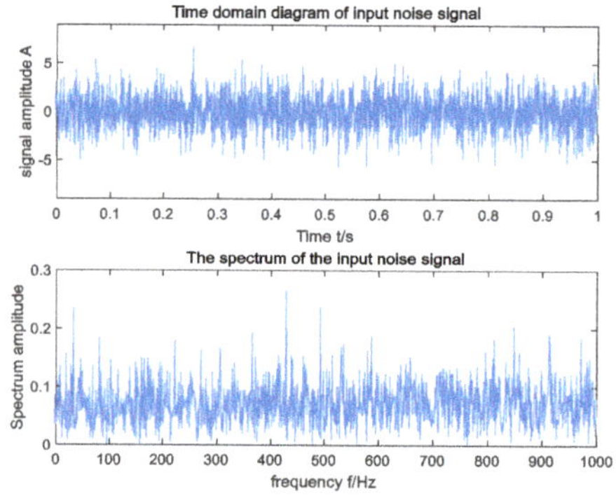

Figure 15. Original vibration signal of crystal rotating motor.

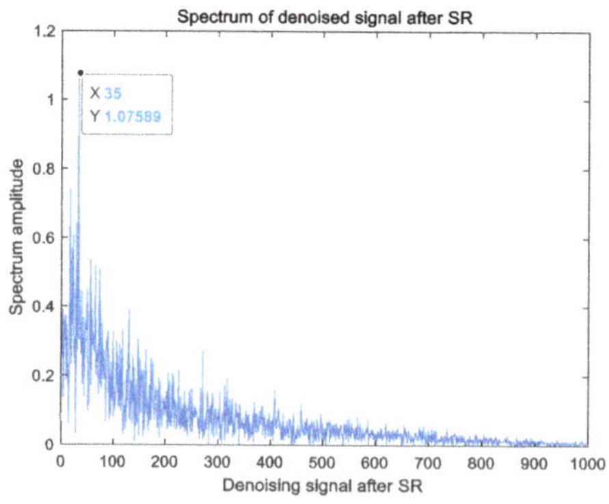

Figure 16. Spectrum amplitude of motor fault.

6. Conclusions

Taking bearing fault-signal detection as the research object, this paper proposes a bearing fault-detection method based on an improved grey wolf algorithm to optimize multistable stochastic resonance parameters, aiming at the problems that multistable stochastic resonance system parameters are difficult to select and basic grey wolf optimization algorithm is prone to local optimization and low convergence accuracy. This method improved the grey wolf optimization algorithm. Firstly, the Sobol sequence was used to initialize the grey wolf population to improve the diversity of the population. Secondly, the exponential rule convergence factor was used to balance the global search and local development stages of the algorithm. At the same time, the adaptive position-update strategy was introduced to improve the accuracy of the algorithm. Additionally, we used Cauchy–Gaussian hybrid variation to improve the ability of the algorithm to escape from the local optimal area. The performance of the proposed algorithm was verified using experiments with 15 benchmark test functions in the CEC23 group of common test functions. The results show that the multi-strategy improved grey wolf optimization algorithm has better optimization performance. Then, the improved grey wolf optimization algorithm was used to optimize the parameters of the multistable stochastic resonance algorithm, so as to realize the detection of bearing fault signals. Finally, the bearing data sets of Case Western Reserve University and the Association for Mechanical Fault Prevention Technology were analyzed and diagnosed with the proposed bearing fault-detection method, and the optimization results were compared with other improved algorithms. At the same time, the method proposed in this paper was used to diagnose the fault of the bearing of the lifting device of a single-crystal furnace. The experimental results show that this method can be used to detect the bearing fault signal and can effectively enhance the fault signal in the noise. Compared with other optimized bearing fault-detection methods based on improved intelligent algorithms, the proposed method has the advantages of fast convergence, high parameter optimization accuracy, and strong robustness.

In the future, this paper will study the following two aspects: Firstly, the MSGWO needs to be further improved to improve its stability due to its poor stability in individual test functions. Secondly, the bearing fault-detection method proposed in this paper will be applied to the bearing fault detection of rotating machinery in different industries, and the corresponding improvement will be made according to the actual detection results, so as to improve the applicability of the bearing fault-detection method proposed in this paper to different industries.

Author Contributions: Conceptualization, W.H.; methodology, W.H.; software, G.Z.; writing—original draft preparation, G.Z.; writing—review and editing, W.H. All authors have read and agreed to the published version of the manuscript.

Funding: This research was supported by the National Natural Science Foundation (NNSF) of China (NOs. 62073258, 62127809, 62003261).

Institutional Review Board Statement: Not applicable.

Informed Consent Statement: Informed consent was obtained from all subjects involved in the study.

Data Availability Statement: Our source code is available on https://github.com/Zfutur1/Code-information.git.

Conflicts of Interest: The authors declare no conflict of interest.

References

1. Li, Y.; Tang, B.; Geng, B.; Jiao, S. Fractional Order Fuzzy Dispersion Entropy and Its Application in Bearing Fault Diagnosis. *Fractal Fract.* **2022**, *6*, 544. [CrossRef]
2. Onufriienko, D.; Taranenko, Y. Filtering and Compression of Signals by the Method of Discrete Wavelet Decomposition into One-Dimensional Series. *Cybern. Syst. Anal.* **2023**, *59*, 331–338. [CrossRef]
3. Grover, C.; Turk, N. Rolling Element Bearing Fault Diagnosis using Empirical Mode Decomposition and Hjorth Parameters. *Procedia Comput. Sci.* **2020**, *167*, 1484–1494. [CrossRef]

4. Li, Y.; Tang, B.; Jiao, S. SO-slope entropy coupled with SVMD: A novel adaptive feature extraction method for ship-radiated noise. *Ocean Eng.* **2023**, *280*, 114677. [CrossRef]
5. You, K.; Qiu, G.; Gu, Y. Rolling Bearing Fault Diagnosis Using Hybrid Neural Network with Principal Component Analysis. *Sensors* **2022**, *22*, 8906. [CrossRef] [PubMed]
6. Markina, A.; Muratov, A.; Petrovskyy, V.; Avetisov, V. Detection of Single Molecules Using Stochastic Resonance of Bistable Oligomers. *Nanomaterials* **2020**, *10*, 2519. [CrossRef] [PubMed]
7. Benzi, R.; Parisi, G.; Sutera, A.; Vulpiani, A. A Theory of Stochastic Resonance in Climatic Change. *SIAM J. Appl. Math.* **1983**, *43*, 565–578. [CrossRef]
8. Zhang, G.; Zeng, Y.; Jiang, Z. A novel two-dimensional exponential potential bi-stable stochastic resonance system and its application in bearing fault diagnosis. *Phys. A Stat. Mech. Appl.* **2022**, *607*, 128223. [CrossRef]
9. Zayed, E.M.E.; Alngar, M.E.M.; Shohib, R.M.A. Dispersive Optical Solitons to Stochastic Resonant NLSE with Both Spatio-Temporal and Inter-Modal Dispersions Having Multiplicative White Noise. *Mathematics* **2022**, *10*, 3197. [CrossRef]
10. Pandey, A.K.; Kaur, G.; Chaudhary, J.; Hemrom, A.; Jaleel, J.; Sharma, P.D.; Patel, C.; Kumar, R. 99m-Tc MDP bone scan image enhancement using pipeline application of dynamic stochastic resonance algorithm and block-matching 3D filter. *Indian J. Nucl. Med.* **2023**, *38*, 8–15. [CrossRef]
11. Huang, W.; Zhang, G.; Jiao, S.; Wang, J. Gray Image Denoising Based on Array Stochastic Resonance and Improved Whale Optimization Algorithm. *Appl. Sci.* **2022**, *12*, 12084. [CrossRef]
12. Ai, H.; Yang, G.; Liu, W.; Wang, Q. A fast search method for optimal parameters of stochastic resonance based on stochastic bifurcation and its application in fault diagnosis of rolling bearings. *Chaos Solitons Fractals* **2023**, *168*, 113211. [CrossRef]
13. Tao, Y.; Luo, B. Monostable stochastic resonance activation unit-based physical reservoir computing. *J. Korean Phys. Soc.* **2023**, *82*, 798–806. [CrossRef]
14. Li, J.M.; Chen, X.F.; He, Z.J. Multi-stable stochastic resonance and itsapplication research on mechanical fault diagnosis. *J. Sound Vib.* **2013**, *332*, 5999–6015. [CrossRef]
15. Meng, Z.; Quan, S.; Li, J.; Cao, L.; Fan, F. A novel coupled array of multi-stable stochastic resonance under asymmetric trichotomous noise and its application in rolling bearing compound fault diagnosis. *Appl. Acoust.* **2023**, *209*, 109405. [CrossRef]
16. Zhang, G.; Shu, Y.; Zhang, T. Piecewise unsaturated multi-stable stochastic resonance under trichotomous noise and its application in bearing fault diagnosis. *Results Phys.* **2021**, *30*, 104907. [CrossRef]
17. Mitaim, S.; Kosko, B. Adaptive Stochastic Resonance in Noisy Neurons Based on Mutual Information. *IEEE Trans. Neural Netw.* **2004**, *15*, 1526–1540. [CrossRef]
18. Huang, W.; Zhang, G.; Jiao, S.; Wang, J. Bearing Fault Diagnosis Based on Stochastic Resonance and Improved Whale Optimization Algorithm. *Electronics* **2022**, *11*, 2185. [CrossRef]
19. Hu, B.; Guo, C.; Wu, J.; Tang, J.; Zhang, J.; Wang, Y. An Adaptive Periodical Stochastic Resonance Method Based on the Grey Wolf Optimizer Algorithm and Its Application in Rolling Bearing Fault Diagnosis. *J. Vib. Acoust.* **2019**, *141*, 041016. [CrossRef]
20. Dong, L.; Yuan, X.; Yan, B.; Song, Y.; Xu, Q.; Yang, X. An Improved Grey Wolf Optimization with Multi-Strategy Ensemble for Robot Path Planning. *Sensors* **2022**, *22*, 6843. [CrossRef]
21. Zhang, L. A local opposition-learning golden-sine grey wolf optimization algorithm for feature selection in data classification. *Appl. Soft Comput. J.* **2023**, *142*, 110319.
22. Vasudha, B.; Anoop, B. Opposition-Based Multi-Tiered Grey Wolf Optimizer for Stochastic Global Optimization Paradigms. *Int. J. Energy Optim. Eng.* **2022**, *11*, 1–26.
23. Rajput, S.S. S-GWO-FH: Sparsity-based grey wolf optimization algorithm for face hallucination. *Soft Comput.* **2022**, *26*, 9323–9338. [CrossRef]
24. Ma, C.; Ren, R.; Luo, M.; Deng, K. Stochastic resonance in an overdamped oscillator with frequency and input signal fluctuation. *Nonlinear Dyn.* **2022**, *110*, 1223–1232. [CrossRef]
25. Li, J.; Chen, X.; He, Z. Adaptive stochastic resonance method for impact signal detection based on sliding window. *Mech. Syst. Signal Process.* **2013**, *36*, 240–255. [CrossRef]
26. Mirjalili, S.; Mirjalili, S.M.; Lewis, A. Grey Wolf Optimizer. *Adv. Eng. Softw.* **2014**, *69*, 46–61. [CrossRef]
27. Sirsant, S.; Hamouda, M.A.; Shaaban, M.F.; Al Bardan, M.S. A Chaotic Sobol Sequence-based multi-objective evolutionary algorithm for optimal design and expansion of water networks. *Sustain. Cities Soc.* **2022**, *87*, 104215. [CrossRef]
28. Digalakis, J.; Margaritis, K. On benchmarking functions for genetic algorithms. *Int. J. Comput. Math.* **2001**, *77*, 481–506. [CrossRef]
29. Arora, S.; Singh, S. Butterfly optimization algorithm: A novel approach for global optimization. *Soft Comput.* **2018**, *23*, 715–734. [CrossRef]
30. Mirjalili, S.; Lewis, A. The Whale Optimization Algorithm. *Adv. Eng. Softw.* **2016**, *95*, 51–67. [CrossRef]
31. Rashedi, E.; Nezamabadi-Pour, H.; Saryazdi, S. GSA: A Gravitational Search Algorithm. *Inf. Sci.* **2009**, *179*, 2232–2248. [CrossRef]
32. Tong, L.; Li, X.; Hu, J.; Ren, L. A PSO Optimization Scale-Transformation Stochastic-Resonance Algorithm With Stability Mutation Operator. *IEEE Access* **2017**, *6*, 1167–1176. [CrossRef]
33. Karaboga, D. Artificial Bee Colony Algorithm. *Scholarpedia* **2010**, *5*, 6915. [CrossRef]
34. Tu, Q.; Chen, X.; Liu, X. Multi-strategy ensemble grey wolf optimizer and its application to feature selection. *Appl. Soft Comput.* **2018**, *76*, 16–30. [CrossRef]

35. Gupta, S.; Deep, K. A memory-based Grey Wolf Optimizer for global optimization tasks. *Appl. Soft Comput.* **2020**, *93*, 106367. [CrossRef]
36. Nadimi-Shahraki, M.H.; Taghian, S.; Mirjalili, S. An improved grey wolf optimizer for solving engineering problems. *Expert Syst. Appl.* **2020**, *166*, 113917. [CrossRef]
37. Liu, H.; Zhang, X.-W.; Tu, L.-P. A modified particle swarm optimization using adaptive strategy. *Expert Syst. Appl.* **2020**, *152*, 113353. [CrossRef]
38. Nadimi-Shahraki, M.H.; Zamani, H. DMDE: Diversity-maintained multi-trial vector differential evolution algorithm for non-decomposition large-scale global optimization. *Expert Syst. Appl.* **2022**, *198*, 116895. [CrossRef]
39. He, B.; Huang, Y.; Wang, D.; Yan, B.; Dong, D. A parameter-adaptive stochastic resonance based on whale optimization algorithm for weak signal detection for rotating machinery. *Measurement* **2019**, *136*, 658–667. [CrossRef]
40. Available online: http://www.eecs.cwru.edu/laboratory/bearing/download.html (accessed on 10 April 2018).
41. Sobie, C.; Freitas, C.; Nicolai, M. Simulation-driven machine learning: Bearing fault classification. *Mech. Syst. Signal Process.* **2018**, *99*, 403–419. [CrossRef]

sensors

MDPI

Article

Intelligent Diagnosis of Rolling Bearings Fault Based on Multisignal Fusion and MTF-ResNet

Kecheng He [1,2,*], Yanwei Xu [1,2,*], Yun Wang [1,2], Junhua Wang [1] and Tancheng Xie [1,2]

1 School of Mechatronics Engineering, Henan University of Science and Technology, Luoyang 471003, China; 13323859953@163.com (Y.W.); wangjh@haust.edu.cn (J.W.); xietc@haust.edu.cn (T.X.)

2 Intelligent Numerical Control Equipment Engineering Laboratory of Henan Province, Luoyang 471003, China

* Correspondence: kecheng_he@163.com (K.H.); xuyanweiluoyang@163.com (Y.X.)

Abstract: Existing diagnosis methods for bearing faults often neglect the temporal correlation of signals, resulting in easy loss of crucial information. Moreover, these methods struggle to adapt to complex working conditions for bearing fault feature extraction. To address these issues, this paper proposes an intelligent diagnosis method for compound faults in metro traction motor bearings. This method combines multisignal fusion, Markov transition field (MTF), and an optimized deep residual network (ResNet) to enhance the accuracy and effectiveness of diagnosis in the presence of complex working conditions. At the outset, the acquired vibration and acoustic emission signals are encoded into two-dimensional color feature images with temporal relevance by Markov transition field. Subsequently, the image features are extracted and fused into a set of comprehensive feature images with the aid of the image fusion framework based on a convolutional neural network (IFCNN). Afterwards, samples representing different fault types are presented as inputs to the optimized ResNet model during the training phase. Through this process, the model's ability to achieve intelligent diagnosis of compound faults in variable working conditions is realized. The results of the experimental analysis verify that the proposed method can effectively extract comprehensive fault features while working in complex conditions, enhancing the efficiency of the detection process and achieving a high accuracy rate for the diagnosis of compound faults.

Keywords: metro traction motor bearings; multisignal fusion; Markov transition field; optimized deep residual network; diagnosis of compound faults

check for updates

Citation: He, K.; Xu, Y.; Wang, Y.; Wang, J.; Xie, T. Intelligent Diagnosis of Rolling Bearings Fault Based on Multisignal Fusion and MTF-ResNet. *Sensors* **2023**, *23*, 6281. https://doi.org/10.3390/s23146281

Academic Editors: Yuxing Li and Luca Fredianelli

Received: 20 June 2023
Revised: 8 July 2023
Accepted: 9 July 2023
Published: 10 July 2023

1. Introduction

As the power source of metro trains, the quality of the traction motor bearings directly affects the normal operation of the motor. The frequent starting and stopping of the metro causes alternating changes in the speed of the traction motor bearings and the loads they are subjected to. With long-term harsh working conditions, the inner and outer rings of bearings and rolling elements will produce varying degrees of pitting, cracking and more complex forms of failure. The adverse vibrations generated by a faulty bearing, when input into the entire system over an extended period, not only damage the traction motor but also pose a risk to other structural components. This poses a serious threat to the safety and reliability of metro trains. The intelligent diagnosis of bearings fault in complex working conditions enables the timely identification of fault types, facilitating early maintenance intervention and providing significant engineering value for practical applications.

Conventional approaches for bearing fault diagnosis predominantly rely on signal processing techniques. To address the issue of noise interference during feature extraction, wavelet thresholding was employed to effectively eliminate significant noise components from the raw data [1,2]. In an effort to enhance the signal-to-noise ratio, ref. [3,4] adopted empirical mode decomposition (EMD) to decompose the signal into multiple intrinsic mode functions. Furthermore, ref. [5] introduced an optimized variational mode decomposition

(VMD) method to facilitate the selection of intrinsic mode functions containing pertinent fault information. Despite the promising outcomes achieved by these traditional methods in bearing fault diagnosis, they are accompanied by inherent limitations. These drawbacks encompass restricted generalization capability, challenges in extracting deep fault features, and complexities associated with parameter optimization. Signal analysis technology, as a research hotspot, has been receiving attention from scholars. Subsequently, the introduction of new methods has successfully addressed many challenges [6,7].

With the development of artificial intelligence technology, machine learning and deep learning [8] have gained significant attention in various fields, and numerous researchers have started extracting deeper features and making notable contributions [9–11]. A convolutional neural network (CNN), as one of their important representatives, possesses a powerful adaptive feature extraction capability. Moreover, CNN has demonstrated remarkable performance in the field of image processing. As such, scholars have increasingly introduced CNN into the field of fault diagnosis and conducted a series of research studies in this area. Ref. [12] has recently proposed a CNN model that utilizes widened convolutional kernels to improve the feature extraction efficiency of the network. Ref. [13] has deployed a CNN to extract features from Mel spectrum generated from the voiceprint signals of motors. Ref. [14] has presented a multiscale CNN model that effectively extracts signal features at different frequencies. This advanced model is further combined with LSTM to identify fault types. In the field of medical imaging, ref. [15] proposed an improved CNN model architecture for the identification of a lung nodule and early-stage cancer diagnosis by comparing multiple photos. In big data environments, to reduce the costs associated with data collection and processing, some researchers have explored unsupervised learning techniques. To synchronously extract local and global structural information from the raw unlabeled industrial data, ref. [16] proposed a new multiple-order graphical deep extreme learning machine (MGDELM) algorithm. Ref. [17] proposed a novel self-training semi-supervised deep learning (SSDL) approach to train a fault diagnosis model together with few labeled and abundant unlabeled samples. The previously discussed research studies have made notable advances in fault diagnosis. However, because of their reliance on single-sensor signals, there may be limitations in accurately characterizing fault information, which could ultimately reduce their overall reliability.

Multisignal fusion technology enables the simultaneous processing of time-series data obtained from multiple sensors, thereby capturing a broader range of system variability while offering heightened complementarity and fault tolerance. In one study, feature extraction was performed on original vibration and acoustic signals, which were subsequently fused using a 1DCNN-based network model [18]. Another approach proposed a frequency-domain multilinear principal component analysis to effectively identify faults by integrating diverse vibration and acoustic signals [19]. Similarly, a two-dimensional matrix was constructed from multi-axial vibration signals, and an enhanced 2DCNN model was employed for fault diagnosis [20]. These methods have demonstrated commendable enhancements in diagnostic accuracy. However, it is worth noting that a limitation common to these approaches is the omission of time correlation among signals, which may result in the loss of crucial fault-related information.

Upon a comprehensive analysis of existing literature, it has been observed that diagnostic approaches leveraging deep learning techniques frequently employ increasing network depths to enhance the model's learning capacity and improve diagnostic performance. Nevertheless, the utilization of progressively deeper networks may give rise to challenges such as the vanishing or exploding gradient problem. To address this issue, deep residual networks were introduced [21], effectively mitigating the aforementioned problem. Furthermore, an innovative activation function named STAC-tanh was proposed by [22], which enables adaptive feature extraction in the bearing system by employing the hyperbolic tangent function with slope and threshold adaptivity. Another compelling approach involved the fusion of Gramian angular field (GAF) with ResNet, leading to notable advancements in bearing fault diagnosis [23]. Additionally, ref. [24] combined transfer learning with ResNet, utilizing a pretrained ResNet model on ImageNet as a fault feature extractor, which yielded remarkably accurate results. These aforementioned studies have demonstrated promising outcomes in the realm of bearing fault diagnosis. However, certain limitations persist, including the sole reliance on a single sensor signal and the absence of experimental verification through the use of a purpose-built platform.

In summary, most of the studies are based on open source datasets with simple working conditions and failure forms, but the actual working conditions of bearings are complex and can present different parts and degrees of failure. To address the challenges faced in compound bearing fault diagnosis under complex working conditions, such as the low reliability of single sensor signals, the tendency for traditional data processing methods to result in important information loss, the degradation of diagnostic models with increasing network depth, and the difficulty of feature extraction, this paper proposes an intelligent diagnosis method for compound bearing faults in metro traction motors by combining MTF-processed acoustic-vibration signals using IFCNN for feature fusion along with an optimized version of ResNet. The main contributions of the paper are expressed as follows:

1. The application of IFCNN in compound bearing fault diagnosis allows for the fusion of multiple signal features, reducing the limitations of single sensor signals and providing more reliable diagnostic results.
2. The optimized ResNet model improves the efficiency of feature extraction by addressing the vanishing gradient problem. Combined with the MTF data processing method, it can effectively extract complex bearing fault features under varying working conditions with good accuracy and stability.
3. The construction of a test platform for metro traction motor bearings was completed, and intelligent diagnosis of composite faults under variable working conditions was conducted, validating the effectiveness of the proposed methods.

The remaining sections of this paper are arranged as follows: In Section 2, the data processing method used in this study and the construction of the dataset are introduced. Section 3 focuses on the multisignal fusion technology used in this study. Section 4 provides a detailed description of the fault diagnosis model and the corresponding diagnostic process. Section 5 explains the specific experimental design, as well as the diagnostic scheme adopted in this study. Section 6 analyzes the experimental results and carries out a series of method comparisons to validate the effectiveness of the proposed approach. Section 7 summarizes the main content of the paper and draws conclusions.

2. Data Preprocessing

In this study, a signal acquisition system will be built to obtain a large amount of raw data using acoustic emission sensors, vibration sensors and PCI acquisition cards. The research focuses on compound faults, with pitting as the main defect. The location of the defect is used as a classification criterion. A total of eight fault types including normal bearings are designed and labeled for subsequent study, using different fault locations as classification indicators. The fault types and labels are shown in Table 1.

Table 1. Label settings for different fault types.

	Fault Types							
	Normal	**Inner Ring**	**Outer Ring**	**Rolling Element**	**Inner Ring + Outer Ring**	**Inner Ring + Rolling Element**	**Outer Ring + Rolling Element**	**Inner Ring + Outer Ring + Rolling Element**
Label	0	1	2	3	4	5	6	7

2.1. Dataset Construction

The vibration and acoustic emission signals were acquired using a PCI data acquisition card with a sampling frequency of 50 kS/s and a sampling time of 10 s, giving a total of 5×10^5 sampling points. In this experiment, the minimum speed of the bearing is determined to be 800 rpm. Based on this speed, the number of sampling points obtained from one cycle of bearing rotation can be calculated to be 3750. In order to ensure the completeness of the sampled fault information, it is recommended that the number of sampling points be at least twice that of the calculated value, resulting in a sampling length of 8192 (2^{13}). With a limited amount of data, the vibration and acoustic emission signals were data augmented using overlapping sampling so that each fault type under each working condition contained 1000 samples for a total of 8000 samples, which were randomly divided into a training set and a testing set at 9:1. Under fixed working conditions, the dataset is divided as shown in Table 2.

Table 2. Dataset partitioning under fixed working conditions.

Data Set	Sample Size for Different Fault Types							
	Normal	Inner Ring	Outer Ring	Rolling Element	Inner Ring + Outer Ring	Inner Ring + Rolling Element	Outer Ring + Rolling Element	Inner Ring + Outer Ring + Rolling Element
Training Set	900	900	900	900	900	900	900	900
Testing Set	100	100	100	100	100	100	100	100

2.2. MTF Image Encoding

In this paper, MTF is used to process vibration signals and acoustic emission signal data, converting the acquired data samples into image samples. MTF is an image encoding method that converts original vibration or acoustic emission signals into time series two-dimensional images through Markov transition probabilities [25].

Suppose a discretized segment of time series data $X = \{x_1, x_2, \cdots, x_n\}$ is partitioned into intervals of its value domain by quantile Q. Each x_t in the sequence can be mapped to the corresponding interval $q_n (n \in [1, Q])$. By calculating the state transfer probabilities through the Markov chain principle, a state transfer probability matrix W of size $Q \times Q$ can be obtained, with an expression, as shown in Equation (1), where w_{ij} denotes the probability that a sample point in interval q_j at moment t is transferred to interval q_i at moment $t+1$ [26].

$$W = \begin{bmatrix} w_{11|P(x_{t+1} \in q_1 | x_t \in q_1)} & \cdots & w_{1Q|P(x_{t+1} \in q_1 | x_t \in q_Q)} \\ w_{21|P(x_{t+1} \in q_2 | x_t \in q_1)} & \cdots & w_{2Q|P(x_{t+1} \in q_2 | x_t \in q_Q)} \\ \vdots & \ddots & \vdots \\ w_{Q1|P(x_{t+1} \in q_Q | x_t \in q_1)} & \cdots & w_{QQ|P(x_{t+1} \in q_Q | x_t \in q_Q)} \end{bmatrix} \quad (1)$$

By incorporating the temporal information into the state transfer probability matrix W and arranging each state transition probability w_{ij} in time sequence, a Markov transition field (MTF) matrix M of size $n \times n$ is obtained as expressed in shown Equation (2) where m_{ij} denotes the transition probability w_{ij} between the intervals $(q_j \to q_i)$ in which the sample points are located in time sequence.

$$M = \begin{bmatrix} m_{11} & m_{12} & \cdots & m_{1n} \\ m_{21} & m_{22} & \cdots & m_{2n} \\ \vdots & \vdots & \ddots & \vdots \\ m_{n1} & m_{n2} & \cdots & m_{nn} \end{bmatrix} = \begin{bmatrix} w_{ij|x_1 \in q_i, x_1 \in q_j} & \cdots & w_{ij|x_1 \in q_i, x_n \in q_j} \\ w_{ij|x_2 \in q_i, x_1 \in q_j} & \cdots & w_{ij|x_2 \in q_i, x_n \in q_j} \\ \vdots & \ddots & \vdots \\ w_{ij|x_n \in q_i, x_1 \in q_j} & \cdots & w_{ij|x_n \in q_i, x_n \in q_j} \end{bmatrix} \quad (2)$$

The elements m_{ij} in the MTF matrix are transformed as pixel points into a two-dimensional feature image with temporal correlation. As the number of sample points selected directly affects the size of the generated coded image, it is clearly inappropriate for an image with too large a size to be used directly as input to the CNN. To improve computational efficiency, a fuzzy kernel $\left\{\frac{1}{m^2}\right\}_{m \times m}$ is used to pixel average each region without overlap. Figure 1 shows images of different fault types after encoding each sample, consisting of 8192 sampling points, using MTF image encoding and subsequently subjecting them to pixel averaging processing. Compared to traditional time domain analysis methods, MTF encoding images preserve time-related information and enable clearer differentiation of various fault types in rolling bearings.

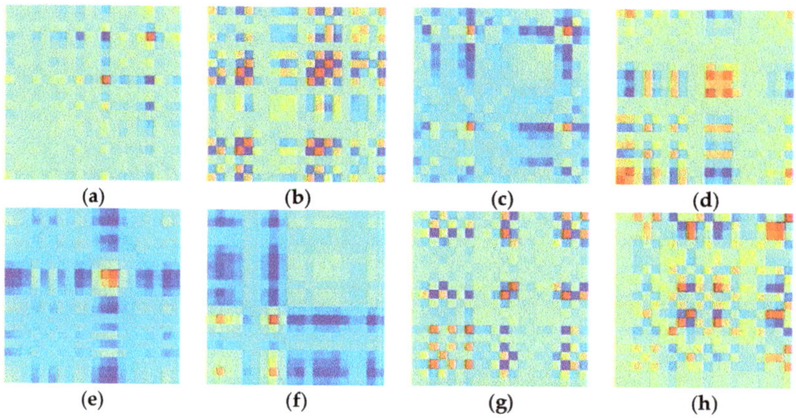

Figure 1. MTF-encoded images of 8 types of fault: (**a**) normal; (**b**) inner ring; (**c**) outer ring; (**d**) rolling element; (**e**) inner ring + outer ring; (**f**) inner ring + rolling element; (**g**) outer ring + rolling element; (**h**) inner ring + outer ring + rolling element.

3. Multisignal Fusion

To enhance system stability and increase diagnostic reliability, this article collected vibration signals and acoustic emission signals and fused them for processing. This fusion processing can establish correlations between multiple signal sources. Usually, information fusion can be divided into three levels: data-level fusion, feature-level fusion, and decision-level fusion. Considering that the sample data in this study consist of MTF encoded images of different fault types, it is advantageous to employ CNN for image processing. Therefore, this paper adopted the IFCNN for feature-level fusion of the data.

IFCNN consists of three modules, namely, the feature extraction module, the feature fusion module and the feature reconstruction module [27], and the structure of this framework is shown in Figure 2.

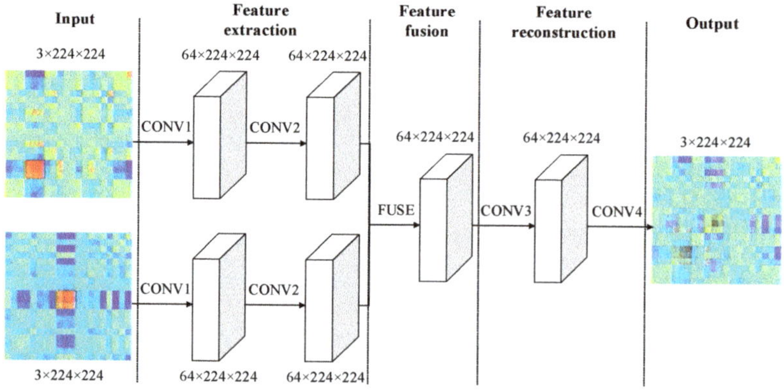

Figure 2. The structure of IFCNN.

The feature extraction module consists of two convolutional layers. The first layer uses the first convolutional layer of the ResNet101 network model, pretrained on the ImageNet dataset. This layer includes 64 convolutional kernels with a size of 7 × 7 and retains the training parameters, enabling effective extraction of image features. The second convolutional layer includes 64 convolutional kernels with a size of 3 × 3, which are used to adjust the features extracted by the first layer in order to adapt to feature fusion. For this study, the feature fusion module adopts an element-wise maximum fusion strategy. The final module is the image reconstruction module, in which the third convolutional layer includes 64 convolutional kernels with a size of 3 × 3. This layer adjusts the fused convolutional features and plays an important role in reconstructing the image. The fourth convolutional layer reconstructs the feature map with three-channel output, and it includes 3 convolutional kernels with a size of 1 × 1.

This framework uses the mean squared error (MSE) as the basic loss function and adds a perceptual loss to optimize the model. The expression for the perceptual loss (P_{loss}) is as follows:

$$P_{loss} = \frac{1}{C_f H_f W_f} \sum_{i,x,y} \left[f_p^i(x,y) - f_g^i(x,y) \right]^2 \tag{3}$$

where f_p and f_g are the feature maps of the predicted fused image and the true fused image, respectively; i is the feature map channel index; C_f, H_f and W_f are the number of channels, height and width of the feature map, respectively. The expression for the basic loss (B_{loss}) is as follows:

$$B_{loss} = \frac{1}{3 H_g W_g} \sum_{i,x,y} \left[I_p^i(x,y) - I_g^i(x,y) \right]^2 \tag{4}$$

where I_p and I_g are the predicted fused image and the true fused image, respectively; i is the RGB image channel index; H_g and W_g are the height and width of the true fused image, respectively. The expression for the total loss (T_{loss}) is as follows:

$$T_{loss} = w_1 B_{loss} + w_2 P_{loss} \tag{5}$$

where w_1 and w_2 are the weighting coefficients. For the fusion of MTF-encoded images in this study, the sums are both set to 1.

4. Fault Diagnosis Method

4.1. Optimized Deep Residual Network

ResNet is built on the basis of CNN and solves the gradient vanishing problem by adding skip connections between the input and output of each convolutional layer. The classic residual module structure is shown in Figure 3.

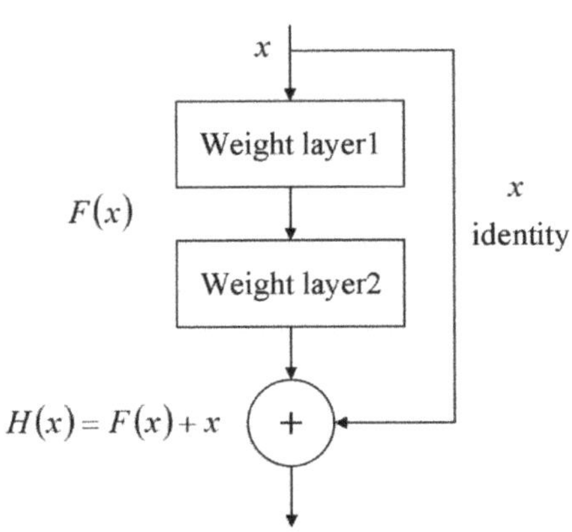

Figure 3. Residual module structure.

The structure contains two mappings, the part of the main path is called the residual mapping and the part of the bypass connection is called the constant mapping. The final output of the residual block is therefore the superposition of the outputs obtained from the two mappings:

$$H(x) = F(x) + x \tag{6}$$

The structure of the residual network model constructed in this study is shown in Table 3. It includes an input layer, a maximum pooling layer, convolutional layers, an average pooling layer, a fully connected layer and a softmax classifier. Conv2, Conv3, Conv4 and Conv5 are residual modules.

Table 3. ResNet model structure.

Layer Name		Kernel Size	Channel	Stride	Padding	Output
Input Image		-	-	-	-	3 × 224 × 224
Conv1		7 × 7	64	2	3	64 × 112 × 112
Maxpool		3 × 3	64	2	1	64 × 56 × 56
Conv2	Conv2_1	3 × 3	64	1	1	64 × 56 × 56
	Conv2_2	3 × 3	64	1	1	64 × 56 × 56
	Conv2_3	3 × 3	64	1	1	64 × 56 × 56
	Conv2_4	3 × 3	64	1	1	64 × 56 × 56
Conv3	Conv3_1	3 × 3	128	2	1	128 × 28 × 28
	Conv3_2	3 × 3	128	1	1	128 × 28 × 28
	Conv3_3	3 × 3	128	1	1	128 × 28 × 28
	Conv3_4	3 × 3	128	1	1	128 × 28 × 28
Conv4	Conv4_1	3 × 3	256	2	1	256 × 14 × 14
	Conv4_2	3 × 3	256	1	1	256 × 14 × 14
	Conv4_3	3 × 3	256	1	1	256 × 14 × 14
	Conv4_4	3 × 3	256	1	1	256 × 14 × 14
Conv5	Conv5_1	3 × 3	512	2	1	512 × 7 × 7
	Conv5_2	3 × 3	512	1	1	512 × 7 × 7
	Conv5_3	3 × 3	512	1	1	512 × 7 × 7
	Conv5_4	3 × 3	512	1	1	512 × 7 × 7
Avgpool		-	-	-	-	512 × 1 × 1
			Fc, Softmax			

Convolutional layers are the core of CNNs, responsible for extracting features from large amounts of input data. Typically, convolutional layers can be described by the following expression:

$$x_j^l = \sigma(\textstyle\sum_{i \in M_j} x_j^{l-1} * k_{ij}^l + b_j^l) \tag{7}$$

where x_j^{l-1} is the input of the $(l-1)$-th layer of the network; x_j^l is the output of the l-th layer of the network; k_{ij}^l is the weight matrix of the convolutional kernel; b_j^l is the bias term; M_j is the set of input feature maps; σ is the nonlinear activation function; and $*$ represents the convolution operation.

Pooling aims to reduce the size of feature maps while retaining the most important feature information. It can effectively reduce computational complexity and improve the model's robustness and generalization capabilities. The pooling process involves four steps: input feature map, sliding window coverage, feature aggregation, and output feature map. The pooling process can be described by the following expression:

$$x_j^l = \sigma(\beta_j^l down(x_j^{l-1}) + b_j^l) \tag{8}$$

where x_j^{l-1} is the input of the $(l-1)$-th layer of the network; x_j^l is the output of the l-th layer of the network; b_j^l is the bias term; σ is the nonlinear activation function; $down(\cdot)$ is the down-sampling function; and β_j^l is the weight.

To improve the efficiency of fault diagnosis, a convolutional block attention module (CBAM) is introduced to optimize the model by focusing it more on important features [28]. CBAM consists of channel attention module, which captures the connections between channels of the feature map, and spatial attention module, which captures the connections between spatial regions of the feature map.

The channel attention module feeds the features F_{avg}^c and $F_{\max}^c$ obtained after using average pooling and max pooling in the channel dimension into the convolutional network, respectively, and sums the results and outputs them. The process is described as:

$$M_c(F) = \sigma(W_1(W_0(F_{avg}^c)) + W_1(W_0(F_{\max}^c))) \quad (9)$$

where σ is a sigmoid function; W_0 and W_1 are convolution operations with a convolution kernel size of 1×1.

The spatial attention module performs a convolution operation on the features F_{avg}^s and $F_{\max}^s$ obtained after stitching using average pooling and max pooling in the channel dimension. The process is described as:

$$M_s(F) = \sigma(f^{7\times7}([F_{avg}^s; F_{\max}^s])) \quad (10)$$

where σ is a sigmoid function; $f^{7\times7}$ is convolution operation with a convolution kernel size of 7×7.

This study introduced CBAM into ResNet without changing the overall structure of the network. The input data are MTF feature images of size 224×224. After passing through the first convolutional layer with a kernel size of 7×7 and a stride of 2, the image size is reduced to 112×112. This is followed by a max pooling layer with a stride of 2, which further reduces the data dimensionality and the image size to 56×56. The channel attention and spatial attention modules are added sequentially after the batch normalization (BN) layer at the end of the residual modules Conv2, Conv3, Conv4 and Conv5, respectively. After passing through the Conv2, which has 64 channels and convolutional kernels of size 3×3 with a stride of 1, deeper features are extracted while maintaining the same image size as the previous layer. The channels in Conv3, Conv4, and Conv5 are doubled successively to 128, 256 and 512. At the same time, down-sampling is implemented in the first convolutional layer with a stride of 2 in each residual module. This results in output image sizes that progressively decrease to 28×28, 14×14 and 7×7, respectively. Afterwards, the network passes through an average pooling layer to reduce the number of parameters and mitigate the occurrence of overfitting. Then, a fully connected layer is used for nonlinear combination of the extracted features, followed by a softmax classifier to produce the final output.

The proposed model uses a cross-entropy loss function to evaluate the error between the predicted and true values, avoiding gradient dispersion, which is defined in the context of a multiclassification problem as:

$$L = \frac{1}{N}\sum_i L_i = -\frac{1}{N}\sum_i \sum_{c=1}^{M} y_{ic} \log(p_{ic}) \quad (11)$$

where M is the number of categories; y_{ic} is the sign function, taking 1 if the true value of sample i is equal to c and 0 otherwise; and p_{ic} is the predicted probability that sample i belongs to category c.

An initial test was carried out with a constant speed of 1600 rpm and a load of 7 kN, the number of epochs was set to 50 and the loss and accuracy (Acc) in training are shown in Figure 4.

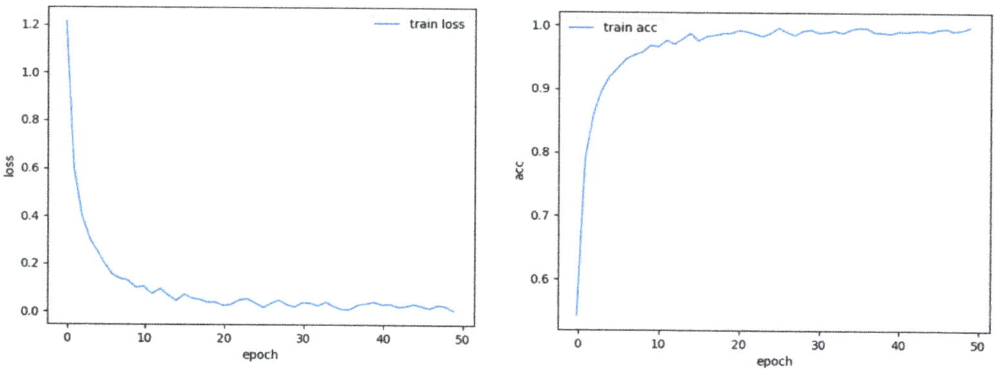

(**a**) (**b**)

Figure 4. Loss and accuracy in training with 50 epochs: (**a**) loss; (**b**) acc.

Overall, from the graph, it can be seen that when the epoch reaches 40, the loss and accuracy have basically converged, and the accuracy has reached nearly 100%. This indicates that the model performs well on the training set and has good generalization ability, which also verifies that the model structure and parameters chosen in this paper are correct. Setting the number of epochs too large can significantly prolong the training time and even cause overfitting, while setting it too small may not find the global optimal solution. After multiple tests, this paper set the learning rate to 0.001 and the number of epochs to 40, which is a good choice. To intuitively demonstrate the advantages of the proposed method in extracting fault features, this paper utilized the uniform manifold approximation and projection (UMAP) algorithm to perform dimensionality reduction on the data and visualize the results. Taking the steady state condition with a speed of 1600 rpm and a load of 7 kN as an example, this paper conducted a layer-by-layer analysis of ResNet models with and without CBAM and extracted the output features of the intermediate layers for calculation. Then, UMAP is utilized to reduce the dimensionality of the extracted features to two dimensions. This paper extracted the fault features from the avgpool layer and visualized the results using a scatter plot where different fault types are marked with different colors. The visualization is shown in Figure 5.

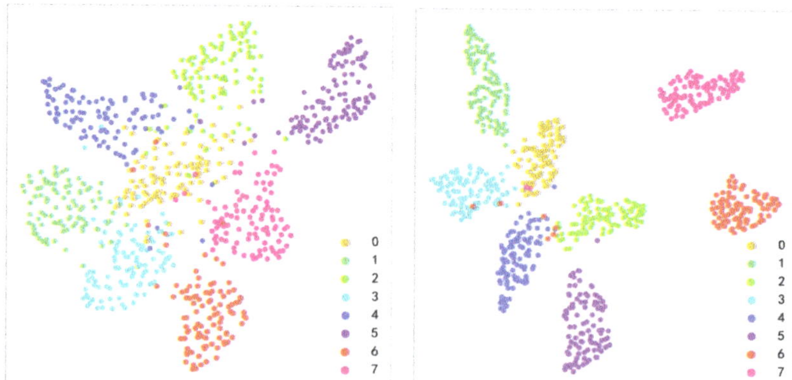

(**a**) (**b**)

Figure 5. Visualization of fault features extracted from the avgpool layer: (**a**) before the introduction of CBAM; (**b**) after the introduction of CBAM.

As can be seen from the figure above, there is a significant difference in the clustering degree of data samples between the two models, and introducing CBAM to ResNet can yield more obvious clustering effect in the avgpool layer. Therefore, it can be concluded that the proposed optimized ResNet has excellent abilities in extracting fault features under complex working conditions.

4.2. Fault Diagnosis Process

This paper proposes a compound fault diagnosis method of rolling bearings based on multisignal fusion and MTF-ResNet. The fused MTF-encoded images are input into the ResNet model for training, and the fault is intelligently diagnosed under different working conditions. The basic process is shown in Figure 6, and the main steps are as follows: (1) acquire vibration and acoustic emission signals; (2) generate feature images of size 224 × 224 by MTF encoding of the original data to build a training set and a test set; (3) fuse the MTF encoded images of the two signals using IFCNN; (4) input the training set into the optimized ResNet model built for training, and save the optimal parameters; and (5) test the test samples and output the results to complete the intelligent fault diagnosis.

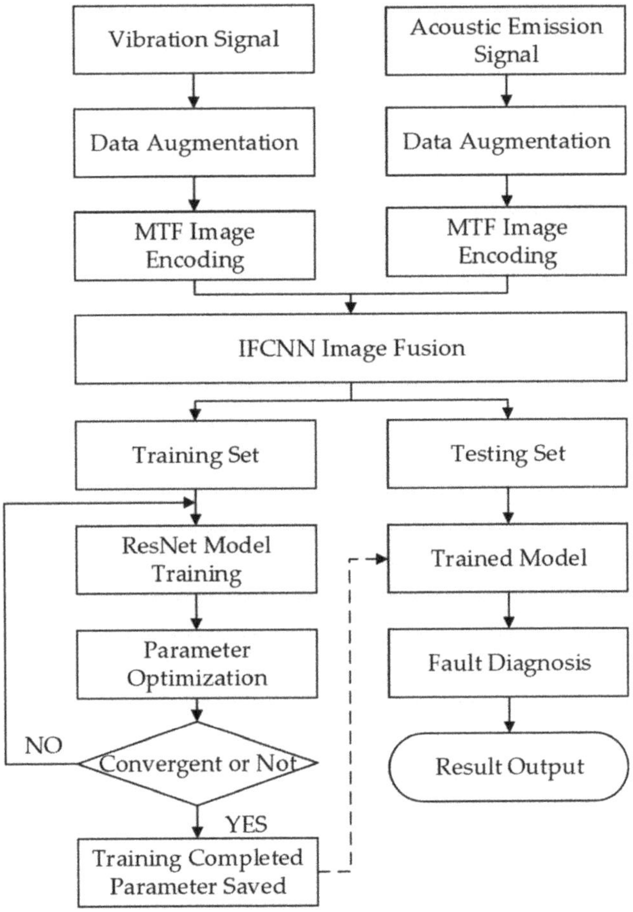

Figure 6. Fault diagnosis process of rolling bearings based on multisignal fusion and MTF-ResNet.

5. Fault Diagnosis Experiment

5.1. Experimental Design

The experimental bearing was selected as NU216 cylindrical roller bearing. Defects were artificially introduced to the inner and outer rings, as well as the rolling elements using a YLP-MDF-152 laser marking machine from Han's Laser. Taking into account the failure mechanism of bearings in actual working environments, alternating loads can cause cracks to form at a certain depth below the surface, which may then propagate to the surface and cause spalling. Fatigue spalling increases vibration and noise during rotation and is usually the main form of rolling bearing failure. Therefore, pitting was produced on the surface of the bearing at different locations to simulate early defects. The pitting diameter was set to 40 μm and the depth was set to 30% of the laser energy. Eight types of faults, as described in Section 2, were designed using different fault positions as classification criterion.

In order to simulate the working conditions of metro traction motors, three additional speeds and three additional loads were included in the experimental design. In consideration of both actual working conditions and minimizing the impact of bearing degradation on the experiment, gradient speeds of 800 rpm (low), 1600 rpm (medium) and 2400 rpm (high) were chosen, along with gradient equivalent dynamic loads of 5 kN (light), 7 kN (medium) and 9 kN (heavy) as the radial loads. There are a total of 72 ($8 \times 3 \times 3$) subexperiments. The experimental arrangement is shown in Table 4.

Table 4. Experimental arrangement.

Speed/rpm	Radial Loads for Different Fault Types/kN							
	Normal	Inner Ring	Outer Ring	Rolling Element	Inner Ring + Outer Ring	Inner Ring + Rolling Element	Outer Ring + Rolling Element	Inner Ring + Outer Ring + Rolling Element
800	5	5	5	5	5	5	5	5
800	7	7	7	7	7	7	7	7
800	9	9	9	9	9	9	9	9
1600	5	5	5	5	5	5	5	5
1600	7	7	7	7	7	7	7	7
1600	9	9	9	9	9	9	9	9
2400	5	5	5	5	5	5	5	5
2400	7	7	7	7	7	7	7	7
2400	9	9	9	9	9	9	9	9

5.2. Construction of the Signal Acquisition System

This study utilized the intelligent testing platform for comprehensive bearing performance, jointly developed by Henan University of Science and Technology, Luoyang Bearing Research Institute, and Intelligent Numerical Control Equipment Henan Provincial Engineering Laboratory, as the signal acquisition system. The testing machine allows for a maximum inner diameter of 120 mm, a maximum speed of 5000 r/min, a maximum radial load of 300 kN, and a maximum axial load of 200 kN for the bearing. The platform is equipped with a PCI-8 acoustic emission transmitter, two R50S-TC acoustic emission sensors, two LC0151T acceleration sensors, two LC0201-5 signal conditioners, and a PCI8510 data acquisition card.

During the experiment, a healthy bearing and a faulty bearing were installed at both ends of the testing machine's spindle, and vibration and acoustic emission signals were collected from both bearings simultaneously. The loading system applies radial loads to the spindle via a pair of NU2218 cylindrical roller bearings, which in turn are transferred to the test bearings at both ends of the spindle. The sensor signals are amplified and conditioned by signal amplifiers, signal conditioners, and input to the computer through a PCI acquisition card. The principle of the signal acquisition system is shown in Figure 7. The physical set-up of the system is shown in Figure 8.

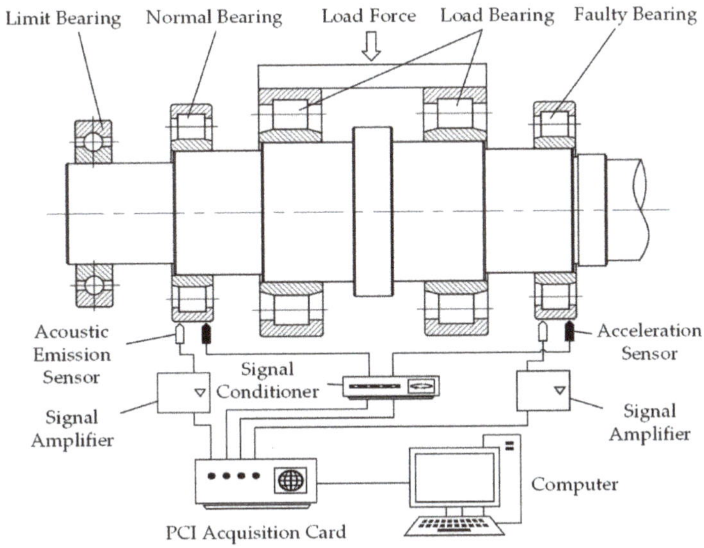

Figure 7. Schematic diagram of the signal acquisition system.

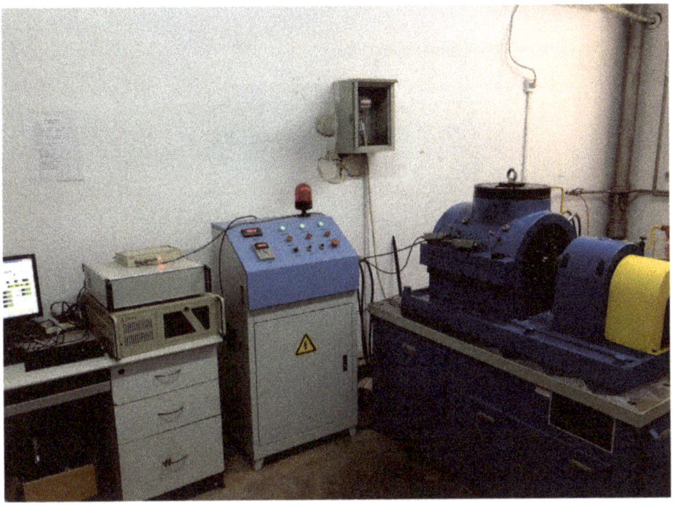

Figure 8. Photograph of the built signal acquisition system.

5.3. Diagnostic Scheme Design

To further validate the effectiveness of the proposed method, three types of diagnostic schemes were designed for single working condition changes, compound working condition changes, and generic working conditions, considering two different factors (speed and load) that affect the test results.

When studying single working condition changes, first control the speed to be constant, put data of two different loads in the training set, and put data of another load in the test set to verify the robustness of the model. When controlling the load to be constant, the method is similar to the above. The specific diagnostic program is shown in Table 5.

Table 5. Diagnostic scheme for single working condition change.

Fixed Variable	Variable	Training Set	Testing Set
Speed/rpm	Load/kN	5, 7	9
		5, 9	7
		7, 9	5
Load/kN	Speed/rpm	800, 1600	2400
		800, 2400	1600
		1600, 2400	800

When studying the change of compound working condition, it is required that the training set contains data with different speeds and loads at the same time. For generic working conditions, it is required that all fault types data under all conditions exist in both the training and testing sets.

6. Experimental Results and Comparison of Methods

During the operational process of a metro system, variations in bearing speed and load are inevitable. While previous steady-state tests have certain limitations, it becomes crucial to analyze the results of variable working condition tests to validate the effectiveness of the proposed method. To further explore the changes in compound working conditions, an additional analysis comparing the fusion of acoustic emission and vibration signals with a single signal was incorporated to emphasize the advantages of the proposed method. In the generic working condition tests, the feature extraction capabilities of four models, namely the proposed model, RepVGG, CBAM-CNN and ResNet, were compared to evaluate their performance.

6.1. Single Working Condition Changes

Based on the fault diagnosis method proposed in Section 5.3, with the control of constant speed and load, the training set was input into the model constructed in this paper, and fault diagnosis was performed on the test set. The diagnostic results are shown in Table 6.

Table 6. The diagnostic results for single working condition changes.

No.	Speed of Training Set/rpm	Load of Training Set/kN	Speed of Testing Set/rpm	Load of Testing Set/kN	Diagnostic Accuracy/%
1	800	5, 7	800	9	99.5
2	800	5, 9	800	7	100
3	800	7, 9	800	5	93
4	1600	5, 7	1600	9	100
5	1600	5, 9	1600	7	99.4
6	1600	7, 9	1600	5	92.4
7	2400	5, 7	2400	9	100
8	2400	5, 9	2400	7	100
9	2400	7, 9	2400	5	94
10	800, 1600	5	2400	5	100
11	800, 2400	5	1600	5	100
12	1600, 2400	5	800	5	78
13	800, 1600	7	2400	7	98.6
14	800, 2400	7	1600	7	90.4
15	1600, 2400	7	800	7	81
16	800, 1600	9	2400	9	98.8
17	800, 2400	9	1600	9	99.4
18	1600, 2400	9	800	9	83

Based on a comprehensive examination of the aforementioned table, it is observed that when maintaining a constant speed while altering the load, the fault diagnosis accuracy reaches nearly 100%. Conversely, in cases where the load remains constant but the speed varies, a decrease in fault diagnosis accuracy is observed, indicating a substantial influence of rotational speed on diagnostic outcomes. Subsequent analysis reveals that the accuracy of items numbered 12, 15 and 18 is significantly low, whereas items numbered 3, 6 and 9 demonstrate accuracy close to 100%, albeit slightly lower than other items within the initial nine numbers. This discrepancy can be attributed to the fact that fault characteristics extracted under medium- to high-speed and medium to heavy load conditions are more discernible compared to those under low-speed and light load conditions.

6.2. Compound Working Condition Changes

Mixed data with different speeds and loads were included in the training set and used to train the model proposed for fault diagnosis on the testing set. Subsequently, a comparison was made between the fusion of acoustic emission and vibration signals and using a single signal. The diagnostic results are shown in Table 7.

Table 7. The diagnostic results for compound working condition changes.

No.	Speed of Training Set/rpm	Load of Training Set/kN	Speed of Testing Set/rpm	Load of Testing Set/kN	Diagnostic Accuracy/%		
					Vibration	Acoustic Emission	Fusion
1	800, 1600	5, 7, 9	2400	5, 7, 9	90	82	94.1
2	800, 2400	5, 7, 9	1600	5, 7, 9	93.4	88.1	97.6
3	1600, 2400	5, 7, 9	800	5, 7, 9	71.4	66	75
4	800, 1600, 2400	5, 7	800, 1600, 2400	9	98	90.5	100
5	800, 1600, 2400	5, 9	800, 1600, 2400	7	96.5	85	99.4
6	800, 1600, 2400	7, 9	800, 1600, 2400	5	97.1	83.4	98.6

The table clearly indicates that the diagnostic results of items numbered 4 to 6 surpass those of items numbered 1 to 3. Notably, the training and testing sets for items numbered 1 to 3 encompass varying rotation speeds, whereas items numbered 4 to 6 involve different loads. It is observed that the diagnostic accuracy of items numbered 4 to 6 remains relatively stable, whereas item numbered 3 exhibits significantly lower accuracy compared to items numbered 1 and 2. The underlying reason behind this phenomenon aligns with the findings presented in Section 6.1 of this paper.

From the standpoint of signal acquisition, the fusion of acoustic emission and vibration signals yields higher diagnostic accuracy in fault diagnosis compared to utilizing a single signal. This finding provides further substantiation that the application of multisignal fusion technology can effectively enhance system stability and diagnostic accuracy. Furthermore, it is evident that employing a single vibration signal for diagnostics yields superior results in comparison to employing a single acoustic emission signal. This can be attributed to the fact that the acoustic emission acquisition system exhibits heightened sensitivity to environmental noise, primarily stemming from the operational testing equipment, which poses challenges in noise elimination.

6.3. Generic Working Conditions

To evaluate the performance of the proposed fault diagnosis model, all fault samples involving three different speeds and three different loads were included in both the training and testing sets. The sample ratio between the two sets was set to 9:1 to ensure the training set was large enough to enable the model to effectively learn the fault data while still reserving an adequate number of samples for testing. Subsequently, the model was applied to diagnose faults on the testing set. To visualize the diagnostic results, a confusion matrix was employed, providing an intuitive and reliable representation of classifications made by the model. The confusion matrix is presented in Figure 9.

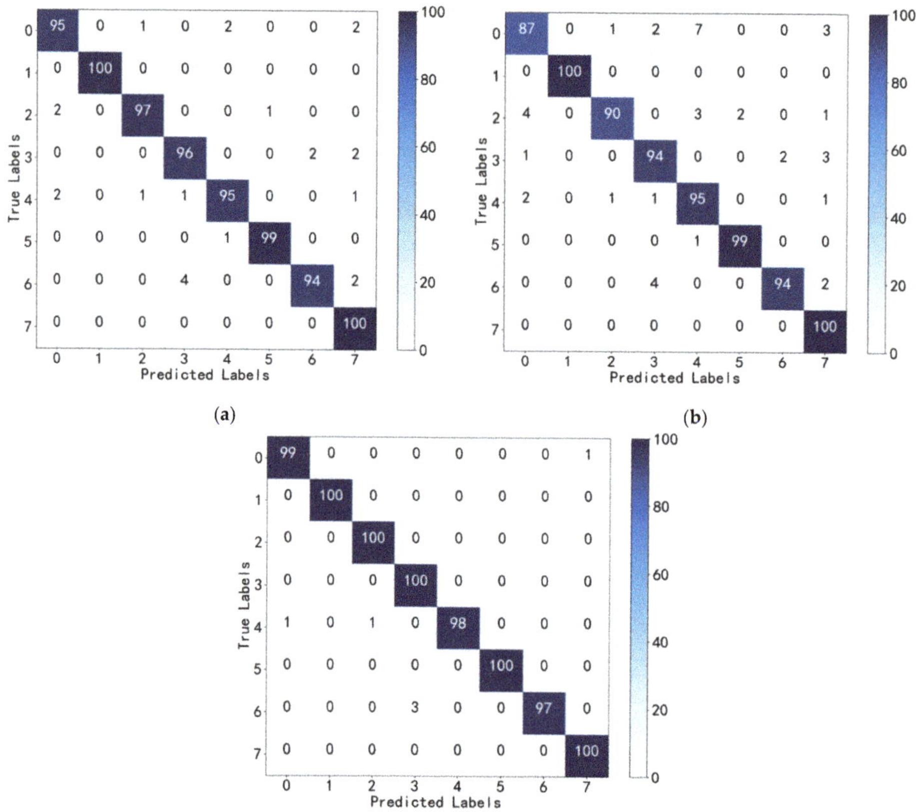

Figure 9. The diagnostic results for generic working condition: (**a**) based on vibration signal (with an accuracy rate of 97%); (**b**) based on acoustic emission signal (with an accuracy rate of 94.88%); (**c**) based on the fusion of acoustic emission and vibration signals (with an accuracy rate of 99.25%).

The confusion matrix provides a clear and intuitive visualization of the model's misclassifications and the types of errors. It can be seen that the overall diagnostic performance is good, and the accuracy rate for the fusion of acoustic emission and vibration signals is almost 100%. However, the diagnosis accuracy rate for label 6, which corresponds to the "outer Ring + rolling element pitting" fault type, is relatively low. The model misclassified three test samples as "rolling element pitting". Further analysis revealed that the two types of faults have similar features, making it difficult to extract differences between them. By comparing (a–c) in Figure 9, the results further confirm that multisignal fusion technology has higher reliability and accuracy compared to a single signal, especially under changing working conditions.

To compare the feature extraction capabilities of different models, the training and testing sets samples of above-mentioned generic working conditions were respectively input into RepVGG, CBAM-CNN and ResNet models for diagnosis. Two types of faults were selected as examples: label 1 (corresponding to "inner ring pitting") with better diagnostic results and label 6 (corresponding to "outer ring + rolling element pitting") with poorer results. The precision–recall (PR) curves and receiver operating characteristic (ROC) curves were generated for the optimized ResNet, RepVGG, CBAM-CNN and ResNet models and evaluation indicators, such as average precision (AP) and area under the curve (AUC) were introduced.

The precision–recall (PR) curve is a graphical representation of the performance of a binary classification model, with recall on the x-axis and precision on the y-axis. It illustrates the trade-off between precision and recall at various classification thresholds. The relevant theoretical formulas for the PR curve are as follows:

$$Precision = \frac{TP}{TP + FP} \tag{12}$$

$$Recall = \frac{TP}{TP + FN} \tag{13}$$

where TP represents the number of true positive instances; FP represents the number of false positive instances; and FN represents the number of false negative instances.

The principle of average precision (AP) is to summarize the Precision-Recall (PR) curve by calculating the average precision value. It can be obtained by computing the area under the PR curve. It provides a comprehensive assessment of how well the model balances precision and recall across different recall levels.

The receiver operating characteristic (ROC) curve is a tool used to evaluate the performance of binary classification models. It plots the false positive rate (FPR) on the x-axis and the true positive rate (TPR) on the y-axis. The principle of the ROC curve can be described using the following formulas:

$$TPR = \frac{TP}{TP + FN} \tag{14}$$

$$FPR = \frac{FP}{FP + TN} \tag{15}$$

where FP represents the number of negative instances incorrectly classified as positive; TN represents the number of negative instances correctly classified as negative; TP represents the number of positive instances correctly classified as positive; and FN represents the number of positive instances incorrectly classified as negative.

Area under the curve (AUC) is obtained by calculating the area under the ROC curve. The resulting AUC value ranges from 0 to 1, where 0.5 represents a random classifier and 1 represents a perfect classifier. A higher AUC value indicates better classifier performance.

The diagnostic results are presented in the form of PR and ROC curves in Figures 10 and 11. The overall accuracy rate, AP and AUC for all fault types were calculated for the four models, and the weighted average values were recorded in Table 8.

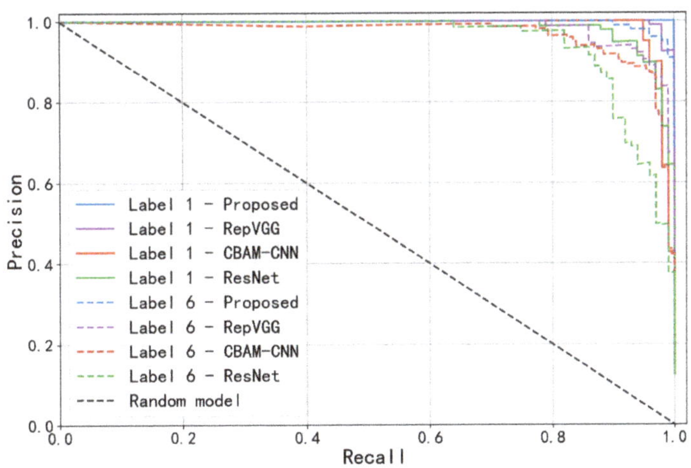

Figure 10. PR curves of four models with two fault types.

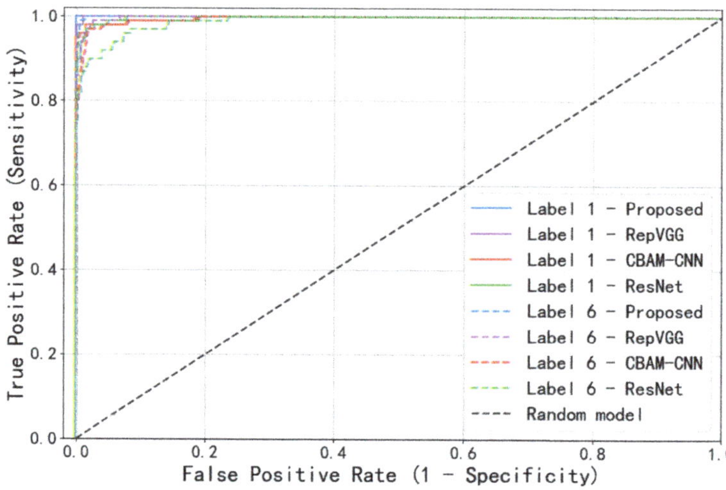

Figure 11. ROC curves of four models with two fault types.

Table 8. The accuracy evaluation indicators of the four models.

Model	Evaluation Indicator		
	Accuracy/%	AP	AUC
Proposed	99.25	0.989	1.000
RepVGG	96.72	0.967	0.996
CBAM-CNN	94.16	0.953	0.993
ResNet	88.35	0.935	0.988

Generally, the closer the PR curve in Figure 10 is to the upper right corner, the larger the AP value, and the better the model performance. The closer the ROC curve in Figure 11 is to the upper left corner, the larger the AUC value, and the better the model performance. Observing the figure above, it can be seen that for the two selected fault types with different diagnostic effects, the PR and ROC curves of proposed model are both closer to the right-angle edge than those of RepVGG, CBAM-CNN and ResNet, indicating better performance. Combined with the data in Table 8, the three accuracy evaluation indicators of the proposed model are higher than those of the compared models, validating the good feature extraction ability of the proposed model.

7. Conclusions

This paper focused on the study of the feature extraction ability of the model for complex working conditions, using the metro traction motor bearings as the research object. On the basis of ResNet, CBAM was introduced to optimize the ResNet model. Nine different working conditions and eight compound fault types were designed for experimentation. In addition, a dataset was constructed using MTF image encoding and IFCNN image fusion technology. During the model training process, UMAP was used for visualization to intuitively demonstrate the feature extraction effect of the proposed model. After the experiment, three evaluation indicators were used for objective evaluation of the feature extraction ability of the optimized ResNet, RepVGG, CBAM-CNN and ResNet models.

The results of the experiment show that the MTF-ResNet model with multisignal fusion performs well under complex working conditions, with a diagnostic accuracy rate of up to 99.25%. Based on the results, some important conclusions can be drawn. Specifically, in terms of sensors, using only vibration signals produces better diagnostic results than using only acoustic emission signals. In addition, compared with a single signal, using acoustic emission and vibration signal fusion can provide more comprehensive and integrated information, while reducing misclassifications caused by the limitations of a single signal, thereby improving fault diagnosis accuracy and making the diagnosis result more reliable. In terms of data processing, MTF image encoding technology is a simple data processing method that retains the time correlation of the data, making it easier for the model to extract more comprehensive fault features. For feature extraction models, introducing CBAM after the batch normalization layers of the ResNet model can make the model more focused on capturing important features, quickly distinguishing different types of fault features, and improving diagnostic efficiency. Furthermore, the ResNet structure can effectively alleviate the gradient disappearance phenomenon that occurs as the network deepens, thereby preventing model degradation.

Undoubtedly, this study presents several avenues for future research in the proposed methodologies. Firstly, the inclusion of additional sensors or exploration of different sensor types holds promise. For instance, incorporating multidirectional vibration sensors or temperature sensors could offer a more comprehensive spectrum of fault information, thereby enhancing diagnostic fault tolerance. Secondly, exploring more advanced data processing techniques warrants investigation to enhance the quality of input signals. The acoustic emission signals acquired in this study exhibited significant levels of environmental noise that proved challenging to eliminate. Therefore, employing sophisticated techniques may substantially improve the value derived from these acoustic emission signals. Moreover, conducting model testing on larger datasets utilizing more complex compound faults can effectively confirm the feature extraction capabilities and generalization of the model. This approach will serve as a more robust means of validation. Furthermore, future research focusing on feature extraction models should prioritize the development of lightweight and efficient models to facilitate practical implementation.

Despite the inherent limitations of the methods proposed in this paper, they exhibit commendable feature extraction capabilities within intricate operational scenarios. Consequently, these methods hold potential for application in fault diagnosis tasks related to metro traction motor bearings, thereby possessing appreciable value in engineering applications.

Author Contributions: Conceptualization, K.H. and Y.X.; methodology, K.H. and Y.X.; validation, K.H., Y.X. and Y.W.; formal analysis, K.H.; investigation, K.H.; data curation, K.H. and Y.W.; writing—original draft preparation, K.H.; writing—review and editing, K.H., Y.X. and J.W.; visualization, K.H.; supervision, T.X.; funding acquisition, Y.X. All authors have read and agreed to the published version of the manuscript.

Funding: This research was funded by the National Natural Science Foundation of China (grant number: 51805151) and the Key Scientific Research Project of the University of Henan Province of China (grant number: 21B460004).

Institutional Review Board Statement: Not applicable.

Informed Consent Statement: Not applicable.

Data Availability Statement: The data presented in this study are available on request from the author.

Conflicts of Interest: The authors declare no conflict of interest.

References

1. Liu, Z.; Zhang, L.; Carrasco, J. Vibration analysis for large-scale wind turbine blade bearing fault detection with an empirical wavelet thresholding method. *Renew. Energy* **2020**, *146*, 99–110. [CrossRef]
2. Chen, W.; Li, J.; Wang, Q.; Han, K. Fault feature extraction and diagnosis of rolling bearings based on wavelet thresholding denoising with CEEMDAN energy entropy and PSO-LSSVM. *Measurement* **2021**, *172*, 108901. [CrossRef]
3. Sun, Y.; Li, S.; Wang, X. Bearing fault diagnosis based on EMD and improved Chebyshev distance in SDP image. *Measurement* **2021**, *176*, 109100. [CrossRef]
4. Zhao, X.; Qin, Y.; He, C.; Jia, L. Underdetermined blind source extraction of early vehicle bearing faults based on EMD and kernelized correlation maximization. *J. Intell. Manuf.* **2022**, *33*, 185–201. [CrossRef]

5. Li, H.; Liu, T.; Wu, X.; Chen, Q. An optimized VMD method and its applications in bearing fault diagnosis. *Measurement* **2020**, *166*, 108185. [CrossRef]
6. Li, Y.; Tang, B.; Geng, B.; Jiao, S. Fractional order fuzzy dispersion entropy and its application in bearing fault diagnosis. *Fractal Fract.* **2022**, *6*, 544. [CrossRef]
7. Li, Y.; Geng, B.; Tang, B. Simplified coded dispersion entropy: A nonlinear metric for signal analysis. *Nonlinear Dyn.* **2023**, *111*, 9327–9344. [CrossRef]
8. LeCun, Y.; Bengio, Y.; Hinton, G. Deep learning. *Nature* **2015**, *521*, 436–444. [CrossRef]
9. Mazhar, T.; Irfan, H.M.; Khan, S.; Haq, I.; Ullah, I.; Iqbal, M.; Hamam, H. Analysis of Cyber Security Attacks and Its Solutions for the Smart Grid Using Machine Learning and Blockchain Methods. *Future Internet* **2023**, *15*, 83. [CrossRef]
10. Naqvi, R.A.; Hussain, D.; Loh, W.-K. Artificial Intelligence-Based Semantic Segmentation of Ocular Regions for Biometrics and Healthcare Applications. *Comput. Mater. Contin.* **2021**, *66*, 715–732. [CrossRef]
11. Rahmani, A.M.; Ali, S.; Yousefpoor, M.S.; Yousefpoor, E.; Naqvi, R.A.; Siddique, K.; Hosseinzadeh, M. An area coverage scheme based on fuzzy logic and shuffled frog-leaping algorithm (sfla) in heterogeneous wireless sensor networks. *Mathematics* **2021**, *9*, 2251. [CrossRef]
12. Song, X.; Cong, Y.; Song, Y.; Chen, Y.; Liang, P. A bearing fault diagnosis model based on CNN with wide convolution kernels. *J. Ambient Intell. Humaniz. Comput.* **2021**, *13*, 4041–4056. [CrossRef]
13. Shan, S.; Liu, J.; Wu, S.; Shao, Y.; Li, H. A motor bearing fault voiceprint recognition method based on Mel-CNN model. *Measurement* **2023**, *207*, 112408. [CrossRef]
14. Chen, X.; Zhang, B.; Gao, D. Bearing fault diagnosis base on multi-scale CNN and LSTM model. *J. Intell. Manuf.* **2021**, *32*, 971–987. [CrossRef]
15. Haq, I.; Mazhar, T.; Malik, M.A.; Kamal, M.M.; Ullah, I.; Kim, T.; Hamdi, M.; Hamam, H. Lung Nodules Localization and Report Analysis from Computerized Tomography (CT) Scan Using a Novel Machine Learning Approach. *Appl. Sci.* **2022**, *12*, 12614. [CrossRef]
16. Zhao, X.; Jia, M.; Bin, J.; Wang, T.; Liu, Z. Multiple-order graphical deep extreme learning machine for unsupervised fault diagnosis of rolling bearing. *IEEE Trans. Instrum. Meas.* **2021**, *70*, 1–12. [CrossRef]
17. Long, J.; Chen, Y.; Yang, Z.; Huang, Y.; Li, C. A novel self-training semi-supervised deep learning approach for machinery fault diagnosis. *Int. J. Prod. Res.* **2022**, 1–14. [CrossRef]
18. Wang, X.; Mao, D.; Li, X. Bearing fault diagnosis based on vibro-acoustic data fusion and 1D-CNN network. *Measurement* **2021**, *173*, 108518. [CrossRef]
19. Al Mamun, A.; Bappy, M.M.; Mudiyanselage, A.S.; Li, J.; Jiang, Z.; Tian, Z.; Fuller, S.; Falls, T.; Bian, L.; Tian, W. Multi-channel sensor fusion for real-time bearing fault diagnosis by frequency-domain multilinear principal component analysis. *Int. J. Adv. Manuf. Technol.* **2023**, *124*, 1321–1334. [CrossRef]
20. Wang, J.; Wang, D.; Wang, S.; Li, W.; Song, K. Fault diagnosis of bearings based on multi-sensor information fusion and 2D convolutional neural network. *IEEE Access* **2021**, *9*, 23717–23725. [CrossRef]
21. He, K.; Zhang, X.; Ren, S.; Sun, J. Deep residual learning for image recognition. In Proceedings of the IEEE Conference on Computer Vision and Pattern Recognition, Las Vegas, NV, USA, 27–30 June 2016; pp. 770–778.
22. Zhang, T.; Liu, S.; Wei, Y.; Zhang, H. A novel feature adaptive extraction method based on deep learning for bearing fault diagnosis. *Measurement* **2021**, *185*, 110030. [CrossRef]
23. Yu, H.; Miao, X.; Wang, H. Bearing fault reconstruction diagnosis method based on ResNet-152 with multi-scale stacked receptive field. *Sensors* **2022**, *22*, 1705. [CrossRef] [PubMed]
24. Wen, L.; Li, X.; Gao, L. A transfer convolutional neural network for fault diagnosis based on ResNet-50. *Neural Comput. Appl.* **2020**, *32*, 6111–6124. [CrossRef]
25. Wang, Z.; Oates, T. Encoding time series as images for visual inspection and classification using tiled convolutional neural networks. In Proceedings of the Workshops at the Twenty-Ninth AAAI Conference on Artificial Intelligence, Austin, TX, USA, 25–30 January 2015.
26. Liu, L.; Wang, Z. Encoding temporal markov dynamics in graph for time series visualization. *arXiv* **2016**, arXiv:1610.07273.
27. Zhang, Y.; Liu, Y.; Sun, P.; Yan, H.; Zhao, X.; Zhang, L. IFCNN: A general image fusion framework based on convolutional neural network. *Inf. Fusion* **2020**, *54*, 99–118. [CrossRef]
28. Woo, S.; Park, J.; Lee, J.-Y.; Kweon, I.S. Cbam: Convolutional block attention module. In Proceedings of the European Conference on Computer Vision (ECCV), Munich, Germany, 8–14 September 2018; pp. 3–19.

sensors

MDPI

Article

A Dual-Optimization Fault Diagnosis Method for Rolling Bearings Based on Hierarchical Slope Entropy and SVM Synergized with Shark Optimization Algorithm

Yuxing Li [1,2,*], Bingzhao Tang [1], Bo Huang [1] and Xiaohui Xue [1]

1 School of Automation and Information Engineering, Xi'an University of Technology, Xi'an 710048, China; 2210321205@stu.xaut.edu.cn (B.T.)

2 Shaanxi Key Laboratory of Complex System Control and Intelligent Information Processing, Xi'an University of Technology, Xi'an 710048, China

* Correspondence: liyuxing@xaut.edu.cn

Abstract: Slope entropy (SlopEn) has been widely applied in fault diagnosis and has exhibited excellent performance, while SlopEn suffers from the problem of threshold selection. Aiming to further enhance the identifying capability of SlopEn in fault diagnosis, on the basis of SlopEn, the concept of hierarchy is introduced, and a new complexity feature, namely hierarchical slope entropy (HSlopEn), is proposed. Meanwhile, to address the problems of the threshold selection of HSlopEn and a support vector machine (SVM), the white shark optimizer (WSO) is applied to optimize both HSlopEn and an SVM, and WSO-HSlopEn and WSO-SVM are proposed, respectively. Then, a dual-optimization fault diagnosis method for rolling bearings based on WSO-HSlopEn and WSO-SVM is put forward. We conducted measured experiments on single- and multi-feature scenarios, and the experimental results demonstrated that whether single-feature or multi-feature, the WSO-HSlopEn and WSO-SVM fault diagnosis method has the highest recognition rate compared to other hierarchical entropies; moreover, under multi-features, the recognition rates are all higher than 97.5%, and the more features we select, the better the recognition effect. When five nodes are selected, the highest recognition rate reaches 100%.

Keywords: fault diagnosis; hierarchical slope entropy; white shark optimizer; optimized support vector machine; bearing signals

Citation: Li, Y.; Tang, B.; Huang, B.; Xue, X. A Dual-Optimization Fault Diagnosis Method for Rolling Bearings Based on Hierarchical Slope Entropy and SVM Synergized with Shark Optimization Algorithm. *Sensors* **2023**, *23*, 5630. https://doi.org/10.3390/s23125630

Academic Editor: Jongmyon Kim

Received: 2 June 2023
Revised: 11 June 2023
Accepted: 13 June 2023
Published: 16 June 2023

1. Introduction

Rolling bearings, as a key component in rotating machinery, serve a very significant role in modern industry. However, because of the increasingly sophisticated and complex structure of bearings and their common use in harsh working environments, rolling bearings are very prone to failures, which can lead to economic losses and even endanger personal safety [1–3]. Therefore, aiming to ensure the normal work of rotating machinery and reduce maintenance costs, it is of great importance to carry out fault diagnoses of rolling bearings [4–6].

Since bearing vibration signals contain rich state information about the bearing during operation, a vibration analysis method is broadly applied to rolling bearing faults [7,8]. In general, the method mainly consists of two steps: feature extraction and fault classification, in which valid feature extraction is crucial for accurate fault diagnosis. As the bearing vibration signal has nonlinear dynamic characteristics, traditional feature extraction methods based on Fourier transform and statistical analysis only characterize features from the time domain or frequency domain, and they cannot detect potential faults through changes in the complexity of the system to achieve effective and accurate extractions of fault features [9,10].

In recent years, nonlinear dynamic methods, such as sample entropy (SE) [11], permutation entropy (PE) [12], and dispersion entropy (DE) [13], have been widely used in the feature extraction of bearing signals and have presented superior performance. Han et al., used SE to extract bearing fault feature information effectively [14], but SE calculation is complicated and not suitable for real-time monitoring. As PE has the strengths of fast calculation and good stability, Xue et al., proposed a bearing fault diagnosis method based on PE and further improved the effectiveness of fault diagnosis [15]. While PE does not consider amplitude information, Dhandapani et al., applied DE to the feature extraction osf rolling bearing faults and considered the amplitude information of bearing signals [16]. Unlike the above entropies, slope entropy (SlopEn) is a new entropy estimator proposed based on symbolic patterns and magnitude information [17] and has been applied in the underwater acoustic field and medical field many times [18–21]. In 2022, SlopEn was introduced into the field of bearing fault diagnosis for the first time, and experimental results showed that, compared with PE and DE, SlopEn could better extract fault information [22]. However, all the above-mentioned entropy-based bearing fault diagnosis methods suffer from two defects: (i) the methods extract only the fault information of the low-frequency component for the bearing signal, and (ii) there is the problem of threshold selection for SlopEn, and the thresholds usually need to be optimized using an optimization algorithm.

Aiming at extracting the bearing fault information more comprehensively, some scholars have proposed fault diagnosis methods based on hierarchical entropy [23–25]. The authors of [26] proposed the concept of hierarchical permutation entropy (HPE) and employed it successfully for fault diagnosis. Moreover, Ref. [27] used hierarchical permutation entropy (HDE) to extract the fault information in both high- and low-frequency components. Diagnosis methods based on hierarchical entropy have proved that they can obtain diagnosis-related information of the whole frequency band and have strong noise resistance and stability; in addition, no scholars have introduced the concept of hierarchy to the SlopEn and used optimization algorithms to optimize the thresholds.

After feature extraction, the next step is fault classification. Commonly used fault classification methods mainly include k-nearest neighbor (KNN) [28], random forest (RF) [29], and the support vector machine (SVM) [30]. The SVM has been widely used in fault diagnosis because of its suitability for small sample classification and its simple structure [31]. Yet, since the parameter penalty factors and kernel functions of SVM models have an impact on fault diagnosis, some existing optimization algorithms have optimized the parameters of SVMs and improved the performance of fault classification, such as the genetic algorithm (GA) [32], particle swarm algorithm (PSO) [33], and whale optimization algorithm (WOA) [34]. Compared to common optimization algorithms, the white shark optimizer (WSO) is a new meta-heuristic optimization algorithm based on deep-sea foraging by great white sharks, proposed in 2022 for solving optimization problems on continuous search spaces [35]; in addition, the results of the basis function tests show that WSO is better than the common optimization algorithm in terms of optimization and has not yet been applied to optimize SVMs.

Based on the analysis above, a dual-optimization fault diagnosis method for rolling bearings is put forward, and the main novelties and contributions of this paper are presented as follows:

(1) To extract the fault information of bearing signals more comprehensively, on the basis of SlopEn, this paper adds the concept of hierarchy and is the first to propose hierarchical slope entropy (HSlopEn).
(2) Since the thresholds of HSlopEn have a relatively large impact on the entropy value and the selection of suitable parameters of an SVM is particularly important for the classification, this paper applies the WSO to optimize the parameters of HSlopEn and an SVM and proposes WSO-HSlopEn and WSO-SVM, respectively.
(3) Targeting the application of bearing fault diagnosis under different operating conditions, this paper proposes a dual-optimization fault diagnosis method for rolling bearings based on HSlopEn and an SVM synergized with the WSO.

The remaining parts of this paper are structured as follows. Section 2 presents the basic concepts of algorithms. Section 3 introduces the steps of the proposed fault diagnosis method. Section 4 carries out the single-feature and multi-feature extraction experiments for bearing signals, and Section 5 summarizes the conclusions of this study.

2. Methodology

2.1. Slope Entropy

Slope entropy (SlopEn) is an algorithm proposed in 2019 to calculate the complexity of time series. It is based on symbolic patterns and magnitude information. The main calculation steps are listed below:

(1) For a given time series $X = \{x_1, x_2, \cdots, x_N\}$, according to the embedding dimension m, extract the subsequences:$X_1 = \{x_1, x_2, \cdots, x_m\}$, $X_2 = \{x_2, x_3, \ldots, x_{m-1}\}, \ldots,$ $X_k = \{x_k, x_{k+1}, \ldots, x_N\}$, of which $k = N - m + 1$.

(2) Two threshold parameters, γ and δ, are used to delimit the symbolic patterns including (+2, +1, 0, −1, −2) between two consecutive samples $x_{i+1} - x_i$ of the subsequence $X_1, X_2, \cdots, X_k$. When $|x_{i+1} - x_i| \leq \delta$, the symbol is 0; when $\delta < x_{i+1} - x_i < \gamma$, the symbol is +1; when $-\gamma < x_{i+1} - x_i < -\delta$, the symbol is −1; when $\gamma < x_{i+1} - x_i$, when symbol is +2; when $x_{i+1} - x_i < -\gamma$, the symbol is −2, where $\gamma > \delta > 0$. Figure 1 displays the symbol allocation of SlopEn.

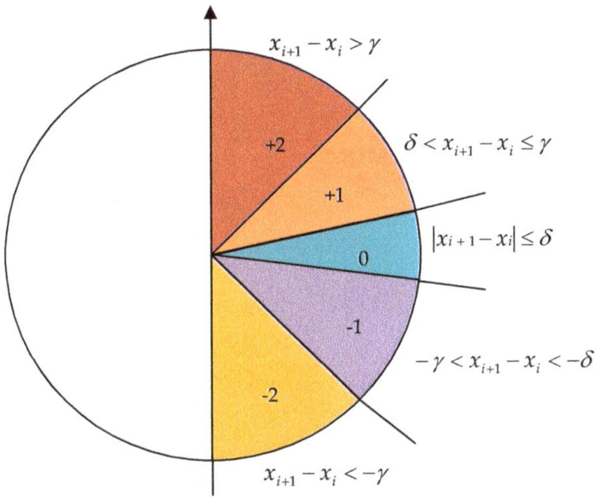

Figure 1. The symbol allocation of SlopEn.

(3) According to the symbol pattern of step (2), divide all the subsequences $X_1, X_2, \cdots, X_k$, and obtain the symbol pattern sequences $Y_1 = \{y_1, y_2, \ldots, y_{m-1}\}$, $Y_2 = \{y_2, y_3, \cdots, y_m\}$, $\ldots, Y_k = \{y_k, y_{k+1}, \cdots, y_{N-1}\}$, where y_1 is obtained from $x_2 - x_1$ through step (2), y_2 is obtained from $x_3 - x_2$ through step (2), … , and y_{N-1} is obtained from $x_N - x_{N-1}$ through step (2).

(4) The total number of types of symbol pattern sequences is recorded as $r = 5^{m-1}$, and the number of occurrences of each symbol pattern sequence is $s_1, s_2, \cdots, s_r$, respectively. Hence, the probability of each symbol pattern sequence appearing is $P_1 = \frac{s_1}{r}, P_2 = \frac{s_2}{r}, \ldots, P_r = \frac{s_r}{r}$, respectively.

(5) On the basis of Shannon information entropy, the definition formula of SlopEn is

$$SlopEn(X, m, \gamma, \delta) = -\sum_{i=1}^{r} P_r \ln P_r \tag{1}$$

2.2. Hierarchical Slope Entropy

Since SlopEn only considers the low-frequency components of the time series, aiming to describe the time series more comprehensively, on the basis of SlopEn and combined with the concept of hierarchy, this paper proposes a new complexity feature, namely hierarchical slope entropy (HSlopEn). The specific process of HSlopEn is as follows:

(1) First, given a time series $X = \{x(i), i = 0, 1, \cdots, N, N = 2^n\}$ of length N, define an average operator $Q_0(x)$ and a difference operator $Q_1(x)$, which can be expressed as

$$Q_0 = \frac{x(2j) + x(2j+1)}{2}, j = 0, 1, 2, \cdots, 2^{n-1} \tag{2}$$

$$Q_1 = \frac{x(2j) - x(2j+1)}{2}, j = 0, 1, 2, \cdots, 2^{n-1} \tag{3}$$

where the $Q_0(x)$ and $Q_1(x)$ operators are the low-frequency part and high-frequency part, respectively, of the original given time series after hierarchical decomposition and n is a positive integer.

(2) The operators $Q_j(j = 0 \text{ or } 1)$ in matrix form is defined as

$$Q_j = \begin{bmatrix} \frac{1}{2} & \left(\frac{-1}{2}\right)^j & 0 & 0 & \cdots & 0 & 0 \\ 0 & 0 & \frac{1}{2} & \left(\frac{-1}{2}\right)^j & \cdots & 0 & 0 \\ 0 & 0 & 0 & 0 & \cdots & \frac{1}{2} & \left(\frac{-1}{2}\right)^j \end{bmatrix}_{2^{n-1} \times 2^n} \tag{4}$$

(3) The l-dimension vectors $[u_1, u_2, \ldots, u_l] \in \{0, 1\}(l \epsilon N)$ are constructed, and an integer e can be expressed:

$$e = \sum_{j=1}^{l} u_j 2^{l-j} \tag{5}$$

where, for a positive e, there is a unique set of l-dimension vectors $[u_1, u_2, \ldots, u_l] \in \{0, 1\}$ and the positive integer e represents the sequence number of the node at each layer, where $0 \leqslant e \leqslant 2^{n-1}$.

(4) The hierarchical decomposition of a given time series X yields a hierarchical component corresponding to the node e at the Kth level, defined as

$$X_{K,e} = Q_{u_l} * Q_{u_{l-1}} * \cdots * Q_{u_1}(x) * X \tag{6}$$

(5) By calculating the SlopEn of nodes on different layers, the HSlopEn can be expressed as

$$HSlopEn(X, m, K, \gamma, \delta) = SlopEn(X_{n,e}, m, \gamma, \delta) \tag{7}$$

where K is the number of layers of decomposition, γ and δ are the two thresholds of SlopEn, and m is the embedding dimension.

As displayed in Figure 2, the hierarchical decomposition structure diagram when K = 3 is shown. SlopEn is calculated on each node after the hierarchical decomposition.

In Figure 2, X indicates the original time series, $x_{1,1}$ is the first node of the first layer, $x_{2,1}$ is the first node of the second layer, and so on.

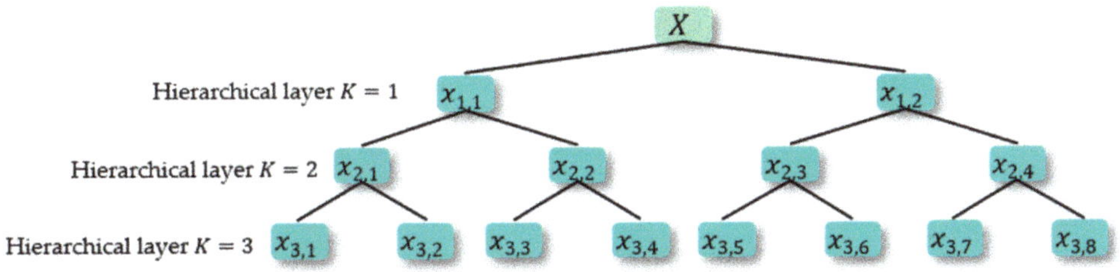

Figure 2. Hierarchical decomposition structure diagram when $K = 3$.

2.3. *Analysis of the Parameters for HSlopEn*

The main parameters of HSlopEn include the number of decomposition layers K, embedding dimension m, two threshold parameters γ and δ, and time delay d. First, the number of decomposition layers K determines the number of nodes in the hierarchical decomposition. When the number of decomposition layers is too large, the number of nodes decomposed is too large, resulting in a large number of calculations for SlopEn values of all nodes; when the value is too small, resulting in a small number of decomposed nodes, there are insufficient frequency bands for the given time series. Referring to other references, the default number of decomposition layers K is 3 in this paper. Then, the embedding dimension m is used to extract the subsequence of a given time series. If it is too small, it is difficult to determine the dynamic changes of the time series; if it is too large, it is difficult to capture the subtle changes in the time series. After that, the two threshold parameters γ and δ are used to divide the symbol pattern of a given subsequence, which affects the change in entropy value. Lastly, the default time delay d is 1, as important information about frequency may be lost at that time if $d > 1$. The effect of embedding dimension and thresholds on the performance of the HSlopEn is investigated below by analyzing the noisy signals.

To investigate the effect of embedding dimension on the entropy value of hierarchical slope entropy, 50 sets of white Gaussian noise (WGN) of signal length 2048 are used, with the embedding dimension m varying from 2 to 5 and the two threshold parameters γ and δ defaulting to 0.1 and 0.001, respectively. Figure 3 shows the mean and standard deviation (SD) of the HSlopEn values for different embedding dimensions in every node.

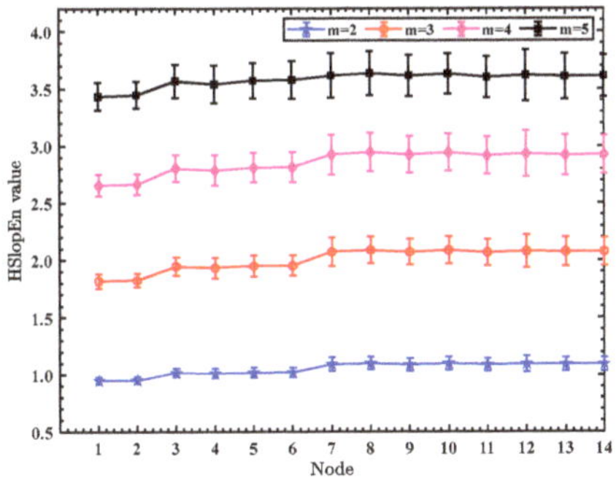

Figure 3. The mean and standard deviation (SD) of HSlopEn values for different embedding dimensions in every node.

As shown in Figure 3, as the embedding dimension m becomes larger, the entropy value of the HSlopEn also becomes larger, but the entropy value of each node for HSlopEn is close to others at different embedding dimensions, and the difference between the mean and SD is small, which indicates that the change in the embedding dimension affects the size of the entropy value, but the stability of the HSlopEn hardly changes. The embedding dimension m is set to 3 in this paper.

In addition, to further study the effect of thresholds γ and δ on the entropy of the HSlopEn, 50 independent pink noise (PN) and WGN signals are selected, where each noise is sampled at 2048 Hz and the embedding dimension m is 3. The three sets of thresholds (γ, δ) for HSlopEn are manually set, which are (0.1, 001), (0.3, 0.1), and (0.8, 0.3), and the mean and standard deviation (SD) of the HSlopEn values for the three sets of thresholds in every node are displayed in Figure 4.

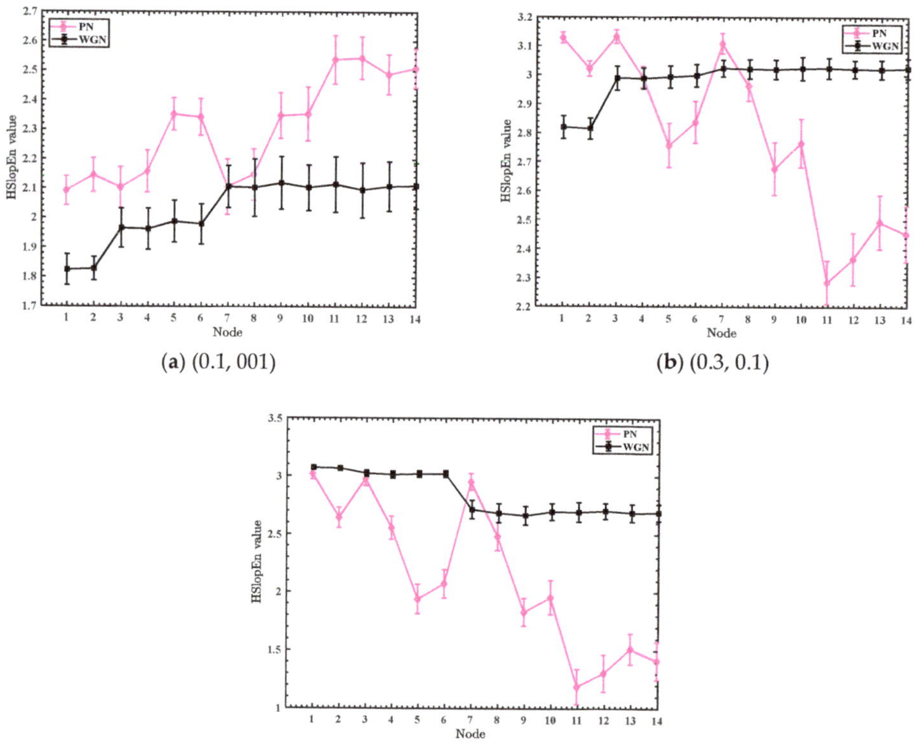

(**a**) (0.1, 001)

(**b**) (0.3, 0.1)

(**c**) (0.8, 0.3)

Figure 4. The mean and SD of HSlopEn values for three sets of thresholds in every node.

It can be seen in Figure 4 that, as the threshold changes, the entropy values of the two types of noise signals change; at the same time, the ability to discriminate between the noise signals is constantly changing, so the threshold has a significant effect on the entropy of the HSlopEn. The WSO is used in the paper to optimize the thresholds to avoid taking values based on artificial experience and further improve the fault diagnosis.

2.4. *WSO-HSlopEn and WSO-SVM*

Following the principle of the HSlopEn algorithm, the two threshold parameters γ and δ of the HSlopEn are used to divide the sign pattern of a given time sequence subsequence. Thus, the two threshold parameters have a great influence on the HSlopEn value. At the same time, the classification effect of the support vector machine (SVM) mainly depends on the selection of the penalty factor (C) and kernel function parameters (g), and it is generally difficult to take the values based on manual experience. Hence, the selection of an appropriate penalty factor and kernel function parameters is also particularly important for the classification and recognition accuracy of the SVM.

To enhance the performance of the fault diagnosis effect, in this paper, taking the recognition rate as the fitness function, the white shark optimizer (WSO) is used to optimize the parameters of HSlopEn and the SVM, and WSO-HSlopEn and WSO-SVM are proposed, respectively, where the WSO is a new meta-heuristic optimization algorithm based on deep-sea foraging by great white sharks, proposed in 2022 for solving optimization problems on continuous search spaces. The main process of optimizing the parameters of HSlopEn and the SVM is shown in Figure 5, and the specific process is as follows:

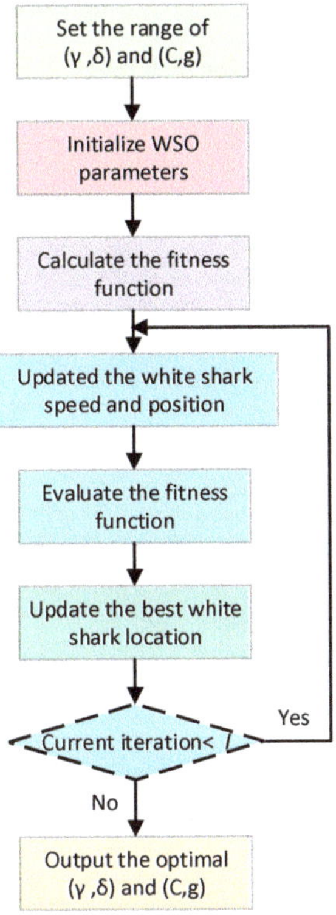

Figure 5. The main process of optimizing the parameters of the HSlopEn and SVM.

(1) Set the initial parameter ranges of HSlopEn (γ, δ) and the SVM (C, g).
(2) Initialize the WSO parameters, such as population size, number of iterations I, position, and speed of white sharks.
(3) Calculate the fitness function, and update the white sharks' position and speed.
(4) Evaluate the fitness function, and update the optimal white shark position
(5) Update the position and speed of the white shark.
(6) Judge whether the current iteration number reaches the maximum iteration number. If so, return to update the speed and position of the white shark and repeat the above steps; otherwise, output the best-optimized parameters (γ, δ) and (C, g).

3. The Proposed Method for Fault Diagnosis of Rolling Bearing

Combining the concept of hierarchical structure, the new complexity feature HSlopEn is proposed, and the parameters of both HSlopEn and the SVM are optimized using the WSO algorithm, and WSO-HSlopEn and WSO-SVM are proposed, respectively. Then, a dual-optimization fault diagnosis method for rolling bearings based on WSO-HSlopEn and WSO-SVM is proposed. Figure 6 presents the flowchart of the proposed fault diagnosis method, and the method mainly includes the following steps:

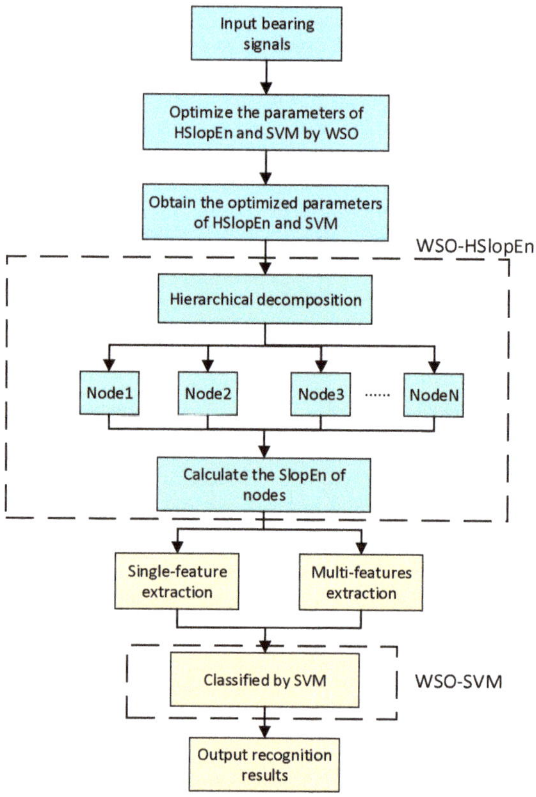

Figure 6. The flowchart of proposed dual-optimization fault diagnosis method for rolling bearings, based on WSO-HSlopEn and WSO-SVM.

(1) The different bearing signals are input. In this paper, each type of bearing signal has 100 samples with 1024 data points.
(2) The WSO algorithm is applied to optimize the parameters of HSlopEn (γ, δ) and the SVM (C, g) by taking the final recognition rate as the fitness function, and the optimized parameters are obtained. At the same time, other optimization algorithms, including SO, marine predator algorithm (MPA), and sparrow search algorithm (SSA), are used for comparison.
(3) Different types of bearing signals are decomposed into several layers, and the nodes are obtained. In this paper, bearing signals are decomposed into three layers.
(4) The nodes of WSO-HSlopEn are calculated, and then single-feature and multi-feature extraction experiments for bearing signals are carried out. Meanwhile, comparisons with some classical entropies, such as HFE, HPE, HSE, and HRDE, are conducted.
(5) WSO-SVM is applied to classify bearing signals, and the recognition results are output. In this paper, for each type, select 25 sample signals as training samples and 75 sample signals as test samples.

4. Experiments and Results

In this chapter, two comparative experiments are implemented to examine the effectiveness of the proposed method in fault diagnosis: (1) In optimizing both HSlopEn and SVM parameters using the WSO, we compare different optimization algorithms, including SSA, MPA, and SO. (2) In extracting the WSO-HSlopEn of nodes, we compare classical hierarchical entropy metrics, including HPE, HSE, HFE, and HRDE.

4.1. Fault Diagnosis of Rolling Bearing Signal

The dataset used in this section was derived from the Bearing Data Center of Case Western Reserve University [36], which is an internationally recognized standard dataset for fault diagnosis of rolling bearings. The schematic of the test rig (Cleveland, USA) is shown in Figure 7.

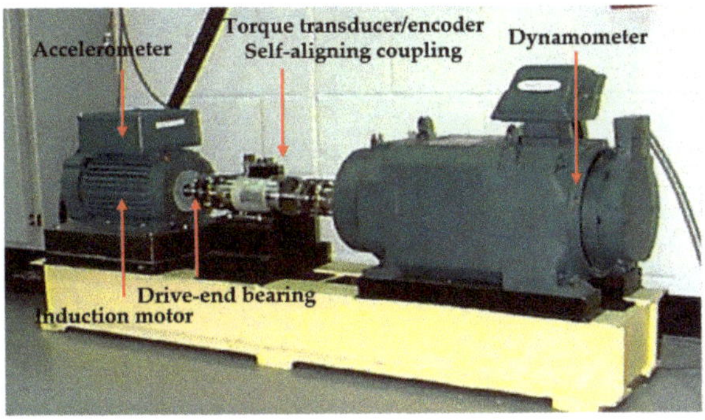

Figure 7. The schematic of test rig.

As shown in Figure 7, the test rig consisted of an induction motor, drive-end bearing, self-aligning coupling, and accelerometer dynamometer. An accelerometer was installed on the base of the motor, which was used to detect the vibration acceleration of the faulty bearing at a sampling frequency of 12 kHz. The dataset divided the fault data into four categories: normal data (NOR), ball faults (BFs), outer race faults (ORFs), and inner race faults (IRFs). Among them, BFs, ORFs, and IRFs were simulated faults with single-point damage as an electric spark. The damage diameters were divided into 0.007, 0.014, and 0.021 inches. At the same time, the processed faulty bearing was reloaded into the test motor, and the vibration acceleration signal data were recorded under the load working conditions of 0, 1, 2, and 3 horsepower.

In this section, bearing signals with ten conditions were collected from the drive-end bearings, including rolling bearings in normal condition and those with damage to the inner race, the outer race, and the ball element. Bearings with various damage diameters were considered under a speed of 1730 rpm with a load of 3 horsepower. Table 1 illustrates the fault diagnosis sample collection of bearing signals. Each fault signal was divided into three types according to the fault diameter. We sampled from point 1001, and each condition had 100 samples with 1024 sampling points. Time-domain waveforms for each state bearing signals are displayed in Figure 8.

Table 1. Fault diagnosis sample collection of bearing signals.

Bearing Condition	Defect Size (Inches)	Sample Label
NOR	-	1
IRF1	0.007	2
IRF2	0.014	3
IRF3	0.021	4
BF1	0.007	5
BF2	0.014	6
BF3	0.021	7
ORF1	0.007	8
ORF2	0.014	9
ORF3	0.021	10

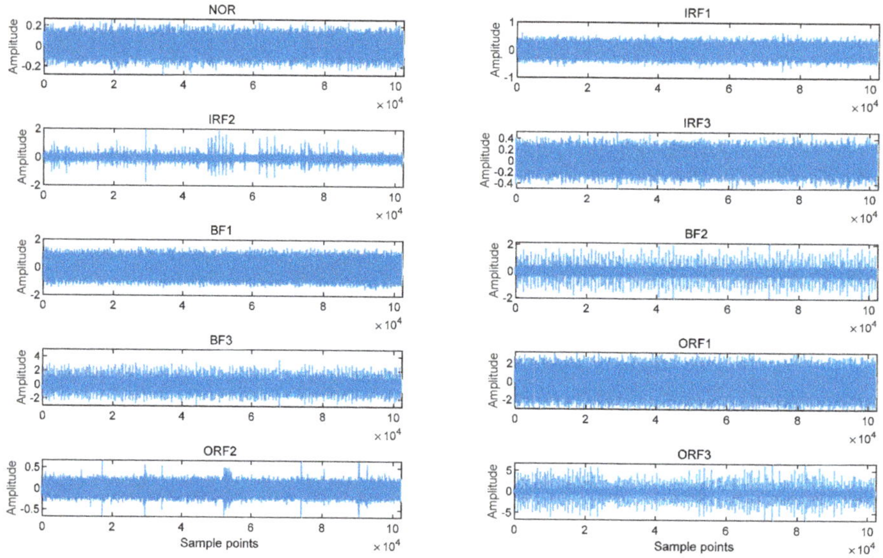

Figure 8. Time-domain waveforms for each state bearing signals.

4.2. Comparison of Different Optimization Algorithms

Designed to verify the performance advantages of the WSO in optimizing HSlopEn and the SVM, this section introduces different optimization algorithms to optimize the parameters of HSlopEn and the SVM, and compares recognition rates of single-feature and multi-feature extractions with those of other optimization algorithms [37–39]. In this experiment, 10 different bearing signal conditions were sampled from the 1001 point as the starting point, and 100 samples were selected. Each sample had 1024 data points. First, the parameters of HSlopEn were set as follows: hierarchical layer $K = 3$, embedding dimension $m = 3$, and threshold parameters γ and δ were adaptively determined using different optimization algorithms. HSlopEn with optimized parameters of bearing signals was extracted. Then, the sample set was divided into the training set and test set, and the select single feature or multi-features were input to optimize the SVM. The penalty factor and kernel function parameters of the SVM were also adaptively determined using the WSO algorithm. Figure 9 presents the fitness iteration curves of different optimization algorithms to optimize HSlopEn and the SVM. These are the fitness iteration curves of different optimization algorithms in the case of extracting five nodes.

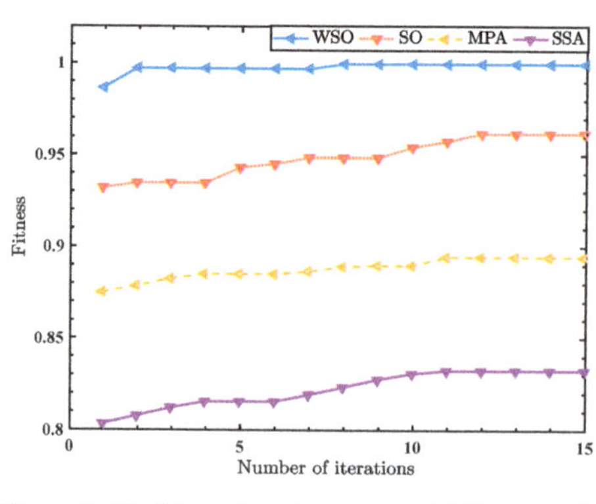

Figure 9. The fitness iteration curve of different optimization algorithms.

It can be found in Figure 9 that, in the condition of extracting five nodes, the highest recognition rate of these ten types of bearing signals reached 100% using the WSO to optimize HSlopEn. At the same time, the convergence speed of the WSO was relatively faster than other optimization algorithms. In addition, the early convergence of the WSO is quick. Its fitness curve eventually converged to a bigger value. To further demonstrate the significant advantages of WSO, we calculated the recognition rate of bearing signals based on using different optimization algorithms to optimize HSlopEn and the SVM under the situation of extracting the single feature and multi-features. The recognition rates of HSlopEn for all single-feature nodes are shown in Tables 2 and 3, and the highest recognition rates of HSlopEn for multi-features for the four types of optimization algorithms are shown in Table 4.

Table 2. The recognition rate of HSlopEn for each single-feature node (nodes 1–7).

Optimization Algorithm	Recognition Rate for Each Node (%)						
	1	2	3	4	5	6	7
WSO	78.40	79.20	70.13	78.00	73.47	79.33	45.73
SO	77.47	74.80	67.87	76.32	72.27	75.43	43.28
MPA	74.00	75.87	62.93	71.20	66.93	76.53	40.40
SSA	74.67	73.07	68.53	71.87	70.67	75.60	45.07

Table 3. The recognition rate of HSlopEn for each single-feature node (nodes 8–14).

Optimization Algorithms	Recognition Rate for Each Node (%)						
	8	9	10	11	12	13	14
WSO	51.60	72.27	59.60	66.93	62.80	69.07	70.93
SO	48.37	70.26	58.23	65.27	60.80	68.27	69.73
MPA	39.87	70.80	56.53	60.13	56.13	67.87	65.47
SSA	50.40	69.73	57.73	63.87	60.93	68.42	69.57

Table 4. The highest recognition rate of HSlopEn for multi-features for four types of optimization algorithms.

Optimization Algorithms	Parameter	Number of Extracted Features			
		2	3	4	5
WSO	Highest recognition rate (%)	97.87	99.60	99.87	100
SO	Highest recognition rate (%)	87.60	90.80	93.87	96.20
MPA	Highest recognition rate (%)	67.87	84.37	86.67	89.47
SSA	Highest recognition rate (%)	64.13	70.40	74.67	83.27

According to the recognition rate of different types of bearing signals, we can find that no matter how many features are extracted, the advantages of the WSO algorithm are obvious. In the case of extracting a single feature, the recognition rate of the WSO reaches 79.33% on node 6, which is much higher than that of other optimization algorithms. Under the circumstances of extracting multi-features, as the number of selected nodes increases, the recognition rate also improves. When we select five features, it realizes the correct identification of all samples. The recognition rate of other optimization algorithms, including SO, MPA, and SSA, is, respectively, 3.8%, 10.53%, and 17.73% lower than that of WSO. Above all, we prove that using the WSO to optimize HSlopEn and the SVM is feasible. Therefore, in this paper, the WSO is used to optimize HSlopEn and SVM parameters.

4.3. Comparison of Different Hierarchical Entropies

Aiming to demonstrate the superiority of WSO-HSlopEn in fault diagnosis, we compared it to other classical hierarchical entropies, including HSE, HFE, HPE, and HRDE.

The single-feature approach was first used to extract the fault feature and compare it with HFE, HSE, HPE, and HRDE. The parameters of HSlopEn were as follows: hierarchical layer $K = 3$, embedding dimension $m = 3$, time delay $d = 1$, and threshold parameter γ and δ were adaptively determined using the WSO algorithm. For a fair comparison, the parameter settings of other hierarchical entropies were the same as those in the HSlopEn method. Among them, the similarity tolerances of HSE and HFE were set as $r = 0.2$, and the category number of HRDE was set as $c = 3$. The entropy distributions of an optimal node for the single-feature extraction of bearing signals are shown in Figure 10.

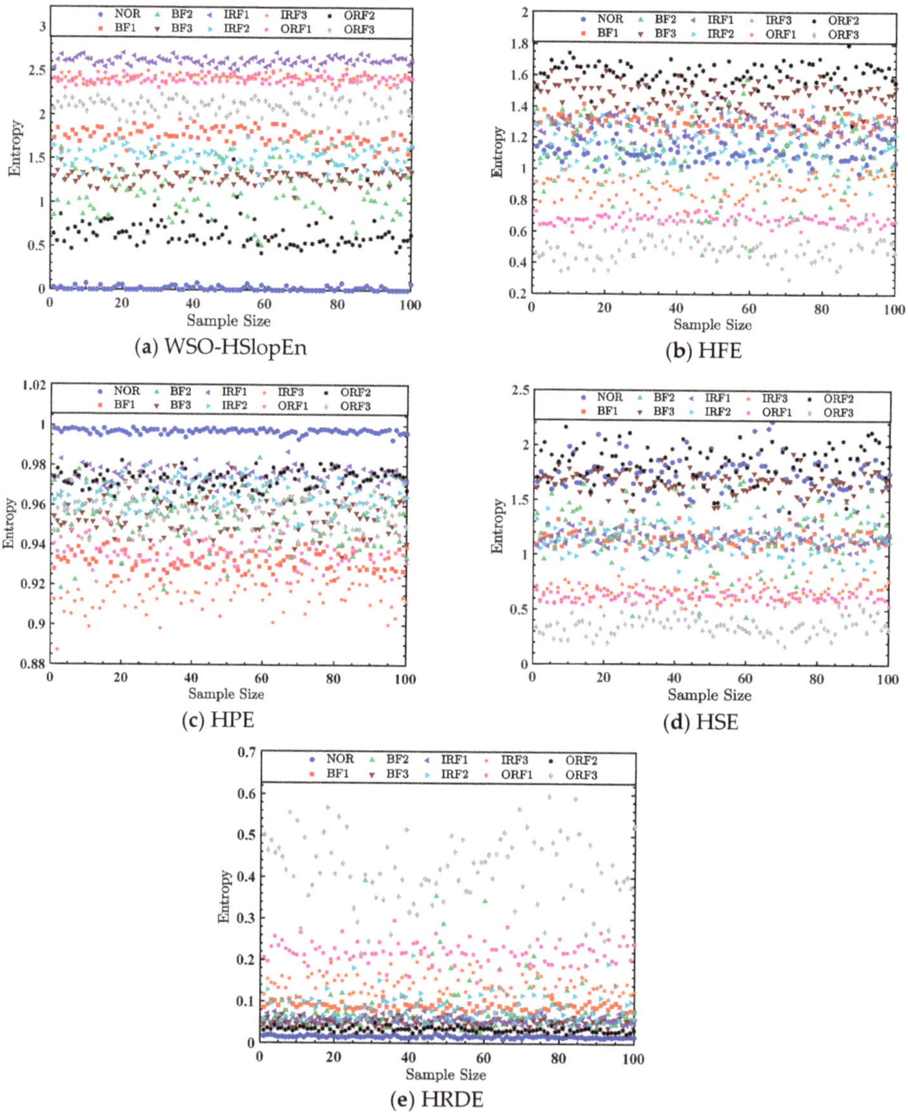

Figure 10. Entropy distribution of optimal node for single-feature extraction of bearing signals.

Figure 10a presents that there was no aliasing phenomenon between the features of the three fault types of NOR, ORF3, and IRF1 and other fault types in the entropy distribution of WSO-HSlopEn. ORF2 only had a few samples of entropy close to BF2, and the entropy values of the samples of the other fault types show a severe overlap. Based on the single-feature extraction shown in Figure 10b–e, compared with other hierarchical entropies, WSO-HSlopEn is not as serious as the aliasing of the distributions of other hierarchical entropies. The entropy values of several types of samples in the ten conditions of bearing signals are quite different from the other types. Relatively speaking, the distance between the various types of samples in the distribution of the entropy of WSO-HSlopEn is relatively large.

After using the WSO-HSlopEn as the fault feature of the bearing signal, the bearing fault diagnosis sample set was divided into a training set and a test set, and the training set was input into the WSO-SVM to train the model, and then the test set was input into the model to finish the fault diagnosis of bearings. The Gaussian kernel function was selected as the kernel function of the SVM. The penalty factor and kernel function parameters of the SVM were also adaptively determined by the WSO algorithm. Recognition rates of single features for the five types of hierarchical entropies are displayed in Tables 5 and 6.

Table 5. Recognition rates of single features for the five types of hierarchical entropies (nodes 1–7).

Entropies	Recognition Rates of Each Node (%)						
	1	2	3	4	5	6	7
WSo-HSlopEn	78.40	79.20	70.13	78.00	73.47	79.33	45.73
HFE	57.07	55.06	37.73	55.33	42.67	59.73	32.53
HPE	43.73	52.53	24.80	40.93	38.13	46.53	18.80
HSE	39.07	47.60	23.87	36.27	21.07	46.67	18.40
HRDE	55.73	65.47	33.87	53.87	42.40	54.53	30.67

Table 6. Recognition rates of single features for the five types of hierarchical entropies (nodes 8–14).

Entropies	Recognition Rates of Each Node (%)						
	8	9	10	11	12	13	14
WSo-HSlopEn	51.60	72.27	59.60	66.93	62.80	69.07	70.93
HFE	27.07	43.73	36.40	38.93	36.40	45.73	40.27
HPE	14.80	29.60	22.27	29.07	20.80	32.27	26.40
HSE	20.40	28.67	20.40	21.60	20.53	37.33	25.87
HRDE	25.07	45.60	37.87	42.93	37.07	50.13	48.93

Tables 5 and 6 illustrate that, when using WSO-HSlopEn, the recognition rate of node 6 was the highest, which was 79.33%. Compared with other hierarchical entropies, under each node, the recognition rate based on WSO-HSlopEn was always the highest, which shows the effectiveness of WSO-HSlopEn as a fault diagnosis feature of bearing signals.

Through observation, when single-feature extraction is used to extract the fault feature, there is still overlap between the features of different conditions of the bearing signals. Furthermore, the recognition rate of the best node was low, and there were many misclassified samples based on single-feature extraction. Aiming to further improve the recognition rate of different conditions of the bearing signals, double features were used to extract the bearing signals. All parameters used in the experiments were the same as those listed in the single-feature extraction. The entropy distribution on the optimal node for double-feature extractions of bearing signals is shown in Figure 11, where the abscissa and ordinate are the entropy values of the two nodes, respectively. For example, in Figure 11a, the abscissa is the SlopEn of node 1, and the ordinate is the SlopEn of node 5.

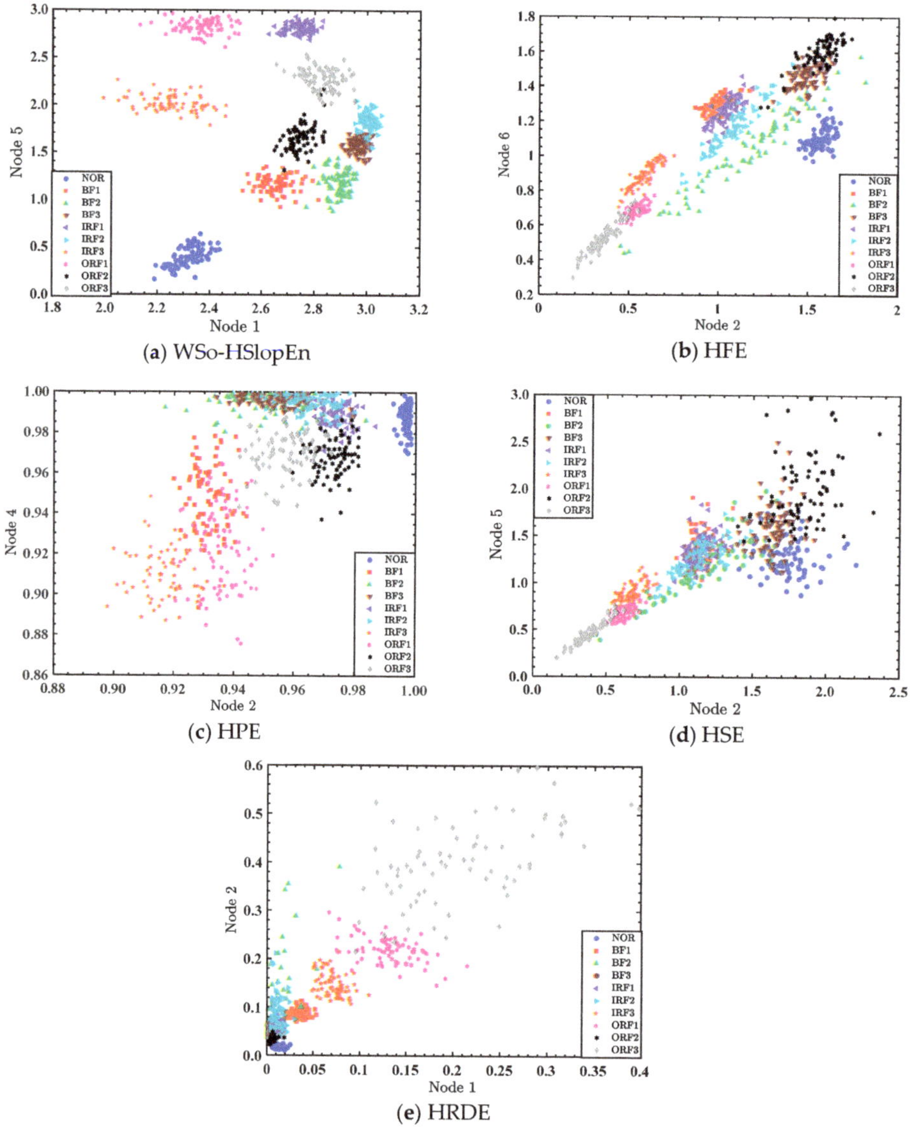

Figure 11. Double-features distribution of ten types of bearing signals.

As can be observed from Figure 11, in the case of double-feature extraction, the WSO-HSlopEn distribution of sample signals belonging to the same type is relatively concentrated compared to other hierarchical entropies; for the other four types of hierarchical entropies, the bearing signals between different types are more divergent, and the entropy values of different types of bearing signals are very close.

To further improve the recognition performance, triple features were used to extract bearing fault features on various hierarchical entropies. The parameters for calculating various hierarchical entropies were the same as those of double features. Figure 12 presents the triple-feature distributions of ten types of bearing signals for different hierarchical entropies.

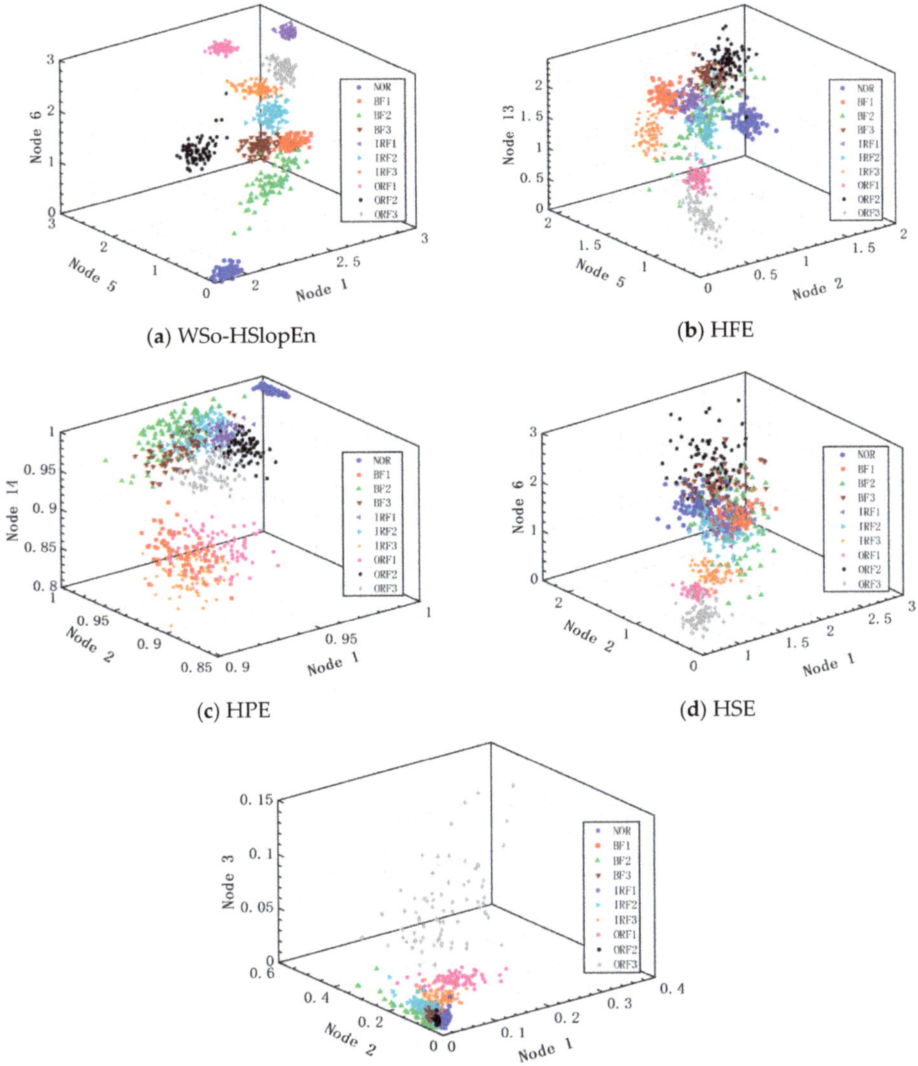

(**a**) WSo-HSlopEn (**b**) HFE

(**c**) HPE (**d**) HSE

(**e**) HRDE

Figure 12. Triple-features distribution of ten types of bearing signals for different entropies.

It can be seen from Figure 12 that there is almost no overlap based on the WSO-HSlopEn, but the feature distributions of the BF2 and IRF2 samples are relatively low in clustering; for the other hierarchical entropies, the clustering of the feature distributions of the samples are very poor because of their approximate entropy distributions. Nevertheless, the entropy distribution of WSO-HSlopEn is more dispersed, and WSO-HSlopEn of different fault types are quite different, which effectively verifies the validity of WSO-HSlopEn as a feature extraction method for ten types of bearing signals.

Next, WSO-SVM is used to construct a fault diagnosis model. The highest recognition rate is calculated for the five types of hierarchical entropies under multi-feature extraction, as shown in Table 7, where (1,5) indicates the combination of nodes with the highest recognition rate for two features are node 1 and node 5, (1,5,6) indicates the combination of nodes with the highest recognition rate for three features are node 1, node 5 and node 6, and so on.

Table 7. Highest recognition rate for the five types of hierarchical entropies for multi-features.

Entropy	Parameter	Number of Extracted Nodes			
		2	3	4	5
WSo-HSlopEn	Highest recognition rate (%)	97.87	99.60	99.87	100
	Choose the node	(1,5)	(1,5,6)	(1,5,6,7)	(1,5,6,7,11)
HFE	Highest recognition rate (%)	87.60	90.80	93.87	96.20
	Choose the node	(2,6)	(2,5,13)	(1,3,4,6)	(2,5,6,12,14)
HPE	Highest recognition rate (%)	67.87	84.37	86.67	89.47
	Choose the node	(2,4)	(1,2,14)	(2,3,5,10)	(1,5,7,12,14)
HSE	Highest recognition rate (%)	64.13	70.40	74.67	83.27
	Choose the node	(2,5)	(1,2,6)	(1,2,3,6)	(2,3,7,12,14)
HRDE	Highest recognition rate (%)	80.60	89.87	93.87	95.87
	Choose the node	(1,2)	(1,2,3)	(1,2,4,6)	(2,5,6,11,12)

Table 7 shows that no matter how many features are extracted, the recognition rate of these ten types of bearing signals using WSO-HSlopEn is higher than that of other hierarchical entropies; additionally, the more features we select, the better the recognition effect we obtain; in the circumstances of multi-features, the recognition rates of WSO-HSlopEn are all higher than 97.5%, yet the highest recognition rates of other hierarchical entropies are all significantly below 97.5%; for WSO-HSlopEn, when five nodes are selected, that is, choosing nodes (1,5,6,7,11), the highest recognition rate of these ten types of bearing signals reaches 100%; however, the highest recognition rate of other entropies is, respectively, 3.80%, 10.53%, 16.73%, and 4.13% lower than that of WSO-HSlopEn. Through the above comparison, we can clearly find the significant advantages of the proposed method based on WSO-HSlopEn, and the recognition results applied to diagnose faults of rolling bearings are higher than those of classic methods.

5. Conclusions

This paper puts forward a dual-optimization fault diagnosis method for rolling bearings based on WSO-HSlopEn and WSO-SVM. The effectiveness of the proposed methods is verified by comparing them with the classical methods. The main innovations and conclusions are as follows:

(1) On the basis of SlopEn, combined with the idea of hierarchical decomposition, HSlopEn is proposed and introduced into the feature extraction of bearing signals for the first time; at the same time, WSO is used to optimize both HSlopEn and the SVM, and WSO-HSlopEn and WSO-SVM are proposed.

(2) In the case of single-feature extraction, the proposed method based on WSO-HSlopEn has the highest recognition rate of 79.33% on node 6, which is, respectively, 19.47%, 26.67%, 31.60%, and 13.73% higher than those of HFE, HPE, HSE, and HRDE.

(3) In the case of extracting multi-features, the recognition rates are higher than 97.5%, which is a significant improvement compared with the single-feature extraction method; moreover, with the different number of features, the recognition rate based on WSO-HSlopEn is always high than the other hierarchical entropies.

(4) For the proposed dual-optimization fault diagnosis method for rolling bearings, based on WSO-HSlopEn and WSO-SVM, the more features we select, the better the recognition effect we obtain. When five nodes are selected, the highest recognition rate reaches 100%.

The proposed WSO-HSlopEn and WSO-SVM solve the problem of dependent parameter settings for SlopEn and the SVM, respectively, and their superiority has been confirmed in fault diagnosis. Therefore, WSO-HSlopEn and WSO-SVM are expected to be applied to other fields in future work, such as underwater acoustic signal processing and medical signal classification.

Author Contributions: Conceptualization, Y.L.; Methodology, Y.L.; Validation, B.T.; Data curation, B.H.; Writing—original draft, B.T.; Visualization, B.T., B.H. and X.X.; Supervision, Y.L. All authors have read and agreed to the published version of the manuscript.

Funding: This research was funded by [Natural Science Foundation of Shaanxi Province] grant number [2022JM-337], [Xi'an University of Technology Excellent Seed Fund] grant number [252082220].

Institutional Review Board Statement: Not applicable.

Informed Consent Statement: Not applicable.

Data Availability Statement: The data used to support the findings of this study are available from the corresponding author upon request.

Conflicts of Interest: The authors declare no conflict of interest.

References

1. Yan, X.; Jia, M. Intelligent fault diagnosis of rotating machinery using improved multiscale dispersion entropy and mRMR feature selection. *Knowl. Based Syst.* **2019**, *163*, 450–471. [CrossRef]
2. Li, X.; Xu, Y.; Li, N.; Yang, B.; Lei, Y. Remaining Useful Life Prediction with Partial Sensor Malfunctions Using Deep Adversarial Networks. *IEEE CAA J. Autom. Sin.* **2023**, *10*, 121–134. [CrossRef]
3. Saucedo-Dorantes, J.J.; Arellano-Espitia, F.; Delgado-Prieto, M.; Osornio-Rios, R.A. Diagnosis Methodology Based on Deep Feature Learning for Fault Identification in Metallic, Hybrid and Ceramic Bearings. *Sensors* **2021**, *21*, 5832. [CrossRef]
4. Li, Y.; Tang, B.; Geng, B.; Jiao, S. Fractional Order Fuzzy Dispersion Entropy and Its Application in Bearing Fault Diagnosis. *Fractal Fract.* **2022**, *6*, 544. [CrossRef]
5. Li, X.; Yu, S.; Lei, Y.; Li, N.; Yang, B. Intelligent Machinery Fault Diagnosis with Event-Based Camera. *IEEE Trans. Ind. Inform.* **2023**, 1–10. [CrossRef]
6. Yadav, E.; Chawla, V.K. An explicit literature review on bearing materials and their defect detection techniques. *Mater. Today Proc.* **2022**, *50*, 1637–1643. [CrossRef]
7. Hu, Q.; Li, Y.; Sun, X.; Chen, M.; Bu, Q.; Gong, B. Integrating test device and method for creep failure and ultrasonic response of methane hydrate-bearing sediments. *Rev. Sci. Instrum.* **2023**, *94*, 025105. [CrossRef]
8. Li, Y.; Tang, B.; Jiang, X.; Yi, Y. Bearing Fault Feature Extraction Method Based on GA-VMD and Center Frequency. *Math. Probl. Eng.* **2022**, *2022*, 2058258. [CrossRef]
9. Li, Y.; Jiao, S.; Geng, B. Refined composite multiscale fluctuation-based dispersion Lempel–Ziv complexity for signal analysis. *ISA Trans.* **2023**, *133*, 273–284. [CrossRef]
10. Li, Y.; Geng, B.; Tang, B. Simplified coded dispersion entropy: A nonlinear metric for signal analysis. *Nonlinear Dyn.* **2023**, *111*, 9327–9344. [CrossRef]
11. Richman, J.S.; Moorman, J.R. Physiological time-series analysis using approximate entropy and sample entropy. *Am. J. Physiol. Heart Circ. Physiol.* **2000**, *6*, 2039–2049. [CrossRef]
12. Bandt, C.; Pompe, B. Permutation entropy: A natural complexity measure for time series. *Phys. Rev. Lett.* **2002**, *88*, 174102. [CrossRef]
13. Rostaghi, M.; Azami, H. Dispersion Entropy: A Measure for Time Series Analysis. *IEEE Signal Process. Lett.* **2016**, *23*, 610–614. [CrossRef]
14. Han, M.; Pan, J. A fault diagnosis method combined with LMD, sample entropy and energy ratio for roller bearings. *Measurement* **2015**, *76*, 7–19. [CrossRef]
15. Xue, X.; Li, C.; Cao, S.; Sun, J.; Liu, L. Fault Diagnosis of Rolling Element Bearings with a Two-Step Scheme Based on Permutation Entropy and Random Forests. *Entropy* **2019**, *21*, 96. [CrossRef]
16. Dhandapani, R.; Mitiche, I.; McMeekin, S.; Mallela, V.S.; Morison, G. Enhanced Partial Discharge Signal Denoising Using Dispersion Entropy Optimized Variational Mode Decomposition. *Entropy* **2021**, *23*, 1567. [CrossRef] [PubMed]
17. Cuesta-Frau, D. Slope Entropy: A New Time Series Complexity Estimator Based on Both Symbolic Patterns and Amplitude Information. *Entropy* **2019**, *21*, 1167. [CrossRef]
18. Li, Y.; Tang, B.; Jiao, S. SO-slope entropy coupled with SVMD: A novel adaptive feature extraction method for ship-radiated noise. *Ocean Eng.* **2023**, *280*, 114677. [CrossRef]
19. Li, Y.; Tang, B.; Yi, Y. A novel complexity-based mode feature representation for feature extraction of ship-radiated noise using VMD and slope entropy. *Appl. Acoust.* **2022**, *196*, 108899. [CrossRef]
20. Cuesta-Frau, D.; Dakappa, P.H.; Mahabala, C.; Gupta, A.R. Fever Time Series Analysis Using Slope Entropy. Application to Early Unobtrusive Differential Diagnosis. *Entropy* **2020**, *22*, 1034. [CrossRef]

21. Cuesta-Frau, D.; Schneider, J.; Bakštein, E.; Vostatek, P.; Spaniel, F.; Novák, D. Classification of Actigraphy Records from Bipolar Disorder Patients Using Slope Entropy: A Feasibility Study. *Entropy* **2020**, *22*, 1243. [CrossRef]
22. Shi, E. Single Feature Extraction Method of Bearing Fault Signals Based on Slope Entropy. *Shock. Vib.* **2022**, *2022*, 6808641. [CrossRef]
23. Jiang, Y.; Peng, C.K.; Xu, Y. Hierarchical entropy analysis for biological signals. *J. Comput. Appl. Math.* **2011**, *236*, 728–742. [CrossRef]
24. Xing, J.; Xu, J. An Improved Incipient Fault Diagnosis Method of Bearing Damage Based on Hierarchical Multi-Scale Reverse Dispersion Entropy. *Entropy* **2022**, *24*, 770. [CrossRef]
25. Peng, C.; Zhao, X.; Jiang, H. A New Method of Fault Feature Extraction Based on Hierarchical Dispersion Entropy. *Shock. Vib.* **2021**, *2021*, 8824901.
26. Wang, X.; Si, S.; Li, Y. An integrated method based on refined composite multivariate hierarchical permutation entropy and random forest and its application in rotating machinery. *J. Vib. Control* **2020**, *26*, 3–4. [CrossRef]
27. Xue, Q.; Xu, B.; He, C.; Liu, F. Feature Extraction Using Hierarchical Dispersion Entropy for Rolling Bearing Fault Diagnosis. *IEEE Trans. Instrum. Meas.* **2021**, *70*, 3521311. [CrossRef]
28. Keller, J.M.; Gray, M.R.; Givens, J.A. A fuzzy K-nearest neighbor algorithm. *IEEE Trans. Syst. Man Cybern.* **1985**, *SMC-15*, 580–585. [CrossRef]
29. Svetnik, V.; Liaw, A.; Tong, C.; Culberson, J.C.; Sheridan, R.P.; Feuston, B.P. Random Forest: A Classification and Regression Tool for Compound Classification and QSAR Modeling. *J. Chem. Inf. Comput. Sci.* **2003**, *43*, 1947–1958. [CrossRef]
30. Samanta, B.; Al-Balushi, K.R.; Al-Araimi, S.A. Artificial neural networks and support vector machines with genetic algorithm for bearing fault detection. *Eng. Appl. Artif. Intell.* **2003**, *16*, 657–665. [CrossRef]
31. Wang, Z.; Yao, L.; Cai, Y.; Zhang, J. Mahalanobis semi-supervised mapping and beetle antennae search based support vector machine for wind turbine rolling bearings fault diagnosis. *Renew. Energy* **2020**, *155*, 1312–1327. [CrossRef]
32. Tang, H.; Yuan, Z.; Dai, H.; Du, Y. Fault Diagnosis of Rolling Bearing Based on Probability box Theory and GA-SVM. *IEEE Access* **2020**, *8*, 170872–170882. [CrossRef]
33. Ye, M.; Yan, X.; Jia, M. Rolling Bearing Fault Diagnosis Based on VMD-MPE and PSO-SVM. *Entropy* **2021**, *23*, 762. [CrossRef] [PubMed]
34. Jin, Z.; Chen, G.; Yang, Z. Rolling Bearing Fault Diagnosis Based on WOA-VMD-MPE and MPSO-LSSVM. *Entropy* **2022**, *24*, 927. [CrossRef]
35. Braik, M.; Hammouri, A.; Atwan, J. White Shark Optimizer: A novel bioinspired meta-heuristic algorithm for global optimization problems. *Knowl. Based Syst.* **2022**, *243*, 108457. [CrossRef]
36. Case Western Reserve University Bearing Data Center. Available online: https://engineering.case.edu/bearingdatacenter (accessed on 26 February 2023).
37. Fatma, A.H.; Abdelazim, G.H. Snake Optimizer: A novel meta-heuristic optimization algorithm. *Knowl. Based Syst.* **2022**, *242*, 108320.
38. Afshin, F.; Mohammad, H.; Seyedali, M.; Amir, H.G. Marine Predators Algorithm: A nature-inspired metaheuristic. *Expert Syst. Appl.* **2020**, *152*, 113377.
39. Xue, J.; Shen, B. A novel swarm intelligence optimization approach: Sparrow search algorithm. *Syst. Sci. Control Eng.* **2020**, *8*, 22–34. [CrossRef]

sensors

MDPI

Article

Deconvolved Fractional Fourier Domain Beamforming for Linear Frequency Modulation Signals

Zhuoran Liu, Quan Tao, Wanzhong Sun and Xiaomei Fu *

School of Marine Science and Technology, Tianjin University, Tianjin 300072, China
* Correspondence: fuxiaomei@tju.edu.cn

Abstract: To estimate the direction of arrival (DOA) of a linear frequency modulation (LFM) signal in a low signal-to-noise ratio (SNR) hydroacoustic environment by a small aperture array, a novel deconvolved beamforming method based on fractional Fourier domain delay-and-sum beamforming (FrFB) was proposed. Fractional Fourier transform (FrFT) was used to convert the received signal into the fractional Fourier domain, and delay-and-sum beamforming was subsequently performed. Noise resistance was acquired by focusing the energy of the LFM signal distributed in the time–frequency domain. Then, according to the convolution structure of the FrFB complex output, the influence of the fractional Fourier domain complex beam pattern was removed by deconvolution, and the target spatial distribution was restored. Therefore, an improved spatial resolution of DOA estimation was obtained without increasing the array aperture. The simulation and experimental results show that, with a small aperture array at low SNR, the proposed method possesses higher spatial resolution than FrFB and frequency-domain deconvolved conventional beamforming.

Keywords: deconvolved beamforming; fractional Fourier transform; direction of arrival estimation; linear frequency modulation signal

1. Introduction

Linear frequency modulation (LFM) signals are widely used in hydroacoustic communication and detection, and direction of arrival (DOA) estimation is important [1,2]. To estimate received signal DOA, beamforming methods represented by conventional beamforming (CBF) [3] and minimum variance distortionless response (MVDR) [4] were proposed. However, those methods cannot directly estimate the wideband signal DOA due to the steering vector being frequency-dependent [5].

An intuitive approach to extend those narrowband beamforming methods to wideband beamforming is dividing the wideband signal into a set of narrowband signals in the frequency domain. Following this idea, two classical frameworks, the incoherent signal-subspace method (ISM) [6] and the coherent signal-subspace method (CSM) [7], were proposed. The ISM framework processes each subband signal separately and obtains a comprehensive result by combining all subband results. The CSM framework aligns each subband datum to a reference frequency by focusing matrices and obtains a result via the focused data. Those two frameworks solve the problem that a single-frequency steering vector cannot perform beamforming on wideband signals but fail to make full use of the time–frequency characteristics of LFM signals [8]. Various time–frequency analysis tools were introduced to utilize the characteristics of LFM signals. Wigner–Ville distribution (WVD)-based methods [9,10] possess a good estimation performance but suffer from the cross terms between multiple targets [11]. Spatial time–frequency distribution (STFD)-based methods [12,13] can directly exploit the nonstationarity information of LFM signals but are highly dependent on time–frequency point selection [12]. Matched filtering (MF)-based methods [14,15] possess a high noise resistance by treating MF as pre-processing or post-processing of beamforming.

check for updates

Citation: Liu, Z.; Tao, Q.; Sun, W.; Fu, X. Deconvolved Fractional Fourier Domain Beamforming for Linear Frequency Modulation Signals. *Sensors* **2023**, *23*, 3511. https://doi.org/10.3390/s23073511

Academic Editors: Yuxing Li and Luca Fredianelli

Received: 17 February 2023
Revised: 23 March 2023
Accepted: 23 March 2023
Published: 27 March 2023

The fractional Fourier transform (FrFT) [16–19] is a generalization of the classical Fourier transform. FrFT can be regarded as the chirp basis decomposition [20], which means it is an extremely effective time–frequency analysis tool for LFM signals. Fractional Fourier domain (FrFD) beamforming methods [8,21,22] can utilize the energy-focusing characteristic of LFM signals by FrFT. The LFM signals are transformed into narrowband signals in the FrFD, and the DOA can be estimated by narrowband beamforming methods. The performance of FrFD beamforming methods is significantly better than that of ISM-based and CSM-based methods, especially at low SNRs [8,23]. Different from the classical time–frequency domain, there are no cross terms in FrFD. The selection of a time–frequency point can be easily achieved by peak search [24]. FrFT has also shown several advantages over MF regarding robustness against distortion [25]. Therefore, beamforming methods based on FrFT are the best in terms of making full use of the time–frequency characteristics of LFM signals. Among all FrFD beamforming methods, fractional Fourier domain delay-and-sum beamforming (FrFB) is one of the most popular methods. FrFB inherits high robustness from CBF but suffers from wide beamwidth, high sidelobes, and limited spatial resolution [8].

Yang proposed deconvolved conventional beamforming (DCBF) to improve the DOA estimation performance of CBF without increasing the array aperture [26]. By regarding the CBF spatial spectrum as the beam pattern convolution with the target spatial distribution, the Richardson–Lucy (R-L) [27] algorithm can be used to restore a high-resolution spatial spectrum. In recent years, DCBF has been widely used for various arrays [28–38]. Deconvolution is also used as postprocessing for other beamforming methods, such as Chebyshev weighting beamforming [39] and frequency-difference beamforming [40], to optimize their spatial spectrum and to improve resolution.

The abovementioned deconvolved beamforming methods all obtain a spatial spectrum with a narrow beamwidth and low sidelobes, which effectively improves the spatial resolution. However, the beam pattern used in deconvolution is a function of frequency, which is applicable to only narrowband signals. A direct wideband deconvolved beamforming method, named ISM-DCBF in this paper, is to divide the LFM signals into several narrowband signals in the frequency domain following the ISM and to use DCBF for each subband signal. In ISM-DCBF, the deconvolution is repeated multiple times and the computational complexity increases linearly with the number of subbands. In addition, the noise significantly reduces the optimizing ability of the deconvolution for the spatial spectrum when the SNR is low [41]. Therefore, ISM-DCBF suffers from weak DOA estimation performance at low SNRs (as verified by the simulation and experimental results of this paper).

To achieve high-resolution LFM signal DOA estimation with a small aperture array in a low SNR hydroacoustic environment, a novel deconvolved fractional Fourier domain beamforming (DFrFB) method is proposed. The proposed method includes the following three steps. First, the LFM signals are focused in the FrFD by FrFT. Accordingly, time-domain broadband signals with low SNRs are transformed into FrFD narrowband signals with high SNRs. Second, CBF is performed to obtain a robust low-resolution spatial spectrum. Third, the high-resolution spatial spectrum is generated by a deconvolution algorithm.

Due to the coherence of LFM signals that are focused at the same time–frequency point in the FrFD, instead of deconvolving the spatial spectrum by the R-L algorithm, a monotone fast iterative shrinkage thresholding algorithm (MFISTA) [42] is introduced to realize the deconvolution of FrFB complex output. Since the FrFB complex output is not converted into beam power, the effect of cross terms between coherent sources on deconvolution can be avoided. The high-resolution DFrFB spatial spectrum is obtained by calculating the power of the deconvolution result of the FrFB complex output. The simulation and experiment show that the proposed DFrFB possesses a narrower beamwidth, lower sidelobes, and higher spatial resolution compared with FrFB and ISM-DCBF.

The remainder of this paper is organized as follows. Section 2 describes the FrFD array signal model. Section 3 introduces the FrFB and proposes the DFrFB. The convolution structure of the FrFB complex output and the shift-invariance of the complex beam pattern are also proven in Section 3. The performance of the DFrFB is demonstrated by simulations and experiments separately in Sections 4 and 5. Section 6 is the summary of this paper.

2. The Fractional Fourier Domain Array Signal Model

An LFM signal with initial frequency f_0 and frequency modulation rate μ can be expressed as

$$s(t) = \exp\left[j2\pi(f_0 t + 0.5\mu t^2)\right] \quad (1)$$

where $t \in [-T/2, T/2]$ and T denotes the time duration. According to the definition of FrFT in [18], the FrFT of $s(t)$ can be expressed as

$$\begin{aligned} S(\alpha, u) &= \mathcal{F}^p[s(t)](u) \\ &= \sqrt{1 - j\cot\alpha} \cdot \exp(j\pi u^2 \cot\alpha) \cdot \int_{-T/2}^{T/2} \exp\left[j(2\pi(f_0 - u\csc\alpha)t + \pi(\mu + \cot\alpha)t^2)\right] dt \end{aligned} \quad (2)$$

where $p = 2\alpha/\pi$ is the order of FrFT ($\alpha \neq n\pi$ and $n \in \mathbb{Z}$), $\mathcal{F}^p$ denotes the FrFT operator of order p, α is the counterclockwise rotation angle (in radians) of the signal coordinate axis, and u is the peak location in the FrFT spectrum. Assume an LFM signal concentrates at (α_0, u_0) in the FrFD, i.e., $S(\alpha_0, u_0)$ is the maximum of $S(\alpha, u)$, which means $2\pi(f_0 - u\csc\alpha)t + \pi(\mu + \cot\alpha)t^2 = 0$ holds for any t. Thus, α_0 and u_0 can be derived as

$$\begin{cases} \alpha_0 = -\text{arccot}\mu \\ u_0 = f_0 \sin\alpha_0 \end{cases} \quad (3)$$

In this case, Equation (2) can be expressed as

$$\begin{aligned} S(\alpha_0, u_0) &= \mathcal{F}^{p_0}[s(t)](u_0) \\ &= \sqrt{1 - j\cot\alpha_0} \cdot \exp(j\pi u_0^2 \cot\alpha_0) \cdot T \end{aligned} \quad (4)$$

where p_0 denotes the optimal order of FrFT ($\alpha_0 \neq n\pi$ and $n \in \mathbb{Z}$).

Consider a uniform linear array (ULA) with M isotropic hydrophones and an LFM signal with an incidence direction of θ. The signal received at the mth hydrophone is denoted by $s(t - \tau_m)$, where $\tau_m = (m-1)d\cos\theta/c$, $(m = 1, 2, \cdots, M)$ is the time delay for the signal to travel from the reference hydrophone to the mth hydrophone, d represents the distance between adjacent hydrophones, and c is the speed of sound. Then, the received signal at the mth hydrophone can be expressed as

$$\begin{aligned} x_m(t) &= s(t - \tau_m) + n_m(t) \\ &= \exp(j2\pi(0.5\mu\tau_m^2 - f_0\tau_m - \mu\tau_m t)) \cdot s(t) + n_m(t) \end{aligned} \quad (5)$$

In Equation (5), the noise is denoted by $n_m(t)$, which is uncorrelated and independent of the source signal. The orientation vector of an LFM signal is time-variant and frequency-dependent, and the conventional individual frequency steering vector is inapplicable. Therefore, a frequency-independent steering vector used in the FrFD needs to be derived.

Assuming $s(t - \tau_m)$ concentrates at (α_m, u_m), then the FrFT of $x_m(t)$ can be derived as

$$\begin{aligned} X_m(\alpha_m, u_m) &= \mathcal{F}^{p_m}[x_m(t)](u_m) \\ &= \mathcal{F}^{p_m}[s(t - \tau_m)](u_m) + \mathcal{F}^{p_m}[n_m(t)](u_m) \\ &= \sqrt{1 - j\cot\alpha_m} \cdot \exp(j\pi u_m^2 \cot\alpha_m) \cdot T \cdot \exp(j\pi(-2f_0\tau_m + \mu\tau_m^2)) + N_m(\alpha_m, u_m) \end{aligned} \quad (6)$$

where $N_m(\alpha_m, u_m) = \mathcal{F}^{p_m}[n_m(t)](u_m)$ represents the FrFT of $n_m(t)$. When $s(t)$ is delayed, the initial frequency changes from f_0 to $f_0 - \mu\tau_m$, but the frequency modulation rate μ remains the same. Thus, α_m and u_m can be represented by

$$\begin{cases} \alpha_m = \alpha_0 \\ u_m = u_0 + \tau_m \cos\alpha_0 \end{cases} \tag{7}$$

Substituting Equation (7) into Equation (6) and replacing f_0, μ by α_0, u_0, $X_m(\alpha_m, u_m)$ can be derived as

$$\begin{aligned} X_m(\alpha_m, u_m) &= X_m(\alpha_0, u_0 + \tau_m \cos\alpha_0) \\ &= A(\tau_m)S(\alpha_0, u_0) + N_m(\alpha_0, u_m) \end{aligned} \tag{8}$$

where $A(\tau_m) = \exp[-j(2\pi\tau_m u_0 \sin\alpha_0 + \pi\tau_m^2 \sin\alpha_0 \cos\alpha_0)]$ is the frequency-independent orientation coefficient in the FrFD. Then, $A(\tau_m)$ can be expressed as a function of θ:

$$A_m(\theta) = \exp(-j2\pi(m-1)d\cos\theta u_0 \sin\alpha_0/c) \cdot \exp(-j\pi(m-1)^2 d^2 \cos^2\theta \sin\alpha_0 \cos\alpha_0/c^2) \tag{9}$$

3. Deconvolved Beamforming in the Fractional Fourier Domain

In this section, we utilize beamforming and deconvolution in the FrFD. For the simplicity of expression, one LFM target is modeled at the beginning. Then, the multitarget model is discussed when the LFM signals are concentrated on the same point.

3.1. Fractional Fourier Domain Delay-and-Sum Beamforming

According to the formation of Equation (9), the steering vector of an LFM signal in the FrFD can be written as

$$\boldsymbol{v} = [v_1(\vartheta), v_2(\vartheta), \cdots, v_M(\vartheta)]^T \tag{10}$$

$$v_m(\vartheta) = \frac{1}{M}\exp(-j2\pi(m-1)d\cos\vartheta u_0 \sin\alpha_0/c) \cdot \exp(-j\pi(m-1)^2 d^2 \cos^2\vartheta \sin\alpha_0 \cos\alpha_0/c^2) \tag{11}$$

where ϑ represents the beam scanning angle. Based on the aforementioned analysis, the FrFD delay-and-sum beamforming complex output is

$$\begin{aligned} Y(\vartheta) &= \boldsymbol{v}^H \boldsymbol{X} \\ &= Y_S(\vartheta) + Y_N(\vartheta) \end{aligned} \tag{12}$$

In Equation (12), $\boldsymbol{X} = [X_1(\alpha_1, u_1), X_2(\alpha_2, u_2), \cdots, X_M(\alpha_M, u_M)]^T$ is the matrix of the peak value in the FrFD, $Y_N(\vartheta) = \boldsymbol{v}^H[N_1(\alpha_1, u_1), N_2(\alpha_2, u_2), \cdots, N_M(\alpha_M, u_M)]^T$ is the beamforming complex output of noise, superscript $(\cdot)^H$ denotes the conjugate transpose, and the beamforming complex output of the LFM signal can be expressed in detail as

$$\begin{aligned} Y_S(\vartheta) &= \boldsymbol{v}^H[A_1(\theta)\ A_2(\theta)\ \cdots\ A_M(\theta)]^T S(\alpha_0, u_0) \\ &= \frac{S(\alpha_0, u_0)}{M}\sum_{m=1}^{M}[\exp(-j2\pi(m-1)d(\cos\theta - \cos\vartheta)u_0 \sin\alpha_0/c) \cdot \exp(-j\pi(m-1)^2 d^2(\cos^2\theta - \cos^2\vartheta)\sin\alpha_0 \cos\alpha_0/c^2)] \end{aligned} \tag{13}$$

Then, the spatial spectrum of the FrFB can be given by Equation (14), and the ϑ corresponding to the peak of $P_{FrFB}(\vartheta)$ is the estimated DOA.

$$P_{FrFB}(\vartheta) = |Y(\vartheta)|^2 \tag{14}$$

For the multitarget scenario, several LFM signals will focus on different points when they have different time–frequency characteristics. In this circumstance, each target derives a corresponding spatial spectrum. The DOA is given by the unique peak of each spatial spectrum [8]. There is no need to discuss spatial resolution further, which will not be discussed in this paper. If the LFM signals from different targets possess the same time–frequency characteristics, they will focus on the same time–frequency point. Multiple

LFM signals are superimposed onto each other in the FrFD. In this circumstance, the FrFB complex output can be expressed as

$$Y(\vartheta) = \sum_{k=1}^{K} Y_{S,k}(\vartheta) + Y_N(\vartheta) \tag{15}$$

$$Y_{S,k} = \frac{S_k(\alpha_0, u_0)}{M} \sum_{m=1}^{M} [\exp(-j2\pi(m-1)d(\cos\theta_k - \cos\vartheta)u_0 \sin\alpha_0/c) \cdot \exp(-j\pi(m-1)^2 d^2(\cos^2\theta_k - \cos^2\vartheta)\sin\alpha_0\cos\alpha_0/c^2)] \tag{16}$$

In Equation (16), $Y_{S,k}$ denotes the FrFB complex output of the kth LFM signal. The spatial spectrum of FrFB for the multitarget scenario can be expressed as

$$P_{FrFB}(\vartheta) = \left| \sum_{k=1}^{K} Y_{S,k}(\vartheta) + Y_N(\vartheta) \right|^2 \tag{17}$$

Although FrFB can effectively estimate LFM signal DOA at low SNRs, it generates a spatial spectrum with a wide mainlobe and high sidelobes because of the CBF. If two LFM signals with the same time–frequency characteristics are closely located, they will be difficult to be distinguished. A typical method for overcoming this problem is to deconvolve the spatial spectrum by the R-L algorithm. However, due to the coherence of LFM signals that focus on the same time–frequency point, this paper introduces the MFISTA algorithm to deconvolve the complex output so that the influence of cross terms between coherent sources on the spatial spectrum can be avoided.

3.2. Deconvolution of the Beamforming Complex Output

In this section, deconvolution is used as postprocessing for the FrFB complex output to obtain a high-resolution spatial spectrum with narrow beamwidth and low sidelobes. First, the FrFB complex output is shown to have a form of convolution, and the complex beam pattern is proved to be shift-invariant. Then, the MFISTA algorithm is introduced to realize deconvolution.

In Equation (16), the quadratic time delay term $(m-1)^2 d^2(\cos^2\theta - \cos^2\vartheta)/c^2$ is proportional to the inverse of the sound speed, which is small enough to be omitted [22,43,44]. Therefore, $\exp(-j\pi(m-1)^2 d^2(\cos^2\theta - \cos^2\vartheta)\sin\alpha_0\cos\alpha_0/c^2) \approx 1$ and Equation (16) can be derived as

$$\begin{aligned} Y_{S,k}(\vartheta) &= \frac{S_k(\alpha_0, u_0)}{M} \sum_{m=1}^{M} \exp(-j2\pi(m-1)d(\cos\theta_k - \cos\vartheta)u_0\sin\alpha_0/c) \\ &= S_k(\alpha_0, u_0)\frac{\sin(\pi M d(\cos\vartheta - \cos\theta_k)u_0\sin\alpha_0/c)}{M\sin(\pi d(\cos\vartheta - \cos\theta_k)u_0\sin\alpha_0/c)}\exp(j(M-1)d(\cos\vartheta - \cos\theta_k)u_0\sin\alpha_0/c) \end{aligned} \tag{18}$$

For the LFM signals, which concentrate at (α_0, u_0) in the FrFD, the complex beam pattern can be proposed as Equation (19), according to the FrFB complex output given by Equation (18).

$$Y_P(\vartheta|\theta) = \frac{\sin(\pi M d(\cos\vartheta - \cos\theta)u_0\sin\alpha_0/c)}{M\sin(\pi d(\cos\vartheta - \cos\theta)u_0\sin\alpha_0/c)}\exp(j(M-1)d(\cos\vartheta - \cos\theta)u_0\sin\alpha_0/c) \tag{19}$$

Assuming the true DOA is ψ, the source spatial distribution corresponding to a specific LFM signal is defined as follows:

$$D(\theta) \triangleq S(\alpha_0, u_0)\delta(\cos\theta - \cos\psi) \tag{20}$$

$$\delta(\cos\theta - \cos\psi) = 0, \quad \cos\theta \neq \cos\psi, \quad \int \delta(\cos\theta - \cos\psi)d\cos\theta = 1 \tag{21}$$

Since the complex beam pattern $Y_P(\vartheta|\theta)$ is a function of $\cos\vartheta - \cos\theta$, it can be regarded as a shift-invariant function on the angle cosine, i.e., $Y_P(\cos\vartheta|\cos\theta) = Y_P(\cos\vartheta - \cos\theta)$. Therefore, $Y_P(\cos\vartheta|\cos\theta)$ is shift-invariant. Hence, the complex output of LFM signals can be expressed in a convolution formation as

$$Y_S(\cos\vartheta) = \int Y_P(\cos\vartheta - \cos\theta)D(\cos\theta)d\cos\theta \tag{22}$$

Therefore, the deconvolution algorithm can be utilized to restore $D(\cos\theta)$, with the knowledge of $Y_S(\cos\vartheta)$ and $Y_P(\cos\vartheta - \cos\theta)$. Considering that both $Y_S(\cos\vartheta)$ and $Y_P(\cos\vartheta - \cos\theta)$ are complex values, the typical R-L algorithm is no longer applicable. The deconvolution of Equation (22) is converted into a sparse-constrained optimization problem, as shown in Equation (23).

$$\underset{\hat{\boldsymbol{D}}\in\mathbb{C}^G}{\text{argmin}}\frac{1}{2}\left\|\boldsymbol{Y}_S - \boldsymbol{Y}_P\hat{\boldsymbol{D}}\right\|_2^2 + \lambda\left\|\hat{\boldsymbol{D}}\right\|_1 \tag{23}$$

$$\boldsymbol{Y}_P = [Y_P(\theta_{ri}|\theta_{ci})]_{G\times G} \tag{24}$$

In Equation (23), $\boldsymbol{Y}_S, \hat{\boldsymbol{D}} \in \mathbb{C}^G$ are the FrFB output vector and the target signal space distribution vector, respectively. $\boldsymbol{Y}_P \in \mathbb{C}^{G\times G}$ is the complex beam pattern matrix, $\lambda \in \mathbb{R}$ is the regularization coefficient, and G is the number of discrete angles. In Equation (24), $ri, ci \in [1, G] \subset \mathbb{Z}^+$ represent the indices of the row and column in the matrix, respectively, and $[\theta_1, \theta_2 \cdots, \theta_G]$ represents the set of discrete angles. The solution to Equation (23) can be obtained by the MFISTA Algorithm 1, which is given as follows.

Algorithm 1: MFISTA algorithm.

Input: FrFB complex output $\boldsymbol{Y}_S$, complex beam pattern $\boldsymbol{Y}_P$, regularization coefficient λ, iteration times I, and initial value $\hat{\boldsymbol{D}}_0$;
Output: the estimation of source signal distribution $\hat{\boldsymbol{D}}_I$;
1. **set** $l_1 = 1$ and $y_1 = \hat{\boldsymbol{D}}_0$;
2. **for** $i = 1$ to I
3. $z_i = \mathcal{S}_{\lambda/L}(\frac{1}{L}\boldsymbol{Y}_P^H(\boldsymbol{Y}_S - \boldsymbol{Y}_P\boldsymbol{y}_i) + \boldsymbol{y}_i)$;
4. $\hat{\boldsymbol{D}}_i = \underset{\hat{\boldsymbol{D}}\in\{z_i,\hat{\boldsymbol{D}}_{i-1}\}}{\text{argmin}} \frac{1}{2}\|\boldsymbol{Y}_S - \boldsymbol{Y}_P\hat{\boldsymbol{D}}\|_2^2 + \lambda\|\hat{D}\|_1$;
5. $l_{i+1} = (1 + \sqrt{1 + 4l_i^2})/2$;
6. $\boldsymbol{y}_{i+1} = \hat{\boldsymbol{D}}_i + \frac{l_i - 1}{l_{i+1}}(\hat{\boldsymbol{D}}_i - \hat{\boldsymbol{D}}_{i-1}) + \frac{l_i}{l_{i+1}}(z_i - \hat{\boldsymbol{D}}_i)$;
7. **end**

In Step 3 of Algorithm 1, L denotes the step size, and the shrinkage-thresholding function $\mathcal{S}_{\lambda/L}(\boldsymbol{x})$ is defined for arbitrary $\boldsymbol{x} = [x_1, x_2, \cdots, x_G]^T \in \mathbb{C}^G$ by

$$\mathcal{S}_{\lambda/L}(x_g) = \begin{cases} 0 & |x_g| < \frac{\lambda}{L} \\ x_g - \frac{\lambda x_g}{L|x_g|} & |x_g| \geq \frac{\lambda}{L} \end{cases} \tag{25}$$

After obtaining the estimation $\hat{\boldsymbol{D}}_I$ using Algorithm 1, the DFrFB spatial spectrum can be expressed as

$$\boldsymbol{P}_{DFrFB} = \left|\hat{\boldsymbol{D}}_I\right|^2 \tag{26}$$

The flowchart of the proposed DFrFB is shown in Figure 1. A coarse-to-fine searching strategy [24] is utilized to obtain the optimal order of the FrFT. The discrete FrFT is implemented using the algorithm proposed in [16]. In the MFISTA deconvolution algorithm, the step size is set to the maximum eigenvalue of the covariance matrix of $\boldsymbol{Y}_P$, and the regularization coefficient λ is set to 0.5.

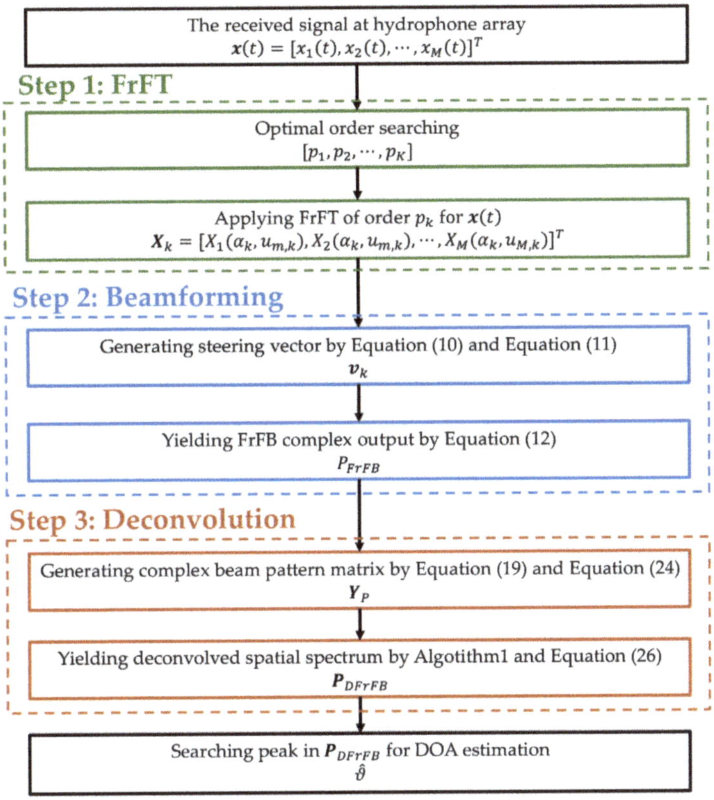

Figure 1. Flowchart of the proposed DFrFB method.

4. Simulation Results

In this section, the performance of DFrFB is verified by five simulations. The same simulations are performed for FrFB and ISM-DCBF for reference. The root mean square error (RMSE) of the DOA estimation and resolution probability (RP) are calculated to evaluate the spatial resolution. The beamwidth and maximum sidelobe level are recorded to evaluate the optimization of the space spectrum. When the estimated DOAs $\hat{\theta_1}, \hat{\theta_2}$ and the real DOAs θ_1, θ_2 satisfy Equation (27), two targets are considered to be successfully discriminated. Assuming N_{total} times Monte Carlo experiments are performed and $N_{success}$ times experiments satisfy Equation (27), the RP is defined as $N_{success}/N_{total}$. The RMSE is calculated only when the RP is 1. The beamwidth is also referred to as the half-power beamwidth, i.e., the width of the mainlobe at $-3\ dB$.

$$\left|\hat{\theta_1} - \theta_1\right| + \left|\hat{\theta_2} - \theta_2\right| < |\theta_1 - \theta_2| \quad (27)$$

The LFM signals emitted by different targets are identical, and the detailed parameters are given in Table 1. The noise signal is Gaussian white noise, and the noise signal received on each hydrophone is independently identically distributed and unrelated to the target signals. The SNR is defined by Equation (28), where $P_{S,k}$ denotes the power of the kth target signal and P_N denotes the power of the noise.

$$SNR = 10\log_{10}\left(\sum_{k=1}^{K} P_{S,k}/P_N\right) \quad (28)$$

Table 1. Parameters of LFM signal.

Parameter	Value
beamwidth	1 kHz
center frequency	7.5 kHz
frequency modulation rate	20 kHz
time duration	0.05 s
sampling rate	62.5 kHz

All time-domain snapshots are processed together by fast Fourier transform (FFT) or FrFT in the corresponding methods. In ISM-DCBF, the received signal is divided into four subbands, and MFISTA is also used as the deconvolution algorithm. Unless stated otherwise, the number of deconvolution iterations is 5000 in ISM-DCBF and DFrFB. The number of iterations of ISM-DCBF refers to that of one subband, and each subband has the same number of iterations. A ULA with 10 hydrophones is used as the received array, whose element spacing is 0.1 m. The sound speed is 1500 m/s. The number of Monte Carlo simulations in Sections 4.2–4.5 is 100. All simulations in this paper are conducted by MATLAB R2020a.

4.1. Spatial Spectrum

The spatial spectra of the FrFB, ISM-DCBF, and DFrFB at 0 dB are given in Figure 2. The cross symbol represents the real DOA of the target, which is 75° and 85°. The estimated DOAs, beamwidth, and sidelobe levels are given in Table 2. As shown in Figure 2, the three methods are all successful in identifying two targets. The estimation error of DFrFB is the smallest, ISM-DCBF is the second smallest, and FrFB is the largest. By deconvolving the complex output of the FrFB, the beamwidth is reduced by 75%, and the sidelobe level is reduced by 17.7 dB, resulting in a much cleaner DFrFB spatial spectrum.

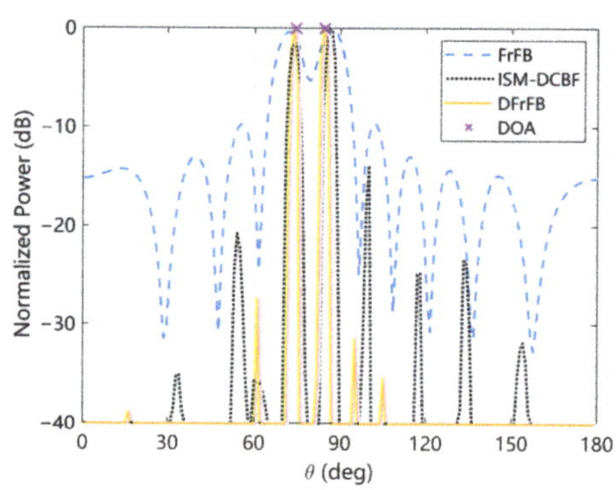

Figure 2. Spatial spectra of the simulation.

Table 2. The performance of the simulation.

Indicators	FrFB	ISM-DCBF	DFrFB
DOA (deg)	72, 87	74, 86	75, 85
beamwidth (deg)	9.3	3.6	2.3
sidelobe level (deg)	−9.6	−13.9	−27.3

Since the energy of the LFM signal is highly concentrated in the FrFD whereas the noise is not [24], noise shows less influence on DFrFB. In contrast, noise shows a large influence on each ISM-DCBF subband signal. Thus, the number and intensity of DFrFB sidelobes are smaller than those of ISM-DCBF. It is demonstrated that DFrFB can effectively improve the spatial spectrum of FrFB and has better performance than ISM-DCBF, which is based on frequency-domain signal processing.

4.2. The Number of Deconvolution Iterations

The influence of the number of iterations in the MFISTA algorithm on DFrFB and ISM-DCBF is discussed in this section. The source direction and SNR are consistent with Section 4.1. The RMSE, beamwidth, and sidelobe level with the number of iterations are given in Figure 3a, Figure 3b and Figure 3c, respectively. Note that FrFB requires no deconvolution, and its data are independent of the number of iterations but still plotted in Figure 3 as a reference. The RP of all three methods is 1 regardless of the number of iterations, so the RP figure with the number of iterations is not given.

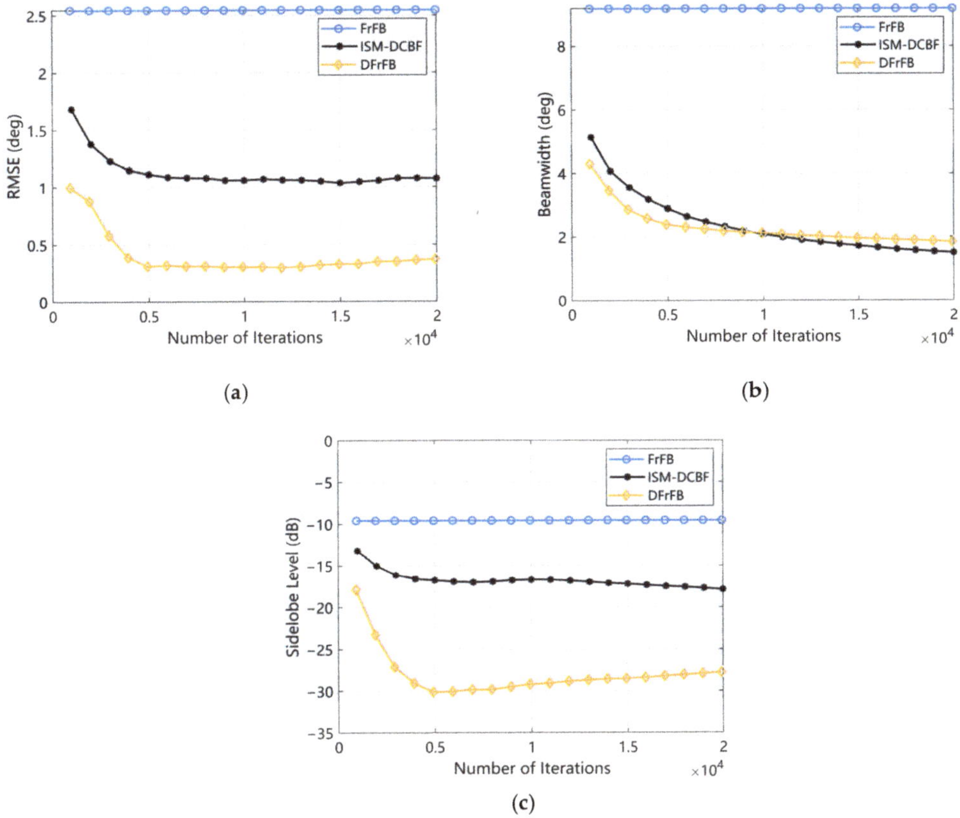

Figure 3. The performances of FrFB, ISM-DCBF, and DFrFB versus the number of iterations. (**a**) The RMSE; (**b**) the beamwidth; (**c**) the sidelobe level.

All the indicators of DFrFB and ISM-DCBF rapidly improve with the increase in iterations when the number is less than 5000. DFrFB performs better than ISM-DCBF for the same number of iterations. However, the effect of noise is magnified by the deconvolution algorithm when the number of iterations exceeds 5000. Only beamwidth can be reduced further. For DFrFB, the RMSE and sidelobe level no longer improve with increasing iterations but deteriorate slightly. For ISM-DCBF, the optimization of RMSE and sidelobe level is minimal and almost stagnant. Figure 3 shows that the DFrFB performs better than ISM-DCBF when the number of iterations is appropriate. However, the number of iterations should not be too large; otherwise, the DFrFB performance will deteriorate in turn.

4.3. SNR

This section investigates the effect of noise on the spatial resolution with the SNR ranging from −20 dB to 20 dB. The real DOAs of the two targets remain 75° and 85°. As shown in Figure 4a, the FrFB is insensitive to noise, and its RMSE fluctuates slightly with the SNR. However, the estimation error is large regardless of the SNR. The RMSEs of DFrFB and ISM-DCBF decrease with increasing SNR. The DFrFB accuracy is better than that of ISM-DCBF, showing better noise resistance. In Figure 4b, benefiting from the FrFT energy-focusing property, the RP of FrFB and DFrFB remains 1 even if the SNR is as low as −20 dB. In contrast, ISM-DCBF reaches an RP of 1 only when the SNR is higher than −12 dB. The DFrFB based on FrFD signal processing demonstrated better spatial resolution at low SNRs compared with ISM-DCBF based on frequency-domain signal processing.

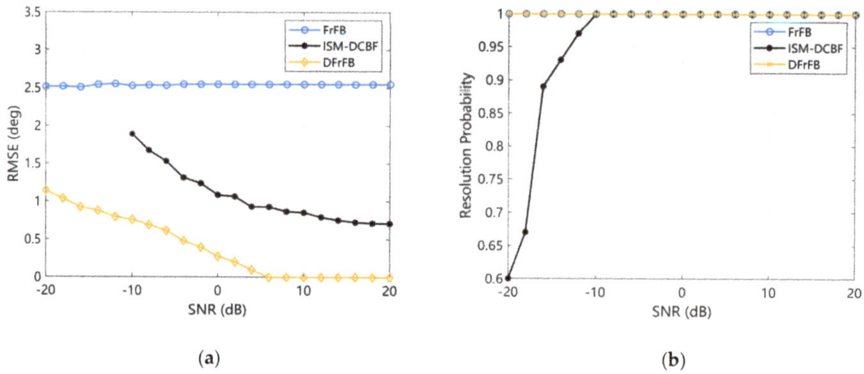

Figure 4. The performances of FrFB, ISM-DCBF, and DFrFB versus the SNR of the received signal. (**a**) The RMSE; (**b**) the RP.

4.4. The Number of Hydrophones

In this section, the effect of the number of hydrophones on the spatial resolution of the DFrFB is studied. The SNR is 0 dB, and the real DOAs of the two targets are the same as in Section 4.3. The RMSE and RP for the number of hydrophones varying from 3 to 15 are shown in Figure 5. DFrFB obtains the smallest DOA estimation error in Figure 5a. As shown in Figure 5b, to reach an RP of 1, DFrFB needs at least six hydrophones, while ISM-DCBF needs eight and FrFB needs nine. DFrFB can realize more accurate DOA estimations with fewer hydrophones. It is shown that DFrFB can obtain a high spatial resolution with a small aperture array.

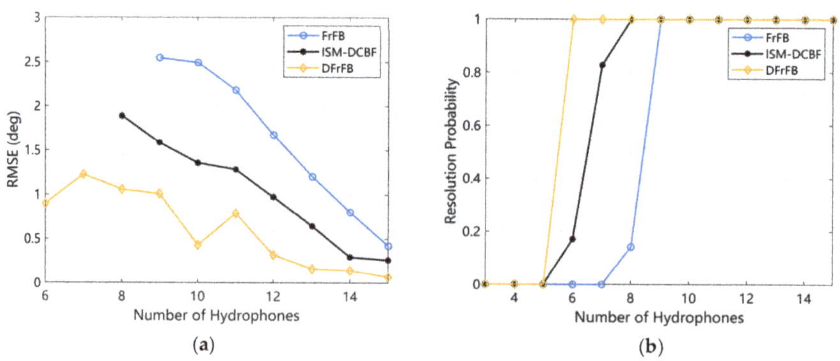

Figure 5. The performances of FrFB, ISM-DCBF, and DFrFB versus the number of hydrophones. (**a**) The RMSE; (**b**) the RP.

Note that the RMSE of DFrFB does not decrease steadily with the increase in hydrophones. The ideal peak location u_m given by Equation (7) is most likely not an integer. However, the peak location selected in practice is an integer due to the received signal being discrete in the time domain. This leads to a roundoff error of u_m and a model mismatch when deconvolution is applied. When the number of hydrophones changes, the roundoff error also changes accordingly. Hence, the roundoff error leads to obvious fluctuations in the RMSE of DFrFB. Nevertheless, DFrFB still obtains the most accurate DOAs.

4.5. Angular Interval

Various angular intervals are simulated in this section to verify the adjacent target discrimination capability of the proposed DFrFB. One target is fixed at 85°, and another target moves from 84° to 70° so that the angular interval gradually changes from 1° to 15°. The number of hydrophones in the ULA is 10. Figure 6a,b show the RMSE and RP with various angular intervals when the SNR is 0 dB. The reason for the fluctuations in the RMSE of DFrFB is similar to that in Section 4.4, which is not redescribed here.

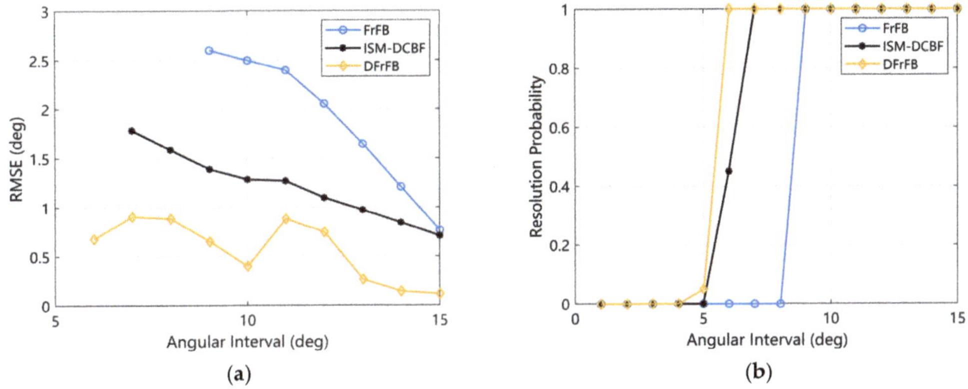

Figure 6. The performances of FrFB, ISM-DCBF, and DFrFB versus different angular intervals. (**a**) The RMSE; (**b**) the resolution probability.

In Figure 6a, despite the apparent fluctuations, DFrFB obtains the most accurate DOA estimations regardless of the angular interval. The minimum discriminable angular interval of the DFrFB is 6°, while that of the ISM-DCBF and FrFB is 7° and 9°, respectively. It is demonstrated that DFrFB possesses the highest adjacent target discrimination capability.

5. Experimental Results

The DFrFB performance is verified by an experiment conducted at Jingye Lake, Tianjin, China. Jingye Lake is broadly rectangular in shape, approximately 126 m long, 80 m wide, and 3 m deep on average. The average sound speed is 1476 m/s, measured by a Valeport miniSVP. In the experiment, the target source continuously emits multiple frames of the same LFM signal, and the time interval between two adjacent frames is 0.25 s, so that no interference between different frames can be guaranteed. The detailed LFM signal parameters can be seen in Table 1. The real DOAs of the two targets are 92.2° and 116.5°, measured by the $NTS - 382R_6$ total station. The receiving ULA consists of eight Brüel & Kjær 8104 hydrophones with an element spacing of 0.8 m. The receiving array is deployed horizontally. Both the receiving and transmitting ends are deployed at 1 m depth to ensure that the target and the array are in the same plane. The received signal is processed by a digital filter with a passband of 7~8 kHz, and the filtered SNR is approximately 4.1 dB. Only the received signal containing the LFM signal is beamformed, and the pure noise received signal is not involved in the calculation. The deconvolution iterations of DFrFB and ISM-DCBF are still 5000.

The spatial spectrum of a one-frame LFM signal is given in Figure 7, where the cross symbols represent the real target DOAs. All three methods yield relatively accurate DOAs for the target located at 116.5°. However, the sidelobe of ISM-DCBF is so significant that it affects the estimation of another target. The sidelobes of DFrFB, in contrast, show no significant effect on detecting the two targets. Due to the complexity of the hydroacoustic channel in the lake and the inaccuracy of the hydrophone installation positions, a slight model mismatch occurs and the spatial spectral optimization capability of both the DFrFB and ISM-DCBF is reduced.

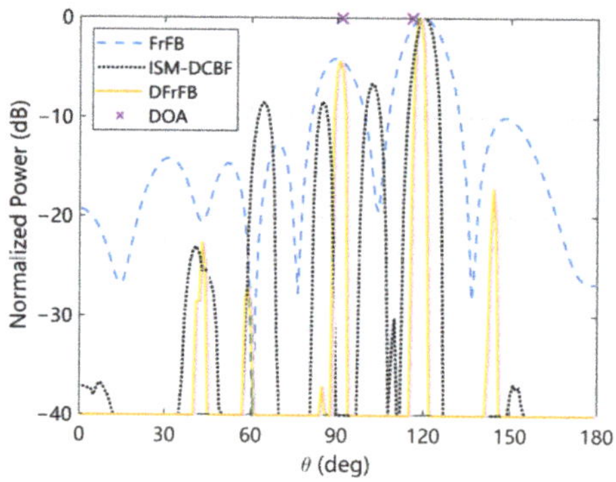

Figure 7. The spatial spectrum of a one-frame LFM signal received in the experiment.

The DOA estimations of nine consecutive frames are given in Figure 8, where the vertical lines represent the real DOAs. The average RMSE, beamwidth, and sidelobe level are listed in Table 3. DFrFB effectively optimizes the spatial spectrum of FrFB under experimental conditions. Compared with FrFB, DFrFB reduces the beamwidth by 77% and the sidelobe by 6.0 dB, resulting in more accurate DOA estimations. Compared with ISM-DCBF, DFrFB yields more stable and accurate DOA estimations, as well as a cleaner spatial spectrum. It is demonstrated that the proposed DFrFB possesses the best performance among the three methods.

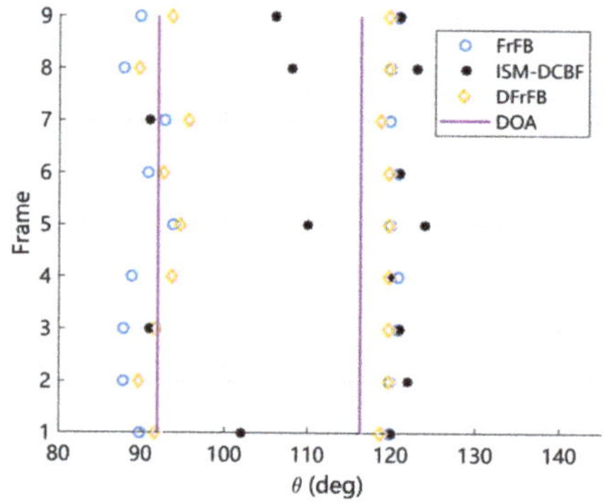

Figure 8. The estimated DOAs of nine frames of LFM signals received in the experiment.

Table 3. The performance of the experiment.

Indicators	FrFB	ISM-DCBF	DFrFB
RMSE (deg)	3.5	15.1	2.8
beamwidth (deg)	14.5	5.7	3.3
sidelobe level (dB)	−10.1	−7.3	−16.1

6. Summary

The proposed DFrFB can achieve high-resolution LFM signal DOA estimation by a small aperture array in a low SNR hydroacoustic environment. In this paper, the FrFD array signal model is established. The FrFB complex output is proven to be the convolution of the shift-invariant complex beam pattern with the source signal spatial distribution. The MFISTA deconvolution algorithm is introduced to solve the deconvolution of complex values. Additionally, a robust high-resolution spatial spectrum is obtained. The results of the simulation and experiment show that the proposed DFrFB can effectively improve the FrFB performance and achieve a high-resolution DOA estimation of multiple LFM signals. In addition, DFrFB possesses better performance compared with directly performing DCBF for the LFM signal frequency-domain subbands.

Author Contributions: Conceptualization, Z.L. and X.F.; formal analysis, Z.L.; resources, W.S.; validation, Z.L. and Q.T.; writing—original draft preparation, Z.L. and Q.T.; writing—review and editing, W.S. and X.F. All authors have read and agreed to the published version of the manuscript.

Funding: This research was funded by Natural Science Foundation of Tianjin, China under Grant 21JCQNJC00650.

Institutional Review Board Statement: Not applicable.

Informed Consent Statement: Not applicable.

Data Availability Statement: Data are available on request.

Conflicts of Interest: The authors declare no conflict of interest.

References

1. Lee, D.-H.; Shin, J.-W.; Do, D.-W.; Choi, S.-M.; Kim, H.-N. Robust LFM target detection in wideband sonar systems. *IEEE Trans. Aerosp. Electron. Syst.* **2017**, *53*, 2399–2412. [CrossRef]
2. Liu, D.L.; Qu, H.R.; Wang, W.; Deng, J.J. Multiple targets detection of linear frequency-modulated continuous wave active sonar using fractional Fourier transform. *Integr. Ferroelectr.* **2020**, *209*, 1–10. [CrossRef]
3. Bartlett, M.S. Smoothing periodograms from time-series with continuous spectra. *Nature* **1948**, *161*, 686–687. [CrossRef]
4. Capon, J. High-resolution frequency-wavenumber spectrum analysis. *Proc. IEEE* **1969**, *57*, 1408–1418. [CrossRef]
5. Chiariotti, P.; Martarelli, M.; Castellini, P. Acoustic beamforming for noise source localization—Reviews, methodology and applications. *Mech. Syst. Signal Process.* **2019**, *120*, 422–448. [CrossRef]
6. Ahmad, Z.; Song, Y.L.; Du, Q. Wideband DOA estimation based on incoherent signal subspace method. *Compel-Int. J. Comp. Math. Electr. Electron. Eng.* **2018**, *37*, 1271–1289. [CrossRef]
7. Wang, H.; Kaveh, M. Coherent signal-subspace processing for the detection and estimation of angles of arrival of multiple wide-band sources. *IEEE Trans. Acoust. Speech Signal Process.* **1985**, *33*, 823–831. [CrossRef]
8. Yin, J.W.; Guo, K.; Han, X.; Yu, G. Fractional Fourier transform based underwater multi-targets direction-of-arrival estimation using wideband linear chirps. *Appl. Acoust.* **2020**, *169*, 7. [CrossRef]
9. Khodja, M.; Belouchrani, A.; Abed-Meraim, K. Performance analysis for time-frequency MUSIC algorithm in presence of both additive noise and array calibration errors. *EURASIP J. Adv. Signal Process.* **2012**, *11*, 94. [CrossRef]
10. Ghofrani, S. Matching pursuit for direction of arrival estimation in the presence of Gaussian noise and impulsive noise. *IET Signal Process.* **2014**, *8*, 540–551. [CrossRef]
11. Cui, K.B.; Wu, W.W.; Huang, J.J.; Chen, X.; Yuan, N.C. DOA estimation of LFM signals based on STFT and multiple invariance ESPRIT. *AEU-Int. J. Electron. Commun.* **2017**, *77*, 10–17. [CrossRef]
12. Han, X.Y.; Liu, M.Q.; Zhang, S.N.; Zheng, R.H.; Lan, J. A passive DOA estimation algorithm of underwater multipath signals via spatial time-frequency distributions. *IEEE Trans. Veh. Technol.* **2021**, *70*, 3439–3455. [CrossRef]
13. Shi, J.; Yang, D.S.; Shi, S.G.; Zhu, Z.R.; Hu, B. Spatial time-frequency DOA estimation based on joint diagonalization using Jacobi rotation. *Appl. Acoust.* **2017**, *116*, 24–32. [CrossRef]
14. Salin, B.M.; Salin, M.B. Methods for measuring bistatic characteristics of sound scattering by the ocean bottom and surface. *Acoust. Phys.* **2016**, *62*, 575–582. [CrossRef]
15. Kim, M.; Lee, S.H.; Choi, I.O.; Kim, K.T. Direction finding for multiple wideband chirp signal sources using blind signal separation and matched filtering. *Signal Process.* **2022**, *200*, 15. [CrossRef]
16. Ozaktas, H.M.; Arikan, O.; Kutay, M.A.; Bozdagt, G. Digital computation of the fractional Fourier transform. *IEEE Trans. Signal Process.* **1996**, *44*, 2141–2150. [CrossRef]
17. Bultheel, A.; Martínez Sulbaran, H.E. Computation of the fractional Fourier transform. *Appl. Comput. Harmon. Anal.* **2004**, *16*, 182–202. [CrossRef]

18. Tao, R.; Deng, B.; Wang, Y. Research progress of the fractional Fourier transform in signal processing. *Sci. China Ser. F* **2006**, *49*, 1–25. [CrossRef]
19. Sejdic, E.; Djurovic, I.; Stankovic, L. Fractional Fourier transform as a signal processing tool: An overview of recent developments. *Signal Process.* **2011**, *91*, 1351–1369. [CrossRef]
20. Coetmellec, S.; Brunel, M.; Lebrun, D.; Lecourt, J.B. Fractional-order Fourier series expansion for the analysis of chirped pulses. *Opt. Commun.* **2005**, *249*, 145–152. [CrossRef]
21. Zhu, Y.C.; Yang, K.D.; Li, H.; Wu, F.Y.; Yang, Q.L.; Xue, R.Z. Research on DOA estimation of nonstationary signal based on fractional Fourier transform. In Proceedings of the OCEANS—MTS/IEEE Kobe Techno-Oceans Conference (OTO), Kobe, Japan, 28–31 May 2018.
22. Wang, X.H.; Li, B.; Yang, H.J.; Lu, W.D. Off-grid DOA estimation for wideband LFM signals in FRFT domain using the sensor arrays. *IEEE Access* **2019**, *7*, 18500–18509. [CrossRef]
23. Sreekumar, G.; Mary, L.; Unnikrishnan, A. Performance analysis of fractional Fourier domain beam-forming methods for sensor arrays. *Smart Sci.* **2019**, *7*, 28–38. [CrossRef]
24. Zhang, H.-x.; Liu, H.-d.; Chen, S.; Zhang, Y.; Wang, X.-y. Parameter estimation of chirp signals based on fractional Fourier transform. *J. China Univ. Posts Telecommun.* **2013**, *20*, 95–100. [CrossRef]
25. Levonen, M.; McLaughlin, S. Fractional Fourier transform techniques applied to active sonar. In Proceedings of the MTS/IEEE Conference on Celebrating the Past—Teaming Toward the Future, San Diego, CA, USA, 22–26 September 2003; pp. 1894–1899.
26. Yang, T.C. Deconvolved conventional beamforming for a horizontal line array. *IEEE J. Ocean. Eng.* **2018**, *43*, 160–172. [CrossRef]
27. Khan, M.K.; Morigi, S.; Reichel, L.; Sgallari, F. Iterative methods of Richardson-Lucy-Type for image deblurring. *Numer. Math.-Theory Methods Appl.* **2013**, *6*, 262–275. [CrossRef]
28. Zhu, Y.; Yang, T.C.; Pan, X. Deconvolved conventional beamforming for a planar array and Cramer Rao bound. *J. Acoust. Soc. Am.* **2018**, *144*, 1971. [CrossRef]
29. Sun, D.; Ma, C.; Mei, J.; Shi, W.; Gao, L. The deconvolved conventional beamforming for non-uniform line arrays. In Proceedings of the 2018 OCEANS—MTS/IEEE Kobe Techno-Oceans (OTO), Kobe, Japan, 28–31 May 2018; pp. 1–4.
30. Huang, J.; Zhou, T.; Du, W.; Shen, J.; Zhang, W. Smart ocean: A new fast deconvolved beamforming algorithm for multibeam sonar. *Sensors* **2018**, *18*, 4013. [CrossRef]
31. Yang, T.C. Performance analysis of superdirectivity of circular arrays and implications for sonar systems. *IEEE J. Ocean. Eng.* **2019**, *44*, 156–166. [CrossRef]
32. Yang, T. Superdirective beamforming applied to SWellEx96 horizontal arrays data for source localization. *J. Acoust. Soc. Am.* **2019**, *145*, EL179–EL184. [CrossRef]
33. Ye, Z.; Yang, T.C. Deconvolved conventional beamforming for a coprime array. In Proceedings of the Oceans 2019—MARSEILLE, Marseille, France, 17–20 June 2019.
34. Sun, D.; Ma, C.; Yang, T.C.; Mei, J.; Shi, W. Improving the performance of a vector sensor line array by deconvolution. *IEEE J. Ocean. Eng.* **2020**, *45*, 1063–1077. [CrossRef]
35. Ma, S.H.; Yang, T.C. The effect of elevation angle on bearing estimation for array beamforming in shallow water. In Proceedings of the Global Oceans 2020: Singapore—U.S. Gulf Coast, Biloxi, MS, USA, 5–30 October 2020.
36. Yang, T.C. Deconvolution of decomposed conventional beamforming. *J. Acoust. Soc. Am.* **2020**, *148*, EL195–EL201. [CrossRef] [PubMed]
37. Zhou, T.; Huang, J.; Du, W.; Shen, J.; Yuan, W. 2-D deconvolved conventional beamforming for a planar array. *Circuits Syst. Signal Process.* **2021**, *40*, 5572–5593. [CrossRef]
38. Wang, F.; Tian, X.; Liu, X.; Gu, B.; Zhou, F.; Chen, Y. Combination complex-valued bayesian compressive sensing method for sparsity constrained deconvolution beamforming. *IEEE Trans. Instrum. Meas.* **2022**, *71*, 9506013. [CrossRef]
39. Wang, S.H.; Lu, M.Y.; Mei, J.D.; Cui, W.T. Deconvolved beamforming using the chebyshev weighting method. *J. Mar. Sci. Appl.* **2022**, *21*, 228–235. [CrossRef]
40. Xie, L.; Sun, C.; Tian, J. Deconvolved frequency-difference beamforming for a linear array. *J. Acoust. Soc. Am.* **2020**, *148*, EL440–EL446. [CrossRef] [PubMed]
41. Prasad, S. Statistical-information-based performance criteria for Richardson-Lucy image deblurring. *J. Opt. Soc. Am. A-Opt. Image Sci. Vis.* **2002**, *19*, 1286–1296. [CrossRef] [PubMed]
42. Zibetti, M.V.W.; Helou, E.S.; Regatte, R.R.; Herman, G.T. Monotone FISTA with variable acceleration for compressed sensing magnetic resonance imaging. *IEEE Trans. Comput. Imaging* **2019**, *5*, 109–119. [CrossRef] [PubMed]
43. Cui, Y.; Wang, J.F. Wideband LFM interference suppression based on fractional Fourier transform and projection techniques. *Circuits Syst. Signal Process.* **2014**, *33*, 613–627. [CrossRef]
44. Cui, Y.; Wang, J.F.; Sun, H.X.; Jiang, H.; Yang, K.; Zhang, J.F. Gridless underdetermined DOA estimation of wideband LFM signals with unknown amplitude distortion based on fractional Fourier transform. *IEEE Internet Things J.* **2020**, *7*, 11612–11625. [CrossRef]

sensors

Article

Leaky Partial Update LMS Algorithms in Application to Structural Active Noise Control

Dariusz Bismor

Department of Measurements and Control Systems, Silesian University of Technology, ul. Akademicka 16, 44-100 Gliwice, Poland; Dariusz.Bismor@polsl.pl

Abstract: Adaptive signal processing algorithms play an important role in many practical applications in diverse fields, such as telecommunication, radar, sonar, multimedia, biomedical engineering and noise control. Recently, a group of adaptive filtering algorithms called partial update adaptive algorithms (partial updates) has gathered considerable attention in both research and practical applications. This paper is a study of the application of PUs to very demanding, structural active noise control (ANC) systems, which are of particular interest due to their ability to provide for a global noise reduction. However, such systems are multichannel, with very high computational power requirements, which may be reduced by the application of partial updates. The paper discusses the modifications necessary to apply PUs in structural ANC systems and the potential computational power savings offered by this application. As a result, leaky versions of the PU LMS algorithms are introduced to the general public. The paper also presents two simulation examples, based on real laboratory setups, confirming high performance of the proposed algorithms.

Keywords: partial updates; least mean squares; Leaky LMS; structural active noise control

Citation: Bismor, D. Leaky Partial Update LMS Algorithms in Application to Structural Active Noise Control. *Sensors* **2023**, *23*, 1169. https://doi.org/10.3390/s23031169

Academic Editors: Yuxing Li and Luca Fredianelli

Received: 19 December 2022
Revised: 17 January 2023
Accepted: 17 January 2023
Published: 19 January 2023

1. Introduction

Many applications of adaptive filtering are computationally too expensive to be implemented in real time, even using modern hardware. Examples of such applications include wireless communication systems [1], radar systems [2], adaptive beamforming of radio signals [3], and active noise control [4,5]. Partial updates are an effective and modern way of reducing this computational effort [6,7]. Partial updates are particularly well-suited for those real-time applications that require a huge number of operations to be performed in a single sampling interval. For example, the Least Mean Square (LMS) algorithm applied to an acoustic echo cancellation and updating a finite impulse response (FIR) filter with L coefficients requires at least $2L$ multiply-and-accumulate (MAC) operations, L signal read, L coefficient read and L coefficient store operations [8]. Other applications, e.g., active noise control, can be even more expensive, computation-wise. For long adaptive FIR filters, these numbers may be too high to fit into one sampling period. For this and other reasons, there is a desire to develop and apply effective algorithms with a smaller number of operations.

Generally, the idea of partial updates can be applied to any iterative descent algorithm, e.g., Recursive Least Squares (RLS). However, in this publication, we will concentrate on the application of partial updates to the family of LMS algorithms. The purpose of this paper is to extend the idea of PU LMS algorithms over the group of LMS algorithms with leakage. To the best of our knowledge, a general analysis of such a combination has not been presented to the general public yet (except for the author's local conference communications [5,9]). In the following chapter, the basic facts about the PU LMS algorithms are presented for the convenience of the reader (for an extended introduction and analysis, we recommend [6]). Then, the PU LMS algorithms with leakage are introduced in Section 3. Section 4 introduces one of the possible applications of such algorithms in the form of a multichannel active noise control system, and presents the results of simulations. Finally, conclusions are presented in Section 5.

2. Partial Update LMS Algorithms

The idea of partial updates is not new, as the first publications within this category are dated back to 1994 [10]. A good summary of scientific achievements concerned with PUs was presented in 2008 [6]. The purpose of this section is to present a short summary of PU LMS algorithms for the convenience of the reader.

There are two main groups of PU LMS algorithms: time-domain algorithms and transform-domain algorithms. This publication is concerned only with the first group. Time-domain PU LMS algorithms can be further divided into data-independent and data-dependent algorithms. An interesting observation is that while application of a data-independent algorithm always results in performance degradation (here we consider the performance to be a combination of the convergence speed and the final error level after a predefined period of time), the application of a data-dependent algorithm does not necessarily result in such degradation. Moreover, under some circumstances, it can even result in performance improvement [8]. On the other hand, data-independent algorithms offer the best computational power savings, requiring no additional operations to analyze the data.

Data-dependent PU algorithms analyze the input data to select those filter coefficients, which will result in the best filter improvement after the update. This operation usually involves sorting of the input vector $u(n)$ (see below). Fortunately, computationally efficient sorting algorithms are now available, especially in cases when the input vector is constructed from a tapped-delay line, where only one vector element is exchanged in each sampling period [6,11].

The only disadvantage of data-independent PU LMS algorithms is slower convergence rate; therefore, those algorithms cannot be used in applications where fast convergence is critical. The level of convergence rate decrease depends on the number of an adaptive filter weights excluded from update. The disadvantage of data-dependent algorithms, on the other hand, is the necessity of implementation of the sorting (or finding the maximum). In this case, the number of parameters excluded from the update must be high enough to overcompensate the computational power of the sorting. This will probably be hard to achieve in case of short adaptive filters.

In this publication, we will focus our attention on the following, popular PU algorithms:

- Data-independent algorithms:
  - Periodic partial update LMS algorithm (periodic LMS),
  - Sequential partial update LMS algorithm (sequential LMS),
  - Sequential partial update NLMS algorithm (sequential NLMS),
  - Stochastic partial update LMS algorithm (stochastic LMS),
- Data-dependent algorithms:
  - M-max partial update LMS algorithm (M-max LMS),
  - M-max partial update normalized LMS algorithm (M-max NLMS),
  - Selective partial update (normalized) LMS algorithm (selective NLMS),
  - One tap update LMS algorithm (OTU LMS).

The periodic LMS algorithm is probably the most intuitive of all: it updates all the filter coefficients once per *several* sampling periods. The properties and behavior of this algorithm are very similar to the full update LMS algorithm (except for the convergence rate); therefore, it will not be considered below.

The one tap update LMS algorithm is in fact the selective NLMS with the extreme selection of an update of a single filter tap only in every sampling period. It was enumerated separately due to a very important property: instead of sorting the whole input vector, this algorithm is based on finding the maximum absolute value of this vector. As searching for a maximum value is computationally more efficient than sorting [11], this algorithm offers the best computational power savings among all the data-dependent algorithms.

Before describing each of the above PU LMS algorithms, we will recall the (full update) Leaky LMS (LLMS) algorithm, which is given by [12]:

$$\mathbf{w}(n+1) = \gamma\mathbf{w}(n) + \mu(n)\,\mathbf{u}(n)\,e(n), \tag{1}$$

where γ is called the leakage factor and $\mu(n)$ is called the step size. The latter can be either constant ($\mu(n) = \tilde{\mu}$, historically the original LMS algorithm [13]) or variable. The filter tap vector and the input vector $\mathbf{u}(n)$ are given by:

$$\mathbf{w}(n) = [w_0(n)\; w_1(n)\; \ldots w_{L-1}(n)]^T, \tag{2}$$

$$\mathbf{u}(n) = [u(n)\; u(n-1)\; \ldots u(n-L)]^T. \tag{3}$$

Thus, the input vector contains samples from a single source of a signal, delayed in time—such a vector is usually referred to as the *tap-delayed vector*. As mentioned above, this specific structure can be utilized to sort the input vector more efficiently. For details, refer, e.g., to [11].

The error signal $e(n)$ is usually calculated as:

$$e(n) = d(n) - y(n) = d(n) - \sum_{l=0}^{L-1} w_l(n)u(n-l), \tag{4}$$

where $d(n)$ is the desired signal, the values of which should be approximated by the filter output $y(n)$ in a way that minimizes the expected value of the squared error.

3. PU LMS Algorithms with Leakage

The general equation describing PU LMS algorithms with leakage discussed in this paper, which applies to all but the periodic LMS, can be given as:

$$\mathbf{w}(n+1) = \mathbf{G}_M(n)\mathbf{w}(n) - \mu(n)\mathbf{I}_M(n)\mathbf{u}(n)e(n), \tag{5}$$

where $\mathbf{G}_M(n)$ is the leakage matrix and $\mathbf{I}_M(n)$ is the coefficient selection matrix.

The coefficient selection matrix $\mathbf{I}_M(n)$ is a diagonal matrix with elements on the diagonal equal to 1, if a corresponding coefficient is selected for update, or to 0, if not:

$$\mathbf{I}_M(n) = \begin{bmatrix} i_0(n) & 0 & \cdots & 0 \\ 0 & i_1(n) & \cdots & 0 \\ \vdots & \vdots & \ddots & \vdots \\ 0 & \cdots & 0 & i_{L-1}(n) \end{bmatrix}, \tag{6}$$

where

$$i_k(n) \in \{0,1\}, \quad \sum_{k=0}^{L-1} i_k(n) \le M. \tag{7}$$

In each iteration, the elements on the diagonal of the selection matrix are selected as equal to 0 or 1, according to the following formula:

$$i_k(n) = \begin{cases} 1 & \text{if } k \in \mathcal{I}_{M(n)} \\ 0 & \text{otherwise} \end{cases}; \tag{8}$$

where $\mathcal{I}_{M(n)}$ denotes a set of the filter weight indexes that define the coefficients to be updated in the n-th iteration; the number of the elements in such set is equal to, or less than M. At least one of the sets has the number of elements equal to M, which is the maximum number of filter weights updated in any iteration.

Similarly, the leakage matrix $\mathbf{G}_M(n)$ is a diagonal matrix and is defined as:

$$\mathbf{G}_M(n) = \begin{bmatrix} g_0(n) & 0 & \cdots & 0 \\ 0 & g_1(n) & \cdots & 0 \\ \vdots & \vdots & \ddots & \vdots \\ 0 & \cdots & 0 & g_{L-1}(n) \end{bmatrix}, \tag{9}$$

where

$$g_k(n) \in \{1, \gamma\}. \tag{10}$$

Specifically, the matrix contains elements equal to 1 in the rows corresponding to 0-s in the $\mathbf{I}_M(n)$ matrix, and a selected value of the leakage γ in the rows corresponding to 1-s in the coefficient selection matrix. In this way, the leaky version of the PU-LMS algorithm in Equation (5) applies leakage in a particular iteration only to those coefficients, which are updated in this iteration. Otherwise, the advantage of processing of only a subset of parameters would be ruined.

The difference between particular PU LMS algorithms is in the way the elements of the coefficient selection matrix (and the leakage matrix) are selected, as described below.

3.1. Sequential LLMS

The sequential LLMS organizes the filter weights into several subsets, which are then updated *sequentially* [10]. The number of subsets the filter is partitioned with can be calculated as:

$$B = \lceil L/M \rceil; \tag{11}$$

where $\lceil \cdot \rceil$ denotes the ceil operation. The way the filter weights are assigned to a specific subset depends on the implementation. For example, consider a simple case with $L = 2M$, e.g., the algorithm updates half of the filter weights in each sampling period. In such a case, it is possible to update the parameters with even indexes in one iteration, and the parameters with odd indexes in the next iteration. The weights selection matrix (6) can than be written as:

$$\begin{aligned} \mathbf{I}_M(n) &= \text{diag}(1,0,1,0,1,0\ldots) \\ \mathbf{I}_M(n+1) &= \text{diag}(0,1,0,1,0,1\ldots). \end{aligned} \tag{12}$$

Another possibility is to update the first half of the weights in one iteration, and the second half in the next iteration:

$$\begin{aligned} \mathbf{I}_M(n) &= \text{diag}(1,1,1\ldots 0,0,0\ldots) \\ \mathbf{I}_M(n+1) &= \text{diag}(0,0,0\ldots 1,1,1\ldots). \end{aligned} \tag{13}$$

Other choices are also possible.

The Sequential LLMS algorithm operates with a constant step size ($\mu(n) = \bar{\mu}$). A property of this LLMS algorithm is that the subsets created by particular partitioning of the weights vector and defined by the weights selection matrix are processed in a sequence. Therefore, after B iterations, all the weights are updated.

3.2. Sequential LNLMS

The sequential LLMS algorithm can also be used with the step size normalization. Similarly to the (full update) NLMS algorithm, the normalization can either be considered as a modification of the LLMS algorithm, with the step size dependent on the energy of the input signal, or may be developed formally as a solution of a constrained optimization problem [14]. Regardless of the approach, the resulting algorithm operates with a variable step size given by:

$$\mu(n) = \frac{\bar{\mu}}{\mathbf{u}^T(n)\mathbf{u}(n)}. \tag{14}$$

Otherwise, the algorithm is identical to the sequential LLMS. We will refer to this algorithm as the sequential Leaky Normalized LMS (sequential LNLMS).

3.3. Stochastic LLMS Algorithm

It may be proved that the sequential LLMS algorithm is permanently unstable (regardless of the step size) for some input signals, e.g., for cyclostationary signals [6]. Therefore, the stochastic LMS algorithm was developed, which selects the subsets of the weights to be updated on a random basis. The random selection should be organized in a way that selects each of the subsets with equal probability. For such algorithm, the set of filter weight indexes from Equation (8) is defined as:

$$\mathcal{P}\big(M(n) = k\big) = \pi_k, \quad k = 1, \ldots B, \quad \sum_{k=1}^{B} \pi_k = 1; \tag{15}$$

where $\mathcal{P}$ denotes the probability density function of an independent random process $M(n)$, and $B = \lceil L/M \rceil$.

Please note that it is necessary to randomize the subset selection rather than the decision if to update every individual parameter. This assures that the complexity reduction is attained during each sampling period. Please note also that the computational cost of the Stochastic LMS algorithm is slightly higher than in case of the Sequential LMS, as the random selection mechanism requires more time than the sequential selection. For further details, refer to [15].

Provided the simulation time is long and the random selection is uniform, the stochastic LLMS algorithm properties and behavior are similar to the sequential LLMS, with the exception of avoiding instability for some signals. Therefore, the algorithm will not be included in the experiments presented below.

3.4. M-Max LLMS Algorithm

One of the simplest ideas about the data-dependent PU algorithms could be to select M those entries of the input vector $\mathbf{u}(n)$, which result in the largest magnitude changes of the filter weights. The algorithm thus constructed is called the M-max LMS algorithm [16], and its weights selection matrix (6) entries can be defined as:

$$i_k(n) = \begin{cases} 1 & \text{if } |u(n-k+1)| \in \max\limits_{1 \le l \le L} (|u(n-l+1)|, M) \\ 0 & \text{otherwise} \end{cases}; \tag{16}$$

where $\max_l(u_l, M)$ denotes a set of M maxima of the elements u_l [6]. In case of this algorithm, the step size is constant: $\mu(n) = \mu$.

The M-max LLMS algorithm requires ranking (sorting) of the input vector elements, based on their absolute values. Therefore, the time savings this algorithm offers are smaller than in case of the data-independent algorithms. On the other hand, computationally efficient sorting algorithms can be applied to lessen the sorting computational burden [11].

3.5. M-Max LNLMS Algorithm

Similarly to the sequential LNLMS, the idea of M-max PU can be applied together with the step size normalization. In this case, the weights selection matrix is defined identically as in the case of M-max LLMS algorithm (Equation (16)), and the step size is normalized as in Equation (14).

3.6. Selective LLMS Algorithm

The M-max LNLMS algorithm has a subtle drawback: it uses the power of the whole input vector to normalize the step size in spite of the fact that only some of its elements are used for the actual weight update. To correct this issue, it is possible to calculate the normalized step size as:

$$\mu(n) = \frac{\bar{\mu}}{\mathbf{u}^T(n)\mathbf{I}_M(n)\mathbf{u}(n)}. \quad (17)$$

This results in the algorithm known as the selective LLMS algorithm.

It is also possible to *derive* the selective LLMS algorithm as a result of instantaneous approximation of the Newton's method, or as an application of the minimum disturbance principle [6]. Both the derivations result in the update Equation (5) and the step-size normalization given by Equation (17).

The difference between the selective LLMS and the M-max LNLMS algorithms may appear to be small, but it will be shown that it results in substantial differences in performances of both the algorithms.

3.7. One Tap Update LLMS Algorithm

As previously mentioned, the selective LLMS algorithm with the extreme choice of $M = 1$ will be considered separately and will be referred to as the One Tap Update LLMS (OTU LLMS). By combining Equations (5) and (17) and using $M = 1$, we notice that the algorithm is given by:

$$\mathbf{w}(n+1) = \mathbf{G}_1(n)\mathbf{w}(n) + \frac{\bar{\mu}}{u_{\max}^2(n)}\mathbf{I}_1(n)\mathbf{u}(n)e(n), \quad (18)$$

where $u_{\max}(n) = \max|\mathbf{u}(n)|$ denotes the maximum absolute value in the input vector at a discrete time n. The maximum absolute value $u_{\max}(n)$ is assumed to be unique, and if it is not, a single element corresponding to one of the maximum absolute values in the input vector is selected for the update at a random basis.

Remembering that $w_k(n)$ denotes the k-th filter coefficient, the algorithm may be expressed as:

$$w_k(n+1) = \begin{cases} \gamma w_k(n) + \frac{\bar{\mu}e(n)}{u(n-k)}, & \text{if } |u(n-k)| = \max|\mathbf{u}(n)| \\ w_k(n), & \text{otherwise} \end{cases}. \quad (19)$$

To conclude, this simple update algorithm developed by Douglas [17], updates only one coefficient—the coefficient that corresponds to the input sample with the maximum absolute value.

Considering the shift structure of the input vector, the *maxline* algorithm [11] can be used for finding its maximum absolute value. The worst case computational complexity of this algorithm is $L + 1$ comparisons, and only one multiplication, one division and one addition per iteration. However, it has been shown in [11] that the average number of comparisons, in case of a random input data with the uniform distribution, is approximately equal to 3 and **does not increase with the filter length**. Of course, calculation of the filter output still requires L multiplications and $L - 1$ additions.

3.8. Computational Power Requirements and Implementation

The computational demands of the discussed algorithms are given in Table 1. To calculate the requirements for the data-dependent algorithms, it was assumed that the input vector is a tap-delayed vector, and the *sortline* algorithm can be used to sort this vector. Therefore, the cost of the sorting was assumed to be $2\lceil\log_2 L\rceil + 2$ comparisons. For further details, refer to [6], but also note that the leakage requires additional multiplications in the number equal to the number of updated weights (L for full update and M for PUs).

Practical implementation of the leaky PU algorithms is similar to the implementation of other LMS-family algorithms. The architecture of modern microprocessors, especially those which are referred to as "digital signal processors", is optimized to perform the MAC operation [18]; therefore, the implementation of the actual filtering (expressed by the summation on the right-hand side of Equation (4)), is usually highly effective. The practical implementation of the adaptation, on the other hand, is quite different from straightforward implementation of Equation (5) in a programming language. The equation suggests the use of two matrices ($\mathbf{I}_M$ and $\mathbf{G}_M$), which are in fact omitted, because they are only used to define indexes of those filter weights, which are selected for updates by a particular PU algorithm. Thus, the algorithm loops through all the selected indexes and performs updates by application of the leakage γ and subtraction of the product of the step size $\mu(n)$, error $e(n)$ and a corresponding value from the input vector $\mathbf{u}(n)$. For more details on the whole digital signal processing path consult, e.g., [19].

As was already mentioned, the data-dependent algorithms require sorting or calculation of a maximum value of a vector. Practical aspects of implementation of such algorithms are discussed in [6].

Table 1. Computational complexity of the discussed PU algorithms.

Algorithm	Multiply	Add	Div	Compare
LLMS	$3L+1$	$2L$		
LNLMS	$3L+2$	$2L+2$	1	
Sequential LLMS	$L+2M+1$	$L+M$		
Sequential LNLMS	$L+2M+2$	$L+M+2$	1	
mMax LLMS	$L+2M+1$	$L+M$		$2\lceil\log_2 L\rceil+2$
mMax LNLMS	$L+2M+2$	$L+M+2$	1	$2\lceil\log_2 L\rceil+2$
Selective LLMS	$L+2M+2$	$L+M+2$	1	$2\lceil\log_2 L\rceil+2$
OTU LNLMS	$L+2$	$L+1$	1	3 *

* The average number of comparisons for uniformly distributed data. In the worst-case scenario, the number of comparisons is the same as with the other data-dependent algorithms.

4. Multichannel Active Noise Control Problem

Noise is one of major civilizational issues and therefore is of great scientific attention [20]. Passive noise control techniques have been used for years and are very effective when it comes to attenuation of high-frequency noise. Unfortunately, low-frequency noise is more difficult to attenuate passively; for this reason, active noise control (ANC) methods were developed [18,21].

Structural ANC systems are the systems that apply active control to the noise-emitting structure rather than to an additional loudspeaker [22,23]. These methods have earned a lot of attention among the scientific community due to the fact that they allow to achieve a global noise reduction, hard to obtain using different methods [24,25]. Unfortunately, structural ANC belongs to the group of the most demanding algorithms, when it comes to the computational power [19]. This is due to the fact that controlling even a simple structure usually requires more than one sensor and actuator, even if the structure is only a single panel or wall. To present a more realistic example, consider a system controlling the noise propagating from a device with a cuboid enclosure. The numbers of walls to control differs from 3 (for the device positioned in a corner of a room) to 5 (for the device standing far from the walls of a room). Suppose the device is positioned close to a wall (but not in a corner), so that 4 devices walls are to be controlled, and each is controlled using 4 actuators. Suppose there are 5 error sensors in the room—the number of active filters to consider will be equal to 80 (4 walls times 4 actuators times 5 error sensors). The hardware demands of such a multichannel system operating with a reasonable sampling frequency and even short filter lengths (e.g., 128) are enormous; therefore, only adaptation algorithms with low computational power demands can be considered for such a case and the alike. Partial updates belong to the group of such algorithms, and we will demonstrate by the means of simulations that leaky PU LMS algorithms can be successfully used.

Several techniques can be used to lower the computational demand. The most intuitive way is to use shorter adaptive filters, but this attitude may result in poor attenuation results for noises more complicated than sinusoidal signals. Another attitude is to use a computational power saving algorithm. There are a few such algorithms, e.g., the switched error algorithm [26]; however, the partial updates can also be used, as reported in previous research [4,7] and this communication. Two experimental setups, based on existing, real laboratory stands, will be used to demonstrate by simulations that PU algorithms with leakage are particularly well-suited for structural ANC.

4.1. Experimental Setup 1: Active Casing

The first experimental setup consists of an active casing with rigid corners and 1 mm thick aluminum plates, with 420 × 420 mm dimensions, as presented in Figure 1. Each plate was equipped with three electrodynamic Monacor EX-1 5 Watt actuators, mounted in carefully selected positions to improve controllability of the system. The casing was placed in the laboratory room a little distance off the walls so five sides of the casing were available and an error sensor (microphone) was positioned in front of each side [22].

Figure 1. The rigid casing used in Experimental Setup 1.

The above setup consists of 15 actuators and 5 error microphones; thus, the number of secondary paths to consider is 75. The models of those secondary paths were identified in the form of finite impulse response (FIR) filters with 256 parameters. The identification experiments were performed using white noise as an excitation and using the sampling frequency of 2 kHz. In addition, a loudspeaker was placed inside the casing and the primary paths were identified as well, in the same form. Finally, a reference signal path was identified between the loudspeaker and a microphone placed inside the casing.

The identified transfer functions (TF) were used to design a simulation system allowing for rapid testing of adaptation algorithms. A part of the block diagram of this system, associated with one of the casing walls, is presented in Figure 2. According to this figure, the noise to be attenuated $u(n)$ is filtered through the reference path TF $X(z^{-1})$, to produce the reference signal, $x(n)$. The reference signal forms the input to the primary path TF $P_j(z^{-1})$, where j is the corresponding error microphone number. This reference signal forms the input signal to the control filters—three such filters, $W_1(z^{-1})$, $W_2(z^{-1})$, and $W_3(z^{-1})$, associated with the front plate are presented in Figure 2. Each control filter produces the output $y_j(n)$, which is then filtered with five different secondary path TFs; one of these TFs for each of the presented filters is visible in the figure. The secondary path TFs are denoted as $S_{i,j}(z^{-1})$, where i is the actuator number ($i \in \{1 \ldots 15\}$), and j is again the error microphone number. Three secondary path TFs, i.e., $S_{1,1}(z^{-1})$, $S_{2,1}(z^{-1})$, and $S_{3,1}(z^{-1})$, are presented in Figure 2, while the remaining twelve belong to the other error signal paths. In consequence, each error signal is a sum of 16 signals: one primary signal and 15 signals coming from the 15 actuators.

All 5 error signals are used by each of the adaptation algorithms, denoted as LMS in Figure 2 [18,27]. In addition, each of the adaptation algorithms uses 5 reference signals, each filtered through a different secondary path TF estimate, e.g., $\hat{S}_{1,1}$–$\hat{S}_{1,5}$ in case of the algorithm associated with the $W_1(z^{-1})$ control filter. The simulation system is then fairly advanced and requires 165 filtrations and 15 adaptations in each simulation iteration.

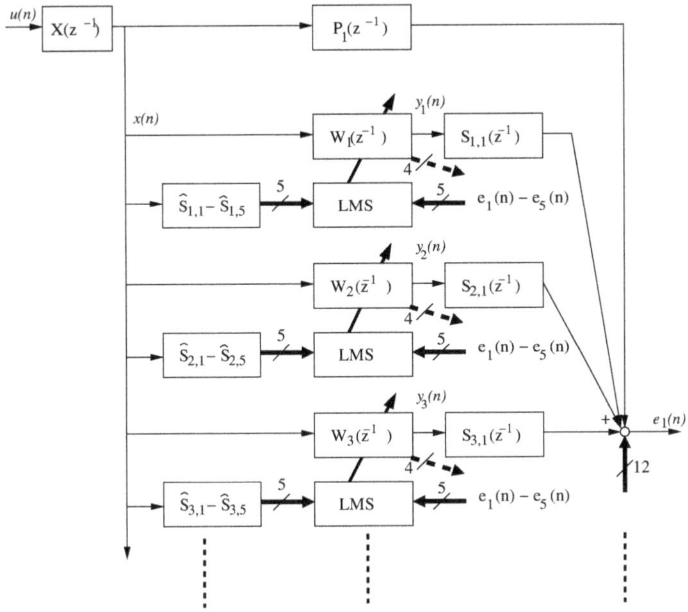

Figure 2. Block diagram of the simulated active casing system.

Exemplary frequency characteristics of the secondary path TFs from one of the actuators mounted on the front plate to all the five error microphones are illustrated in Figure 3. From the figure, it can be noticed that the magnitude of the presented TFs is similar in the average; however, large variations of magnitude for all the five TFs are observed at several particular frequencies. For example, $S_{1,4}(z^{-1})$ and $S_{1,5}(z^{-1})$ have a very low magnitude at 115 Hz, the difference between $S_{1,5}(z^{-1})$ and $S_{1,3}(z^{-1})$ is around 40 dB at this frequency.

Figure 4 presents the frequency characteristics of the secondary path TFs from the three top plate actuators to one of the error microphones. In this case, the magnitudes of the presented TFs differ much more from each other than those presented in Figure 3. Moreover, similar differences were observed for the TFs to the other four error microphones. Special attention should be paid to the $S_{11,y}(z^{-1})$ TF, which has the lowest average magnitude among all the secondary path TFs.

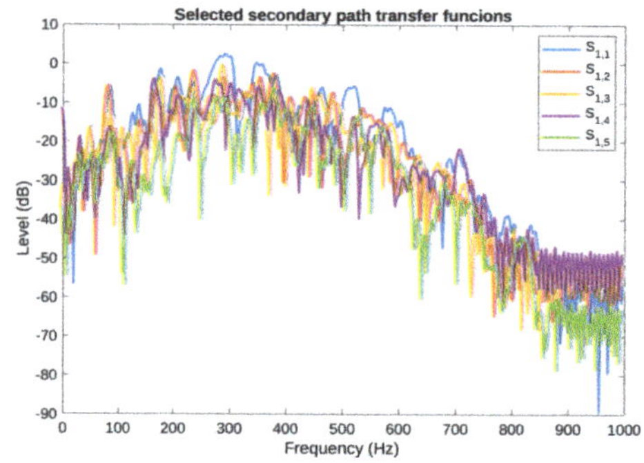

Figure 3. Secondary path transfer functions from the first actuator on the front plate to all the five error microphones.

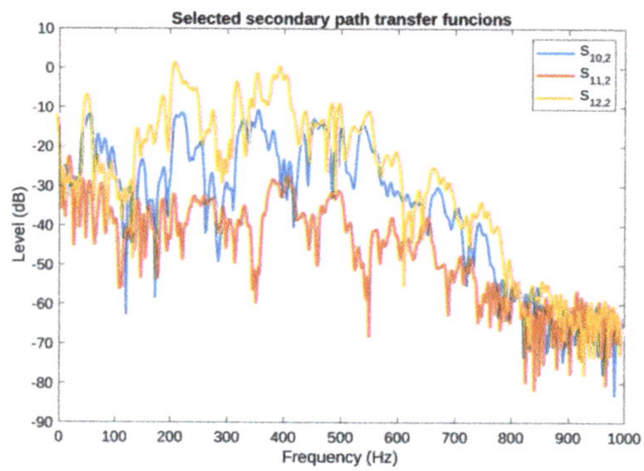

Figure 4. Secondary path transfer functions from the top plate actuators to the left plate error microphone.

Both the above mentioned figures suggest that the identified models are accurate and contain a variety of phenomena present in real environment, contrary to some other, simplified models (e.g., TF of 16 order) used in simulations presented in the literature. Moreover, the secondary path TF estimates ($\hat{S}_{x,y}(z^{-1})$ in Figure 2) used in the simulations presented below were not perfect. Instead, all the models were truncated down to 128 parameters, and each impulse response parameter was disturbed with a small, random value.

The simulations described in this section, unless otherwise specified, were performed using the following setup. The sampling frequency was assumed to be 2 kHz. The noise signal was generated as a sum of two sinusoids, with frequencies 112 and 191 Hz, shifted in phase by $\pi/4$. A white noise sequence was also generated of the same length and was filtered with a band-pass filter with pass band between 100 and 500 Hz, resulting with a sequence for which the variance was equal to 0.4. Such noise was added to the two-sine signals to present a more realistic signal. The active filter length was set at 256. Three different step sizes were used: 1×10^{-5} for the algorithms with a constant step size, 8×10^{-3} for the algorithms with the step size normalization, except for the selective LLMS (including the One Tap Update LLMS), for which the step size was equal to 4×10^{-3}. The leakage factor used was 0.999999, except for the eleventh active filter—the filter associated with the second sensor on the top plate—for which the leakage factor was 0.99999. The step sizes and leakage factors were determined by trial and error, with an intention to achieve fast adaptation and robust performance (the robust performance was confirmed by simulation of combinations of different frequencies, within the band 100–300 Hz). The simulations were repeated 50 times, with different white noise sequences, to achieve smoother mean square error (MSE) curves.

4.2. Simulation Results

The reasons for using the algorithms with leakage are presented in Figures 5 and 6, where the results of simulation of the NLMS algorithm without leakage are presented. The figures present the output signals from the adaptive filters only (the final attenuation towards the end of this simulation was around 37 dB). From the figures, it is clear that some of the output signals have substantially higher levels than the others. For example, y_1 output level reaches ± 15, y_{10} output stays within ± 20, and y_{11} exceeds ± 100. On the other hands, the majority of signals are bounded by ± 5. It can be also noticed that some of the signals did not achieve a steady state even after 500 s of the simulation, and are still growing slowly. Both the facts suggest that the described algorithm without leakage is not suitable for practical applications, because it would probably result in exceeding the digital-to-analog converters limits, and it would probably be unstable after longer periods of operation. Therefore, all the remaining simulations presented in this paper use algorithms with leakage.

The results also suggest that a lower value of the leakage should be applied for those filters that give higher output values. Unfortunately, particular values of the leakage factor are hard to determine in a way other than by trial end error. In practical ANC applications, this is usually done based on observation of the filter parameters, which should be approximately constant in the steady state.

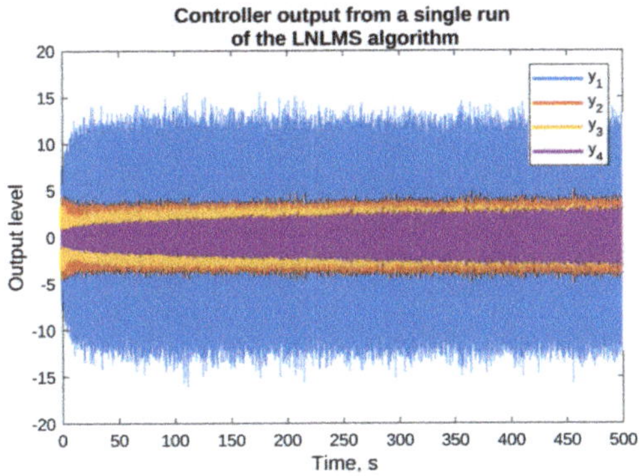

Figure 5. Selected outputs of the adaptive filters for the simulations of the NLMS algorithm without the leakage.

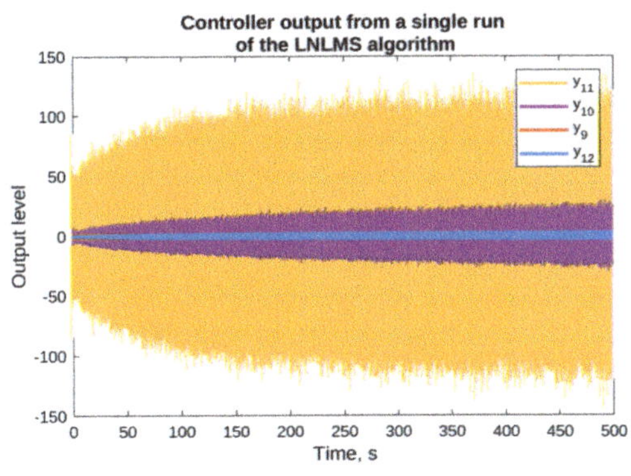

Figure 6. Selected outputs of the adaptive filters for the simulations of the NLMS algorithm without the leakage.

Figures 7 and 8 present the mean squared error (MSE) obtained on the selected error microphones during simulations of algorithms without and with the step size normalization, respectively. The PU algorithms were simulated with $M = 16$, i.e., 16 adaptive filter parameters out of 256 were updated in each simulation step. The full update LLMS and LNLMS algorithms were added to the simulations for a reference. The attenuation level was calculated based on the MSE during the last 50 s of each simulation.

The figures show that all the algorithms are stable and converge at different speeds, reaching noticeable attenuation levels towards the end of the simulation. Among the algorithms without the step size normalization (Figure 7), the full update algorithm is the fastest to converge, and attains the highest level of attenuation; however, the M-max LLMS algorithm converges only a little slower, and attains the same level of attenuation. The sequential LLMS algorithm converges very slowly (approximately 16 times slower); therefore, it achieves only 24 dB of attenuation within the simulation window.

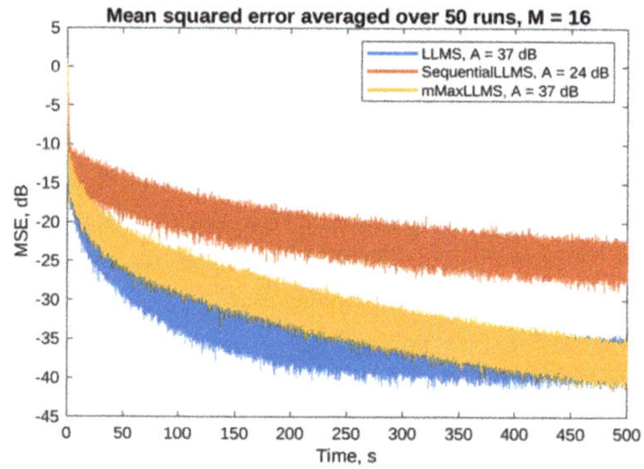

Figure 7. Active noise control results on the third error microphone for the algorithms without the step size normalization and $M = 16$. Final attenuation level for each algorithm is given in the legend.

The algorithms with the step-size normalization (Figure 8) are capable of achieving even better performance: converge faster and achieve higher levels of attenuation. It is interesting to notice that all the PU algorithms except the sequential LNLMS algorithm show similar performance than the full update LNLMS. The selective LNLMS algorithm shows the fastest initial convergence speed and attains an attenuation level only slightly worse than the full update. However, the OTU LNLMS algorithm is the third, reaching an impressive attenuation level of 36 dB despite the fact that it updates only one filter weight in each simulation step.

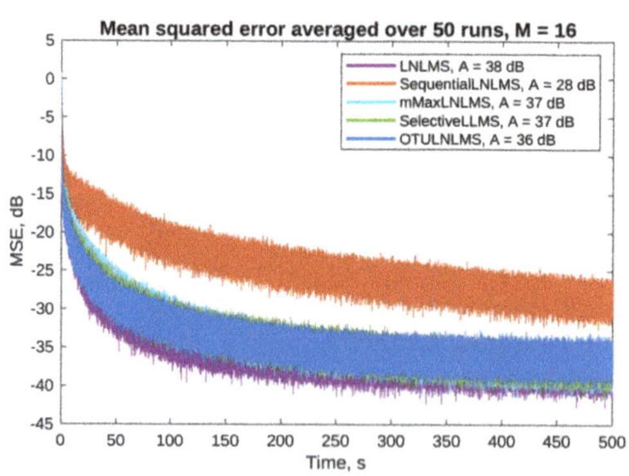

Figure 8. Active noise control results on the first error microphone for the algorithms with the step size normalization and $M = 16$. Final attenuation level for each algorithm is given in the legend.

The final attenuation levels obtained for all five error microphones for all the simulated algorithms are presented in Table 2. From the table, it is clear that the attenuation level calculated based on e_3 was the best, while the attenuation for e_5 was the worst. The latter fact can be explained based on the geometry of the laboratory, where the fifth microphone, corresponding to the back panel, was close to the room wall.

Figure 9 presents selected output signals obtained during simulation of the LNLMS algorithm, corresponding to those in Figure 6. It can be observed that introduction of the leakage prevented the output signals from reaching very high levels—this fact is especially visible in case of y_{11}.

Table 2. Active casing attenuation levels obtained for each error microphone for $M = 16$.

Algorithm	e_1	e_2	e_3	e_4	e_5
LLMS	33.6 dB	29.1 dB	37.2 dB	26.4 dB	24.8 dB
LNLMS	37.8 dB	30.0 dB	40.5 dB	28.1 dB	25.3 dB
Sequential LLMS	21.8 dB	19.8 dB	24.5 dB	18.5 dB	18.7 dB
Sequential LNLMS	28.4 dB	22.2 dB	31.4 dB	20.9 dB	20.5 dB
mMax LLMS	33.6 dB	30.2 dB	37.1 dB	26.0 dB	22.4 dB
mMax LNLMS	37.2 dB	31.3 dB	40.7 dB	28.3 dB	24.0 dB
Selective LLMS	36.9 dB	33.4 dB	39.1 dB	31.5 dB	27.1 dB
OTU LNLMS	36.2 dB	34.0 dB	38.6 dB	32.7 dB	27.7 dB

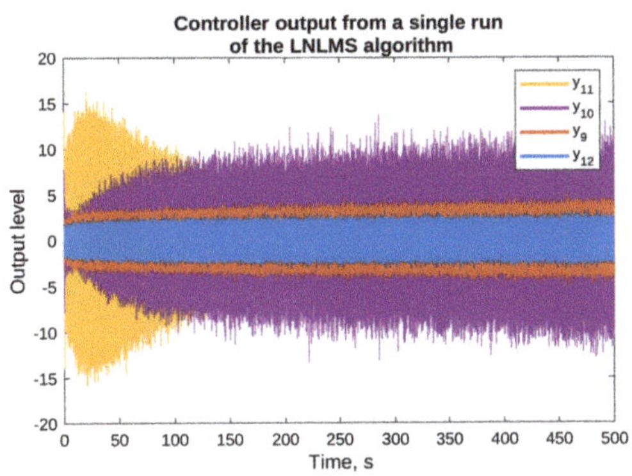

Figure 9. Selected outputs of the adaptive filters for the simulations of the LNLMS algorithm with leakage.

Figure 10 and Table 3 present the results of simulations with $M = 8$ (eight filter taps updated in each simulation step). The figure shows only the algorithms with the step size normalization, which show better performance. The table does not include the algorithms which do not use the M parameter (i.e., full update algorithms and the OTU LNLMS). It can be noticed that updating only eight parameters results in slightly worse performance, compared to simulations with $M = 16$, but the results are still acceptable in the means of both the convergence speed and the attenuation level.

Table 3. Active casing attenuation levels obtained for each error microphone for $M = 8$.

Algorithm	e_1	e_2	e_3	e_4	e_5
Sequential LLMS	19.9 dB	18.0 dB	22.2 dB	16.9 dB	16.7 dB
Sequential LNLMS	24.2 dB	18.8 dB	26.1 dB	18.3 dB	19.0 dB
mMax LLMS	31.9 dB	28.1 dB	34.7 dB	24.1 dB	20.6 dB
mMax LNLMS	35.1 dB	28.4 dB	39.4 dB	26.1 dB	21.8 dB
Selective LLMS	37.7 dB	34.4 dB	41.1 dB	32.0 dB	29.8 dB

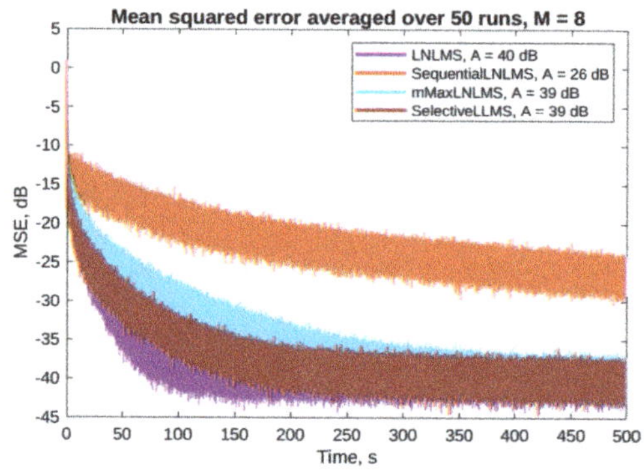

Figure 10. Active noise control results on the third error microphone for the algorithms with the step size normalization and $M = 8$. Final attenuation level for each algorithm is given in the legend.

4.3. Experimental Setup 2: Washing Machine

The second experimental setup uses a commercial washing machine, as presented in Figure 11. The device was placed in the laboratory close to one of the walls; therefore, only four walls of the machine were available. Thirteen 5 W electrodynamic actuators were placed on the machine: three on each of the three side walls and four on the top wall (contrary to the Figure, which was taken before the final mount). The reference sensor was placed inside the washing machine, below the drum. The laboratory room was equipped with eight microphones, positioned in a quarter-sphere arrangement, used as error sensors (the final application will probably use other type of sensors, e.g., accelerometers attached to the device, and using the Virtual Microphone technique [28]). As was described in [22], examination of the results of attenuation on the eight error microphones allows us to determine the global attenuation effect.

Figure 11. The washing machine used in the experiments.

Similar to the case of the first experimental setup, impulse responses of primary and secondary paths were identified, each with 256 parameter. There were eight primary paths from the reference microphone to the error sensors, and 104 secondary paths, from the 13 actuators to each of the error sensors. The same sampling frequency of 2 kHz was used. Again, the simulation system was implemented in the Matlab environment with the block diagram similar to this presented in Figure 2. This system was used to produce the results presented below.

The simulations were performed using a signal recorded during the washing machine spinning cycle, which is the loudest part of each washing. The spectrogram of this signal is presented in Figure 12, where the variety of the spinning signal harmonic components can be observed. For example, the spinning frequency is constant and equal to 1140 rpm between 5th and 8th minute of the cycle; therefore, the 19 Hz component and its products are easily distinguishable. Nevertheless, it should be noted that the signal is nonstationary: the washing machine increases the rotation speed slowly or even sometimes abruptly (as right before the 5th min), maintains a constant speed for an amount of time, and then slowly reduces the speed. The simulations presented below used only a part of this recording, approximately between minute 3 and 8.

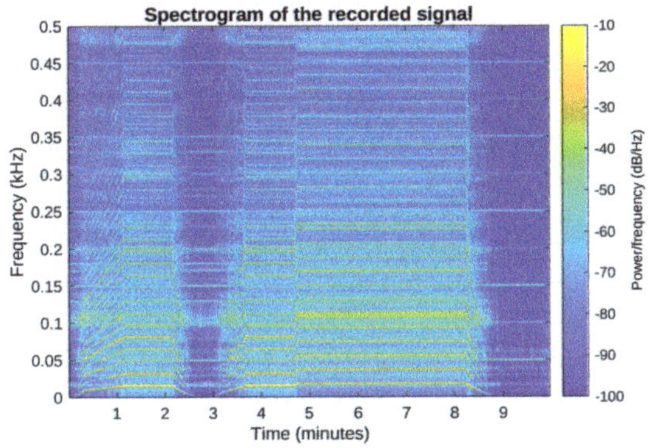

Figure 12. Spectrogram of the signal recorded during the washing machine final spinning phase.

The active filter length used in the following simulations was 256. Based on the results obtained with the rigid casing, it was assumed that it will be sufficient to update eight filter parameters in each iteration ($M = 8$). Three different step sizes were used: 1 for the algorithms with constant step size, 4×10^{-3} for the algorithms with the step size normalization, except for the selective LLMS (including the One Tap Update), for which the step size was equal to 1×10^{-3}. The leakage factor used was 0.999999 for all the filters. Again, the step sizes and leakage factors were determined by trial and error, with an intention to achieve fast adaptation and robust performance. The simulations were repeated five times without resetting the filter weights to avoid zero initial conditions, and the results of only the last run are presented below.

4.4. Simulation Results

The adaptation process of a filter that does not start with zero initial conditions does not produce results similar to those presented in Figures 7 and 8: the MSE level is low from the beginning of the simulation, and increases only a little during the attenuated signal rapid changes. Additionally, the MSE cannot be smoothed by averaging several runs, since each single run uses exactly the same spinning cycle recording. Therefore, Figures 13 and 14 present only time plots of signals simulated on a selected error microphone: the primary noise (blue) and the attenuated signal (orange). The fifth error microphone was selected because it produced the most spectacular results; however, substantial attenuation was recorded on each of the 8 error microphones. Table 4 presents the results of attenuation for every simulated algorithm on each of the error microphones. The attenuation was calculated from the ratio of the primary noise variance and the attenuated signal variance during the last 10% (29 s) of the simulation experiment.

By analyzing the results, we conclude that all the PU Leaky LMS algorithms are capable of achieving good attenuation levels, from 8 dB in the worst-case scenario, up to almost 30 dB. Of course, the sequential algorithms result in the poorest attenuation levels, but it must be remembered that these algorithms offers the best computational power savings. What can be surprising is the fact that two PU algorithms, namely the selective LNLMS and the OTU LNLMS algorithms achieved better results than the full update LNLMS. However, it must be remembered that the step size used with these algorithms was slightly different (0.004 for the full update vs. 0.001 for the PU algorithms), which might have given a little advantage to the PU algorithms. Nevertheless, the results obtained with the algorithm that updates only 1 filter weight out of 256 (OTU) are impressive.

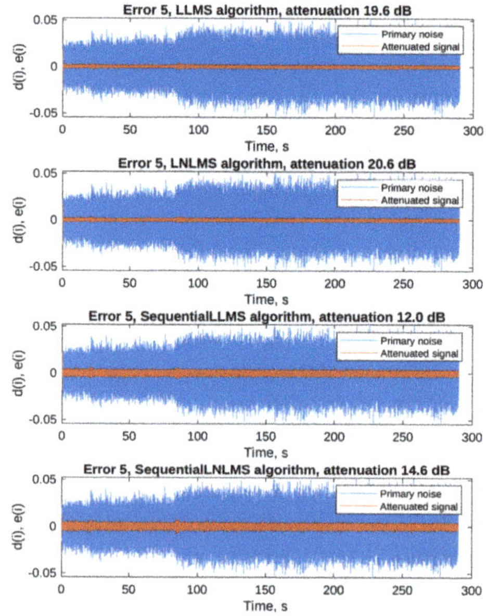

Figure 13. Primary noise and the attenuated signal on the error microphone 5—part 1.

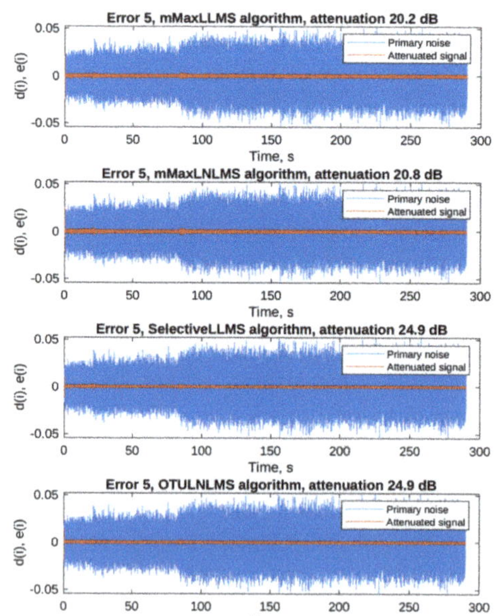

Figure 14. Primary noise and the attenuated signal on the error microphone 5—part 2.

Table 4. Washing machine attenuation levels obtained for each error microphone for $M = 8$.

Algorithm	e_1	e_2	e_3	e_4	e_5	e_6	e_7	e_8
LLMS	14.0 dB	13.6 dB	17.7 dB	12.7 dB	26.5 dB	16.9 dB	21.0 dB	19.6 dB
LNLMS	14.2 dB	13.3 dB	15.4 dB	12.8 dB	26.2 dB	16.3 dB	20.9 dB	20.6 dB
Sequential LLMS	10.4 dB	9.5 dB	14.3 dB	8.3 dB	21.0 dB	12.9 dB	16.4 dB	12.0 dB
Sequential LNLMS	9.8 dB	9.8 dB	11.0 dB	8.7 dB	19.8 dB	12.1 dB	15.1 dB	14.6 dB
M-max LLMS	12.9 dB	13.2 dB	17.2 dB	12.4 dB	26.6 dB	15.9 dB	20.8 dB	20.2 dB
M-max LNLMS	13.5 dB	12.8 dB	15.2 dB	12.5 dB	26.5 dB	15.9 dB	20.7 dB	20.8 dB
Selective LLMS	15.9 dB	16.1 dB	17.2 dB	15.6 dB	29.5 dB	17.4 dB	23.1 dB	24.9 dB
OTU LNLMS	15.8 dB	16.8 dB	17.1 dB	16.0 dB	28.9 dB	17.0 dB	22.8 dB	24.9 dB

Figures 15 and 16 show the spectrograms of the primary noise and the attenuated signal in the most interesting frequency range from 0 to 500 Hz, for the selective LNLMS algorithm and the fifth error microphone. The spectrogram of the attenuated signal confirms that the attenuation is obtained during the full simulated period of time, for all the spectral components of the noise signal.

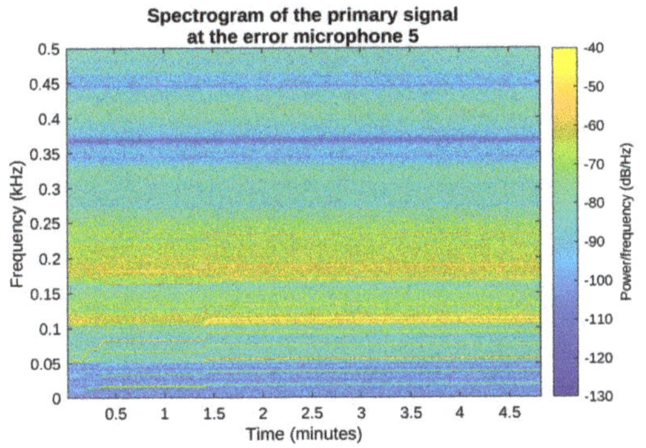

Figure 15. Spectrogram of the primary noise on the error microphone 5, Selective LNLMS algorithm.

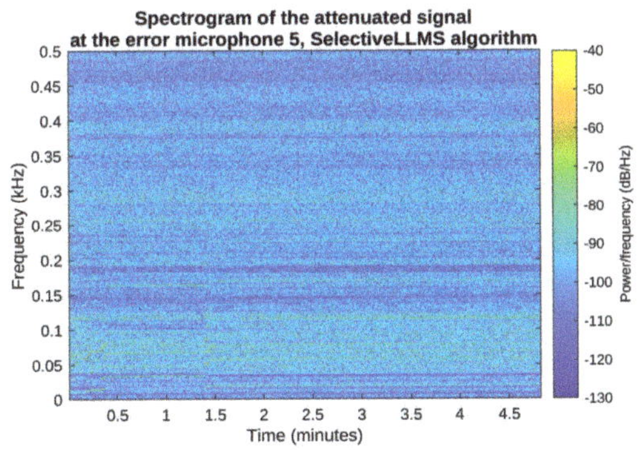

Figure 16. Spectrogram of the attenuated signal on the error microphone 5, Selective LNLMS algorithm.

Finally, Figures 17 and 18 present power spectral densities of the noise signal and the attenuated signal during the last 10% of the simulation time, for the OTU LNLMS algorithm and error microphones where the best and the worst attenuation was obtained, i.e., e_5 and e_4. In Figure 17 we observe very good attenuation of almost all harmonic components, except for the lowest frequency component (17 Hz). Fortunately, such low frequency is already outside the normal human hearing frequency range. The figure also shows a good attenuation of broadband noise present in the recorded signal. On the other hand, Figure 18 shows that some lower frequency harmonic components were not attenuated sufficiently in the error microphone 4 location. Observe, however, that the whole level of the signals from the fourth microphone is lower than in case of the fifth microphone. Considering this, the final levels of both the attenuated signals are similar.

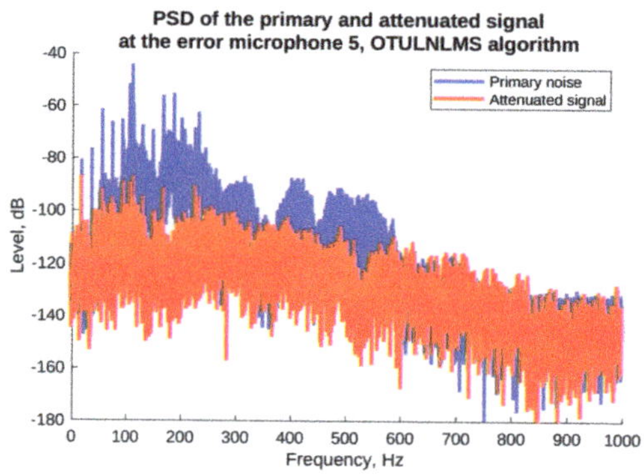

Figure 17. Powers spectral density without and with ANC for the microphone with the highest attenuation, OTU LLMS algorithm.

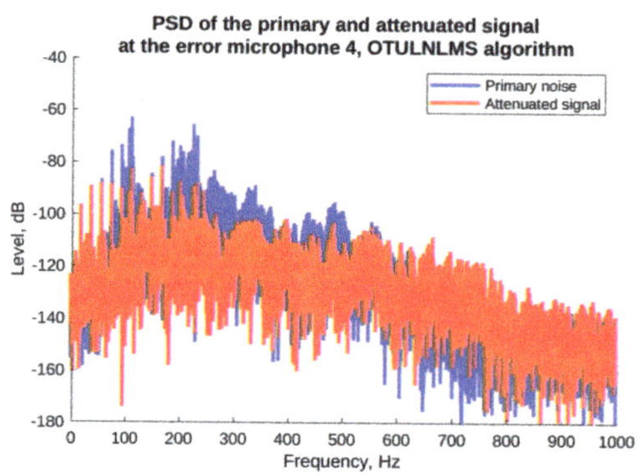

Figure 18. Powers spectral density without and with ANC for the microphone with the poorest attenuation, OTU LLMS algorithm.

5. Conclusions

The goal of the research presented in this paper was to study usefulness of several partial updated LMS algorithms in a very demanding application to structural ANC. From a practical implementation point of view, overestimated multichannel ANC systems usually require adaptation algorithms with leakage, because leakage allows us to equalize individual actuator control efforts, thus resulting in stable adaptation and higher attenuation levels. Therefore, to allow for application of PU LMS algorithms in such systems, leaky versions of the algorithms were introduced. In case of PU algorithms, the leakage must be implemented in a way that does not ruin the computational power savings. The main contribution of this paper is the presentation of the proper attitude towards solution of this problem, and the presentation of the resulting leaky PU LMS algorithms (the sequential LLMS, sequential LNLMS, M-max LLMS, M-max LNLMS, selective LLMS and One Tap Update LLMS). The computational power savings offered by these algorithms with the respect to the full update LLMS algorithm are also presented and discussed.

To test the proposed algorithms in a rapid development environment, two simulation systems were constructed, based on the structural ANC real laboratory setups. The first system was based on an active casing with rigid walls; five panels were controlled by three actuators on each wall. The second was based on a commercial washing machine placed close to the laboratory room wall, so four washing machine walls were controlled with a total of 13 actuators. Both the systems were multichannel, with 75 and 104 secondary paths, respectively. Contrary to many other publications, the secondary path estimates used in these simulations were not perfect. The simulated noises included two sinusoids embedded in noise and a real recording of a washing machine spinning cycle.

The simulations confirmed that the proposed leaky PU algorithms can be used in structural ANC systems. All the leaky PU LMS algorithms tested were capable of providing between 8 and 40 dB of attenuation. It must be emphasized that in case of the simulated structural ANC system, the attenuation effect is global and is not concerned only with the error microphone location. Therefore, the obtained results are satisfactory and prove the leaky PU LMS algorithms have a high potential for real application.

The main disadvantage of PU LMS algorithms is possible degradation of the convergence speed, which is an inherent feature of all data-independent PUs. Therefore, data-independent algorithms should be used with caution when fast convergence is important. Data-dependent algorithms, on the other hand, do not result in the convergence speed decrease, but require the implementation of sorting of the input vector. This in turn lowers the computational power savings these algorithms offer. An interesting compromise may be the leaky One Tap Update algorithm, which instead of sorting utilizes maximum values searching—the algorithms that can be implemented in a very optimal way.

Funding: This research was supported by Polish state budget for science.

Data Availability Statement: The data presented in this study are available on request from the author.

Acknowledgments: The author expresses sincere gratitude to Krzysztof Mazur from the Silesian University of Technology, who performed identification experiments resulting in models used in this paper. The author would also like to thank the Upper Silesian Centre of Scientific and Engineering Calculations (GECONiI) for the hardware support allowing to speed up the simulations.

Conflicts of Interest: The author declares no conflict of interest. The funders had no role in the design of the study; in the collection, analyses, or interpretation of data; in the writing of the manuscript; or in the decision to publish the results.

References

1. Wang, W.; Li, J.; Li, M. Selective Partial Update of NLMS Adaptive Filter Algorithm. *J. Phys. Conf. Ser.* **2021**, *1966*, 012009. [CrossRef]
2. Ma, Y.; Liu, S.; Lu, J. A multi-channel partial-update algorithm for sea clutter suppression in passive bistatic radar. In Proceedings of the IEEE Sensor Array and Multichannel Signal Processing Workshop, Sheffield, UK, 8–11 July 2018; pp. 252–256.
3. Ur Rahman, M.Z.; Aswitha, P.V.S.; Sriprathyusha, D.; Farheen, S.K.S. Beamforming in cognitive radio networks using partial update adaptive learning algorithm. *Acta IMEKO* **2022**, *11*, 8. [CrossRef]
4. Bismor, D. Simulations of Partial Update LMS Algorithms in Application to Active Noise Control. *Procedia Comput. Sci.* **2016**, *80*, 1180–1190. [CrossRef]
5. Bismor, D. Chapter Leaky Partial Updates in Application to Structural Active Noise Control. In *Postępy Akustyki*; Polskie Towarzystwo Akustyczne, Oddział w Krakowie: Kraków, Poland, 2021; pp. 44–53.
6. Doğançay, K. *Partial-Update Adaptive Signal Processing. Design, Analysis and Implementation*; Academic Press: Oxford, UK, 2008.
7. Bismor, D. Partial Update LMS Algorithms in Active Noise Control. In Proceedings of the Forum Acusticum, Krakow, Poland, 7–12 September 2014; pp. 1–7.
8. Douglas, S.C. Adaptive filters employing partial updates. *IEEE Trans. Circuits Syst. II Analog. Digit. Signal Process.* **1997**, *44*, 209–216. [CrossRef]
9. Bismor, D. Leaky Partial Updates to Control a Real Device Casing. *Vib. Phys. Syst.* **2022**, *33*, 1–8. [CrossRef]
10. Douglas, S.C. Simplified stochastic gradient adaptive filters using partial updating. In Proceedings of the 1994 Sixth IEEE Digital Signal Processing Workshop, Yosemite National Park, CA, USA, 2–5 October 1994; pp. 265–268. [CrossRef]
11. Pitas, I. Fast algorithms for running ordering and max/min calculation. *IEEE Trans. Circuits Syst.* **1989**, *36*, 795–804. [CrossRef]
12. Bismor, D.; Pawelczyk, M. Stability conditions for the leaky LMS algorithm based on control theory analysis. *Arch. Acoust.* **2016**, *41*, 731–740. [CrossRef]

13. Widrow, B.; Hoff, M.E. Adaptive Switching Circuits. 1960 IRE Wescon Convention Record, Part 4. 1960; pp. 96–140. Available online: http://isl-www.stanford.edu/~widrow/papers/c1960adaptiveswitching.pdf (accessed on 16 January 2023).
14. Bismor, D. *Stability of a Class of Least Mean Square Algorithms*; Problemy Współczesnej Automatyki i Robotyki, AOW EXIT: Warszawa, Poland, 2016.
15. Godavarti, M.; Hero, A.O. Partial update LMS algorithms. *IEEE Trans. Signal Process.* **2005**, *53*, 2382–2399. [CrossRef]
16. Douglas, S. Analysis and implementation of the max-NLMS adaptive filter. In Proceedings of the Conference Record of the Twenty-Ninth Asilomar Conference on Signals, Systems and Computers, Pacific Grove, CA, USA, 30 October–1 November 1995; Volume 1, pp. 659–663. [CrossRef]
17. Douglas, S. A family of normalized LMS algorithms. *IEEE Signal Process. Lett.* **1994**, *1*, 49–51. [CrossRef] [PubMed]
18. Kuo, S.M.; Morgan, D.R. *Active Noise Control Systems*; John Wiley & Sons: New York, NY, USA, 1996.
19. Mazur, K.; Wrona, S.; Pawelczyk, M. Design and implementation of multichannel global active structural acoustic control for a device casing. *Mech. Syst. Signal Process.* **2018**, *98*, 877–889. [CrossRef]
20. World Health Organization. *Reducing Global Health Risks through Mitigation of Short-Lived Climate Pollutants. Scoping Report for Policy-Makers*; WHO: Geneva, Switzerland, 2015.
21. Bismor, D. Comments on "A New Feedforward Hybrid ANC System". *IEEE Signal Process. Lett.* **2014**, *21*, 635–637. [CrossRef]
22. Mazur, K.; Pawełczyk, M. Active control of noise emitted from a device casing. In Proceedings of the 22nd International Congress on Sound and Vibration, Florence, Italy, 12–16 July 2015.
23. Wrona, S.; Pawelczyk, M. Controllability-oriented placement of actuators for active noise-vibration control of flexible structures using memetic algorithms. *Arch. Acoust.* **2013**, *38*, 529–536. [CrossRef]
24. Wrona, S.; de Diego, M.; Pawelczyk, M. Shaping zones of quiet in a large enclosure generated by an active noise control system. *Control. Eng. Pract.* **2018**, *80*, 1–16. [CrossRef]
25. Wyrwal, J.; Pawelczyk, M.; Liu, L.; Rao, Z. Double-panel active noise reducing casing with noise source enclosed inside—Modelling and simulation study. *Mech. Syst. Signal Process.* **2021**, *152*, 107371. [CrossRef]
26. Mazur, K.; Wrona, S.; Chraponska, A.; Rzepecki, J.; Pawelczyk, M. FXLMS with Multiple Error Switching for Active Noise-Cancelling Casings. *Arch. Acoust.* **2019**, *44*, 775–782. [CrossRef]
27. Elliott, S. *Signal Processing for Active Noise Control*; Academic Press: London, UK, 2001.
28. Mazur, K.; Pawelczyk, M. Virtual microphone control for an active noise-cancelling casing. In *Solid State Phenomena*; Trans Tech Publications Ltd.: Bäch, Switzerland, 2016; Volume 248, pp. 57–66. [CrossRef]

sensors

MDPI

Article

Statistical Pass-By for Unattended Road Traffic Noise Measurement in an Urban Environment

Elena Ascari [1], Mauro Cerchiai [2], Luca Fredianelli [1,*] and Gaetano Licitra [1,2,*]

1 Institute for Chemical-Physical Processes of the Italian Research Council (CNR-IPCF), Via Giuseppe Moruzzi 1, 56124 Pisa, Italy
2 Environmental Protection Agency of Tuscany Region, Pisa Department, Via Vittorio Veneto 27, 56127 Pisa, Italy
* Correspondence: luca.fredianelli@ipcf.cnr.it (L.F.); g.licitra@arpat.toscana.it (G.L.)

Abstract: Low-noise surfaces have become a common mitigation action in the last decade, so much so that different methods for feature extraction have been established to evaluate their efficacy. Among these, the Close Proximity Index (CPX) evaluates the noise emissions by means of multiple runs at different speeds performed with a vehicle equipped with a reference tire and with acoustic sensors close to the wheel. However, signals acquired with CPX make it source oriented, and the analysis does not consider the real traffic flow of the studied site for a receiver-oriented approach. These aspects are remedied by Statistical Pass-By (SPB), a method based on sensor feature extraction with live detection of events; noise and speed acquisitions are performed at the roadside in real case scenarios. Unfortunately, the specific SPB requirements for its measurement setup do not allow an evaluation in urban context unless a special setup is used, but this may alter the acoustical context in which the measurement was performed. The present paper illustrates the testing and validation of a method named Urban Pass-By (U-SPB), developed during the LIFE NEREiDE project. U-SPB originates from standard SPB, exploits unattended measurements and develops an in-lab feature detection and extraction procedure. The U-SPB extends the evaluation in terms of before/after data comparison of the efficiency of low-noise laying in an urban context while combining the estimation of long-term noise levels and traffic parameters for other environmental noise purposes, such as noise mapping and action planning.

Keywords: SPB method; sound pass-by; low-noise surfaces; noise modeling; road traffic noise; unattended noise measurement procedure; traffic measurements; noise emission; environmental noise; sound

Citation: Ascari, E.; Cerchiai, M.; Fredianelli, L.; Licitra, G. Statistical Pass-By for Unattended Road Traffic Noise Measurement in an Urban Environment. *Sensors* **2022**, *22*, 8767. https://doi.org/10.3390/s22228767

Academic Editors: Antonio Lázaro and Mateusz Malanowski

Received: 26 September 2022
Accepted: 11 November 2022
Published: 13 November 2022

1. Introduction

The "Noise in Europe 2020" report [1] by the European Environment Agency confirmed that road traffic is the most dominant source of environmental noise, with an estimated 113 million Europeans affected by noise levels greater than 55 dB(A) of L_{den} (day–evening–night level). Moreover, at least 20% of the EU population lives in areas where road traffic noise levels are high, and it is likely that the inhabitants underestimate the well-known long-term effect on health that prolonged exposure can produce.

Among the many different mitigation actions that can be applied, the laying of new low-noise pavement (LNP) [2,3], such as open-graded pavement, rubber asphalts or poroelastic surfaces, has become popular as it significantly affects a wider area and a greater number of citizens compared to other actions focused on a single building. LNP is defined as a road surface which can reduce sound emissions to some extent compared to a reference pavement. The most widely used definition of LNP is "pavement able to provide a reduction of 3 dB(A) with respect to a Dense Asphalt Concrete (DAC)" [4].

Noise barriers have visive impact that reduces their acceptability [5,6], and the installation of soundproof windows is much more expensive in cases where there are multiple receivers to mitigate [6,7]. Moreover, new windows do not completely solve the problem as citizens leave them open in summertime [8].

The laying of LNP is now a common mitigation action in the EU, and a few methods have been established by technical norms that have emerged over the years in order to evaluate the performances of road surfaces. Among these, the Close Proximity Index (CPX), according to ISO 11819-2:2017 [9], the Statistical Pass-By (SPB-ISO), according to ISO/DIS 11819-1 [10], and the ISO 13472-1:2022 [11], are all non-intrusive methods requiring on-site measurements, unlike ISO 10534-1:1996 [12], which applies in-lab Kundt tube tests to asphalt samples. However, different studies in the past demonstrated that these standard monitoring techniques are not sufficient to reliably evaluate the pavement acoustic efficacy or to directly compare the properties of different ones [13,14].

CPX has been applied and improved over the years, reaching the status of a well-recognized methodology [15–18], and its indicators are mentioned within the minimum environmental levels given by the EU Green Public Procurement Criteria (GPP), which were finally defined in 2016 for Europe [19]. The quality of new LNP should then be guaranteed by a mandatory evaluation of its noise emission through CPX measurements, as it is the method that is more directly focused on evaluating the source emission by measuring noise with microphones next to the reference tire of a lab-moving vehicle.

On the contrary, SPB-ISO is the right method to identify the sound emitted by real flow or for different vehicle categories, as it is based on noise and traffic data measurements performed at the roadside and at a standard distance. It provides an indicator of noise perceived for a single vehicle and may help to verify the noise efficacy of the pavement or to introduce policies for specific category restrictions or speed limits in contrast to methodologies which acquire the overall noise. Moreover, SPB-ISO is also very helpful for the definition of noise models in the noise mapping and action plan phases. This method is also able to characterize new categories, such as, for example, electric vehicles, which are an emerging issue the definition of which is left open even in the noise mapping official method CNOSSOS-EU [20]. In particular, a variation of the SPB-ISO, the Controlled Pass-By (CPB) [21] is being used in the ongoing LIFE Project E-VIA, aiming to assess the noisiness of electric vehicles on specific-noise pavements [22]. In fact, low-noise pavements can reduce rolling noise; thus, their effect is maximal for electric vehicles [23], whilst it is reduced for heavy vehicles and mopeds. Low-noise surfaces were also recently under study in a combined solution with Intelligent Transportation Systems (ITS) in order to maximize their mitigation effect with more focused traffic [24].

Even though both CPX and SPB-ISO are widely performed around the world [25–27], they have been shown to not fit perfectly into the urban environment, especially the standard SPB methodology [13]. In addition, the SILENCE project contributed to the development of the Backing Board variant (ISO 11819-4 [28]), the measurement of which is still influenced by the real world in front of the microphone.

SPB-ISO measurements not only require a free field around the mics, which is rare in an urban context, but also single vehicles passing within a sufficient time spacing. Such a condition almost never happens on urban roads during the daytime, especially on the major and most impactful roads that have high traffic flow. Traffic lights, pedestrian crossing and roundabouts are also urban issues that group vehicles together, especially during daytime, which is the period when attended SPB-ISO should be carried out.

Thus, a method able to determine vehicle noise on road pavements in any urban context was still being looked for when the LIFE NEREiDE [29] project was conceived. The project, among others, also aims to develop innovative measurement protocols and new methods to verify the efficacy of low-impact surfaces to improve soundproofing performances and reduce annoyance in the urban context, such as through the use of P-U sensors [30]. Therefore, the NEREiDE project intends to improve the existing measurement protocol of SPB in order to provide institutions with more reliable data from monitoring campaigns.

The present work presents a methodology to determine average noise levels emitted by vehicles categories in an urban context. The outcome is called Urban Pass-By (U-SPB) and mixes standard long-term noise measurement with the SPB-ISO method. It has the advantages of an easy setup of common instrumentation, and, most importantly, it can be unattended during the data acquisition, thus, sparing man hours. The developed algorithms allow the elaboration of the measurements to obtain a U-SPB index that is comparable with the SPBI set by ISO/DIS 11819-1. Finally, the most important advantage is that this procedure implements statistical pass-by in an urban context, ascribing value to such results for mitigation purposes.

The present paper details the measurement setup and the analysis procedure of U-SPB. The testing and validation phases are based on specific measurement campaigns performed in Tuscany, Italy, and application to monitor the noise performances of new surfaces in an urban context is also shown.

The locally estimated values are not generalizable for all road surfaces, making the U-SPB a method that is not developed enough to label road surfaces over their entire stretch as the CPX does. However, this behavior turned out to be a pro in different contexts, such as the local evaluation of a road noise mitigation action or the evaluation of the effectiveness of traffic reduction measures (times of prohibition of freight transport, bus lines, electric vehicle policies, speed limits, etc.) on the noise impact. This aspect represents an added value for policy makers and allows the correct planning of traffic route policies such as freight-banning times, bus routes, electric vehicle policies, speed limits, etc. The procedure also allows the estimation of long-term L_{den} according to a relation based on measured road traffic flows only. This allows the avoidance of spurious events or unwanted noise sources in the evaluation of people exposed to the noise specifically emitted by the investigated road.

2. Background

In 2006, the Tuscany region planned the LEOPOLDO project with the intent to develop innovative noise mitigation techniques for action plans relating to road infrastructure to find the best surface criteria based on the surroundings of the laying and to develop measurement protocols useful for assessing road surface effectiveness and time stability in terms of both acoustical and safety characteristics. SPB-ISO and CPX were applied within the LEOPOLDO project, and improved protocols for measurements and for data postprocessing were developed [31,32]. An improved version of SPB was developed (SPB-L) using a different noise metric than the original version. The SPB-L procedure was based on measuring the acoustical energy of the passing vehicle; it uses the Sound Exposure Level (SEL or L_E), calculated according to the ISO 1996-1 [33], and the pass-by event is the signal part in which L_{Afmax} exceeds the background noise by more than 10 dB(A), as defined in ISO/DIS 11819-1 [10].

A scientific debate was raised at that time [34–37], and different metrics were proposed for SPB analysis, which led the LEOPOLDO project to adopt SEL instead of L_{Afmax} as a metric in its SPB-L method. Later, the ROSANNE project [38] found good correlation between L_{Afmax} and SEL. Although SEL is considered more sensitive to ground effects, it is still considered as a relevant option for helping the definition of common noise modeling methods [39].

The SPB-L followed the HARMONOISE [35] and IMAGINE [36] projects, which introduced a second measurement position placed at 3.0 m height and 7.5 m distance from the center of the road lane (SPB-HI). This improved the evaluation of the influence of the local context by avoiding the roadside ground influence that was not negligible with only the 1.2 m height position.

Thus, SPB-L measurements include two microphone heights. Thus, the statistical sample of many single passages constitutes the dataset for the logarithmic regression

(Equation (1)) between the measured speed v and the SEL, which estimates the level at the reference speed v_0, in accordance with SPB-HI.

$$SEL = a + b \cdot log\left(\frac{v}{v_0}\right) \quad (1)$$

where a is the SEL at reference speed (typically 50 km/h), and b is a speed-related correction.

A further improvement of SPB-L was the introduction of the use of statistical data binning since the SPB-HI procedure fails when speed data are collected around a single value, such as the speed limit. In fact, the variability due to driver behavior and to vehicle characteristics dominates when the speeds are almost the same. Data then have a cloud-like shape, and those falling outside the cloud influence the fit algorithm. This influence is avoided through statistical data binning applied to the whole dataset and a minimum chi-square fit of central values with their uncertainties.

In more detail, data are grouped in velocity classes, or bins, about 10 km/h in width. The actual width is chosen to minimize the total chi-square of the final fit. The mean and standard deviations of data in each class are computed with the hypothesis of a Poissonian distribution of the vehicles data, with each class represented by the triplet: central velocity of the class and the mean and standard deviation of the SEL data contained in each class. These triplets are used as inputs in the best fit between SEL and speed. In this way, the information of data spread out, and the numerousness in each speed bin is taken into account by means of the uncertainty associated with the central values.

Within the NEREiDE project, a further improvement of SPB was looked for in order to simplify SPB measurement and analysis protocol, especially sparing man time on site. The U-SPB protocol, presented in the present work, intends to derive indexes able to represent pavement efficacy in their implementation context and to derive a method able to identify the contribution that a new pavement might provide in that context, taking into account the fleet composition, average speed and surrounding additional noise sources.

In Table 1, the characteristics of U-SPB are compared to the different adaptations of the SPB-ISO method in order to clarify the different approaches mentioned in the present work.

Table 1. Summary of SPB method adaptations mentioned in the paper and their characteristics.

	SPB-ISO	SPB-HI	SPB-L	U-SPB
Reference	ISO/DIS 11819-1 [10]	HARMONOISE/IMAGINE [35,36]	LEOPOLDO [31,32]	NEREIDE [29]
Parameter	L_{Amax}	SEL	SEL	SEL
Microphone heights	1 mic at 1.2 m height	2 mics at 1.2 m and 3 m height	2 mics at 1.2 m and 3 m height	1 mic at 4 m height
Measurement distance from road lane	7.5 m	7.5 m	7.5 m	3–15 m
Measurement conditions	Attended	Attended	Attended	Unattended
Events triggering	On site	On site	On site	In lab by use of counter traffic
Fitting technique	Linear fit	Linear fit	Binning linear fit	Binning linear fit

3. Materials and Methods

An improved SPB methodology that is able to provide similar indicators but avoids the need for the measurements to be attended is defined on the basis of literature review and previous experiences. The methodology described in the next chapter consists of obtaining the pass-by index directly from the noise monitoring station that is usually installed for law requirement purposes. The microphone used for noise level acquisition is placed on the roadside at 4 m height. The single-vehicle passages, required by SPB-ISO, can be easily identified in the noise time histories acquired during the nighttime, when traffic flows and background noise are lower. After its definition, the methodology is finalized according to the following phases:

- Testing, including a feasibility test, performed to confirm the hypothesis, and a controlled comparison with the SPB-L method;

- Validation: comparison of the results obtained by the application of the U-SPB and SPB-L methods;
- Application: within the NEREiDE project, U-SPB is applied to the sites where new low-noise surfaces have been laid in order to evaluate both ante- and post-operam conditions.

Each phase is based on measurement campaigns specifically performed in Tuscany (Italy), which are summarized in Table 2 together with a summary of their most peculiar features.

Table 2. Summary of measurement campaigns.

Step	Code	Site	Methods	Measurement Conditions	Results
Testing	T0	SR 439 Capannori (LU)	U-SPB	Extra urban road, low-noise surface, data previously acquired	Feasibility confirmed
Testing	T1	SP 224 Marina di Pisa (PI)	U-SPB, SPB-L	Urban road, worked surface, data acquired at roadside day and night, with 6 h controlled situations	Calculated parameter chosen and comparison test performed and passed
Validation	V	Viale Togliatti Sesto Fiorentino (FI)	U-SPB, SPB-L	Urban road, low-noise surface, full measurement protocol applied	Validation completed
Application	A	SR 439 Massarosa (LU)	U-SPB	Extra urban road, ante-/post-operam measurements	Surface evaluated

3.1. Testing

The work started with a preliminary analysis of previously gathered data, performed in order to verify the applicability of the main idea. The dataset was retrieved from a noise measurement campaign carried out in SR 439 (Capannori, LU, Italy) for the monitoring of a one-year-old low-noise pavement (ISO 10844 optimized texture, commonly used by the Tuscany region in its road action plan). In this case, the acquisitions were originally performed with a noise monitoring station at the roadside, 10 m from the center lane, in order to verify law requirements in terms of ante- vs. post-operam noise levels. As these data were not acquired ad hoc, a preliminary procedure was applied in order to link night traffic acquired passages with noise levels acquired with a short time interval (50 ms). The pass-by events were identified, and those surpassing the background noise by at least 10 dB(A) in the noise time history were selected. According to vehicle length, categories were established, setting appropriate thresholds for two wheelers and light and heavy vehicles. This preliminary test was intended to verify the feasibility of deriving a model predicting the noise emitted by categories of vehicles by means of fits for speed vs. the different noise metrics (SEL, L_{Amax}, L_{Aeq}) of a single vehicle passing in front of the measurement station. Distances from the monitoring station to the center lanes were known, and data were corrected for distance using a linear divergence model [37,40] in order to enable comparison with the standard requirements (1.2 m height, 7.5 m distant from center lane). Such a technique is valuable for a comparison at different sites and can provide indexes that are comparable with standard ones. A linear correction is mandatory to compare techniques, i.e., the single moving vehicle passing by has to be represented as a linear source, and divergence rules are applied, as in Equation (2).

$$NM = NM_{meas} + 10 \log \frac{\sqrt{d^2 + 4^2}}{\sqrt{7.5^2 + 1.2^2}} \quad (2)$$

where NM is the selected noise metrics.

Once the feasibility was achieved, a procedure was then established for acquiring data. A comparative analysis with the standard method was also performed in order to:

- Define which sound level metric should be fitted with speed (SEL, L_{Amax}, L_{eq});
- Verify if categorization with a traffic counter is comparable with a man-made one or if, at least, outliers are not relevant;
- Verify the influence of microphone height and position on the results' quality and the ability of the instrumentation to detect both lanes.

A simple, straight road with free field conditions and a standard pavement (dense asphalt concrete) was selected for this phase. The site was chosen for the performance of both standard SPB and U-SPB in a short time. Figures 1 and 2 report the measurement location and measurement setup as an example.

Figure 1. Urban pass-by location in T1 (Marina di Pisa).

Figure 2. Measurement setup in T1 (Marina di Pisa). Near lane is the one close to the traffic counter and monitoring station; far lane is the one in the opposite direction.

The test compared the U-SPB with SPB-L on two different sets of data; the first set compared the morning traffic flow sample acquired with SPB-L and at the monitoring station, while the second set compared the SPB morning measured data with nighttime data on the monitoring station, identified by the traffic counter, i.e., applied the U-SPB procedure as detailed in the following chapter.

This phase was also dedicated to demonstrating the feasibility of estimating noise levels at a roadside from an SEL vs. speed model. In fact, a simulated, hourly, A-weighted L_{eq} ($L_{Aeq,h}$) was derived using flows acquired with the traffic counter, and the noise model for SEL was obtained. Namely, the $L_{Aeq,h}$ was estimated according to the following equation:

$$L_{Aeq,h} = \bigoplus_i \left(10\cdot \log(Q_i) + a_i + b_i\cdot \log\left(\frac{v_i}{50}\right)\right) - 10\cdot \log(3600) \quad (3)$$

where the symbol $\oplus$ means the energetic sum, Q_i is the traffic flow and v_i is the average speed of the ith category at the specific hour.

3.2. Validation

After the positive outcomes of the testing phase, a validation on low-noise pavements was carried out to increase the statistics of the U-SPB application. The objective of the validation was to confirm the validity of the methodology results through a comparison with other standard on-site methods and to provide a definitive protocol for the urban pass-by methodology, including measurements, a to-do list and analysis guidelines. This phase was carried out by applying the complete protocol to a new area where a mitigation was in force and where the local urban configuration also allowed a full standard application of SPB-L. On the chosen sites, SPB-L, U-SPB and noise equivalent level measurement methods were tested. Noise level estimation through the SEL models was performed and compared with a standard measurement technique in the validation phase.

3.3. Application

Finally, the procedure was applied to 4 other pavements in SR 439, corresponding to the NEREiDE project ones. This meant applying the procedure to the data acquired during the ante- and post-operam campaigns and deriving U-SPB values and verifying noise level estimations over a long-term campaign. While the NEREiDE project studied a total of 12 stretches, the present paper only considered the stretches referenced as nos. 1, 3, 5 and 6 in the project [41].

4. U-SPB Procedure

The applied U-SPB measurement procedure is derived from the SPB-L experimental protocol.

The procedure intends to combine data from a traffic counter and a standard monitoring station to catch single passages, as in the SPB-L. Then, the statistical data binning is applied and data fitted according to the procedure followed in the LEOPOLDO project, and the SPB index (SPBI) is derived accordingly.

4.1. Measurement Setup

The monitoring station includes a microwave traffic counter system, a sound level meter, a power system and a weather station to exclude periods with a wind speed higher than 5 m/s or rainy periods. The traffic counter acquires single-vehicle passage data, such as vehicle length, velocity, transit time and time distance from the previous vehicle passage. The devices should be carefully synchronized in time in order to simplify analysis. The sound level meter should fulfill the ISO 1996-2 requirements and should be positioned preferably 7.5 m away from the centerline. If impossible, the distance from the roadside should not be less than 3 m in order to avoid source directivity problems and be no more than 15 m away in order to avoid ground and spurious reflection issues. In this way, the sight angle of the road is maximized without going too far from centerline and avoiding measuring a lower sound signal.

The sound meter level should be placed at 4 m height and should be set to acquire:

- L_{Aeq}, in fast 50 ms;
- Third-octave band spectrum.

The microwave traffic counter should be installed at the roadside, positioned in order to acquire traffic data of all lanes, and should be set to acquire vehicle length accurately in order to identify the different vehicle categories.

The sound level meter and the traffic counter should be positioned as close to each other as possible in order to avoid speed differences between the acquisitions.

A brief on-site observation is suggested at the beginning of the setup to fix category length limits for both lanes, thus, minimizing category mismatching.

A fixed length for the measurement is not fixed and depends on the flow, but a two-night duration is generally enough to obtain a sufficient number of transits.

4.2. Data Analysis

The analysis procedure starts from the raw data obtained in the measurement phase by matching traffic counter flows with noise data (sensor acquisition phase). Measurement periods are filtered according to the national law requirements for weather conditions.

The feature detection phase starts with downloading traffic counter data into a spreadsheet. Then, a time interval between consecutive passages has to be fixed to identify valid transits. The choice of the time interval depends on the source distance and source average speed. For the purpose of this study, a time interval between 3.5 and 4 s is applied since all the roads have 50 km/h as a speed limit. The candidate transits are highlighted by conditional formatting on the spreadsheet. Analogous conditional formatting is performed over vehicle lengths returned by the traffic counter to discriminate between vehicle categories. Then, single candidates are identified in the time history of sound pressure level, and events are selected if L_{Afmax} exceeds the background noise by at least 10 dB(A); otherwise, the event is discarded.

For valid events, the noise average spectrum is acquired in the L_{Afmax}−10 dB(A) time interval. The analysis spreadsheet is designed to be able to count valid events per direction and categories so that it is possible to stop analysis as soon as the minimum requirements are fulfilled. A diagram of the event selection procedure is reported in Figure 3.

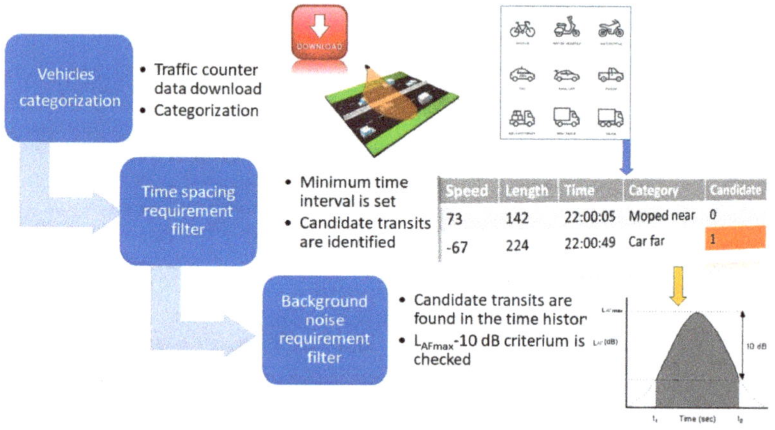

Figure 3. Features extraction phase: event selection process.

The event extraction phase starts by collecting the average spectrum together with speed and direction and storing them in order to collect all the events. The average spectrum is then fitted according to Equation (1).

Thus, the statistical sample of many single passages constitutes the dataset for the statistical data binning developed in SPB-L. In Figure 4, an example of statistical data binning is provided according to Zei et al. [42] and Lafferty et al. [43].

The analysis with data binning provides regression of SEL values and each frequency band as a function of speed. An example is reported in Figure 4. In case of fits with fewer than 10 events, fits are carried out without statistical data binning.

If enough events are acquired, at least 100 light and 30 heavy vehicles, the SPBI can be estimated according to Equation (3), taken from the ISO/DIS 11819-1 for low-speed roads:

$$SPBI = 10\ \log\left[W_1 10^{\left(\frac{L_1}{10}\right)} + W_2\left(\frac{v_1}{v_2}\right)10^{\left(\frac{L_2}{10}\right)} + W_3\left(\frac{v_1}{v_3}\right)10^{\left(\frac{L_3}{10}\right)}\right] \quad (4)$$

where L_i is the vehicle sound levels for the ith vehicle categories; W_i is the weighting factors determined by compositions of traffic flow and are established by the ISO.

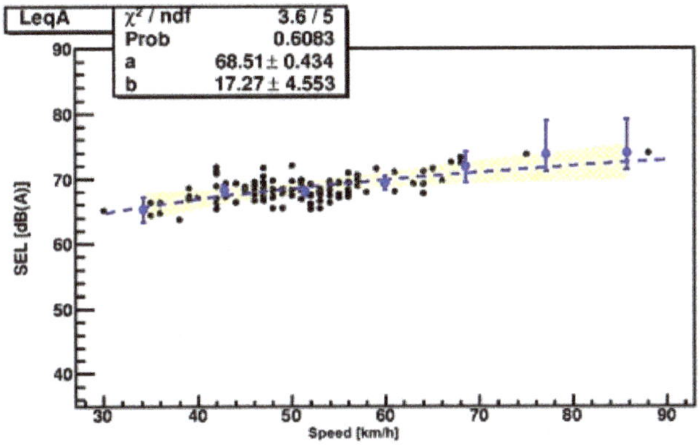

Figure 4. Example of data collected in binning chunks (blue points with error bars) for the best fit between SEL and velocity. The 95% confidence interval from the fit is reported in yellow.

5. Results

The present chapter describes the obtained results in the different phases of testing, validation and application of the U-SPB method.

5.1. Testing

The feasibility test provided results for SEL, L_{Aeq} and L_{Amax}. The pros and cons of each of the metrics are briefly described:

- SEL was chosen by the NORDTEST method, HARMONOISE, IMAGINE and the Italian LEOPOLDO projects for its representativeness in noise models and its ability to estimate the overall noise levels. A potential weakness is represented by its partial dependence on the 10 dB(A) cut made by the operator [41];
- On the contrary, L_{Aeq} is weakly dependent on the 10 dB(A) cut, but it is representative of average vehicles only when their speed variation is small and when the time of passage is almost the same for all vehicles. These conditions are especially difficult to achieve at nighttime, which is when data are intended to be collected, because single-passage events are more frequent;
- L_{Amax} is the ISO/DIS 11819-1 parameter and does not depend on cut, but it is strongly dependent on the local peculiarities of the road (such as potholes, bumps, etc.), which lack homogeneity, and vehicle discrepancies. Thus, its results tend not to be stable even for the same vehicle and pavement.

The data analysis revealed a relevant difference between fits performed by plotting speed against SEL, L_{Aeq} (recomputed by the average spectrum) and L_{Amax} (from spectrum on maximum) for the light vehicle category.

Then, the model in Equation (4), which is a generalization of Equation (2), with f being one of the three tested parameters, was applied to the measured dataset:

$$f = a + b \cdot log\left(\frac{v}{50}\right) \quad (5)$$

Table 3 shows the results obtained in T0 (Capannori) for each lane and fit values for the light vehicle category. As mentioned in the previous chapter, the near lane direction was the lane closest to the noise monitoring station, and the far lane was the opposite one. Distances from microphones to center lanes were known, and data were corrected for distance using a linear divergence model in order to allow comparison with the standard requirements (1.2 m height, 7.5 m distant from center lane) and between lanes.

Table 3. Coefficients of the fit of data measured in T0. a and b units are dB(A).

	Near Lane			Far Lane		
Metric	$a \pm da$	$b \pm db$	χ^2	$a \pm da$	$b \pm db$	χ^2
SEL	69.5 ± 0.4	19.2 ± 4.5	0.42	68.6 ± 0.4	18.0 ± 5.6	0.10
L_{Aeq}	62.8 ± 0.4	25.8 ± 4.6	0.69	60.5 ± 0.4	25.2 ± 5.7	0.23
L_{Amax}	66.8 ± 0.4	25.2 ± 4.5	0.78	64.7 ± 0.4	26.4 ± 5.7	0.42

The SEL slope results were lower than the other metrics, probably because they already considered time/speed effect. The near lane was noisier according to L_{Amax} and L_{Aeq}, while it was not for SEL.

The preliminary test also showed that pass-by parameters can be derived with good fit agreement in terms of χ^2 by starting from data acquired with the monitoring station, and the noisiness of a single vehicle category can be determined as a function of speed.

Then, the testing measurement in T1 (Marina di Pisa) allowed a comparison of the SPB-L and the U-SPB applied over the same sample flows. The test intended to address whether there were differences due to the microphones' positions, regardless of the speed and sample variability. The three parameters were once again tested. U-SPB data were grouped into the morning common sample, which was intended to be the passages collected on the monitoring station at the same time as the attended SPB-L, and the night independent sample.

Table 4 reports the results for the light vehicle category for each indicator in the first U-SPB dataset. Linear correction for distance was applied to the monitoring station and to the 3 m microphone to allow comparisons.

Table 4. Results with the morning common sample acquired in T1 for light vehicles. a and b units are dB(A).

Lane	Coefficient	Monitoring Station	Mic 1.2 m	Mic 3 m
		SEL		
near	$a \pm da$	75.1 ± 0.6	72.6 ± 0.6	75.4 ± 0.6
	$b \pm db$	22.8 ± 4.5	22.3 ± 4.5	21.6 ± 4.5
far	$a \pm da$	76.3 ± 0.5	76.4 ± 0.6	76.4 ± 0.6
	$b \pm db$	18.9 ± 5.4	20.5 ± 5.4	20.2 ± 5.4
		L_{Aeq}		
near	$a \pm da$	68.7 ± 0.6	67.2 ± 0.6	69.6 ± 0.3
	$b \pm db$	31.3 ± 4.5	30.1 ± 4.5	29.6 ± 1.2
far	$a \pm da$	69.2 ± 0.5	71.3 ± 0.5	70.7 ± 0.6
	$b \pm db$	26.5 ± 5.4	28.6 ± 5.5	28.7 ± 5.5
		L_{Amax}		
near	$a \pm da$	72.7 ± 0.6	70.9 ± 0.6	73.1 ± 0.6
	$b \pm db$	30.4 ± 4.5	29.3 ± 4.9	28.6 ± 4.9
far	$a \pm da$	73.3 ± 0.5	75.6 ± 0.5	75.3 ± 0.6
	$b \pm db$	26.3 ± 5.3	29.1 ± 5.4	26.1 ± 5.5

Some relevant differences emerged for each microphone, with lower values for the near lane and higher values for the far lane from the 1.2 m SPB-L microphone. This might be due to the specific location, which presented a small, green hedge at the roadside that might have partially screened noise from the near lane at low heights. The far lane might have been noisier because of some bumps and because of a more reflective pavement than the one in the parking area where the microphones for the near lane were located. Moreover, the levels for all metrics at the 3 m microphone and at the monitoring station were more similar between the lanes, as they were not as affected by ground reflections as the standard microphone at 1.2 m height. As reported in Figure 5, the acquisitions with the microphone at 3 m and with the station were compared in terms of the difference between the two lanes. L_{Amax} was greatly different between lanes for both microphones, confirming that L_{Amax} is more sensitive to local bumps than other indicators and is highly dependent on microphone position. Thus, the L_{Amax} was discarded for its sensitivity to local issues. L_{Aeq} difference highly depends on microphone position due to the duration of the event, which can vary significantly with the sight angle. As expected, the SEL difference was similar at both microphones.

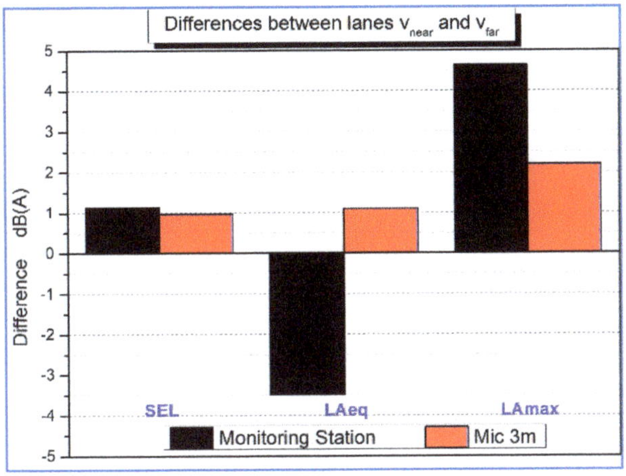

Figure 5. Difference between near and far lanes results with different parameters in T1.

SEL, as from Equation (1), can also be used to estimate the annual averaged A-weighted equivalent sound levels L_{Aeq} from the annual averaged traffic flow, as explained in Equation (2). This represents an additional advantage that should drive the sound pass-by methodology to be optimized for this indicator. On the other hand, SEL depends on time interval above 10 dB(A), while L_{Aeq} is not corrected by the time length of the passage. Thus, as well as being corrected by the distance, the two lanes were differently influenced by the time length for the L_{Aeq} indicator. The greater difference with this indicator resides in the directivity of the source and the geometry of the measurement setup; as the acquisition was next to the source, a shorter time interval was needed to identify a 10 dB(A) cut compared to an acquisition at a greater height from the ground. Thus, the SEL is the one with less influence on position in general, and it can be considered as more reliable. As reported in Table 5, the fits' parameters of the night dataset with SEL were compared with SPB-L for the three vehicle categories. Table 6 reports the number of events considered for the fits during one-night acquisition for light vehicles and two-night acquisition for the other categories.

The differences between SPB-L and U-SPB appear to be mainly due to the sample when statistics were low, but a good agreement was achieved for light vehicles. Some differences could be due to category mismatching, especially mismatching in identification of small trucks (potentially long light vehicles) and short light vehicles (potentially two wheelers).

Even though the measurement period was very short, a model was built to compare measured hourly L_{Aeq} with estimated ones according to Equation (1) (calculated with the night dataset). Figure 6 reports the estimate obtained with U-SPB and measurement over a 24 h monitoring period. The results show that the model is in very good agreement with the measurement, confirming the possibility of estimating noise levels with U-SPB. In particular, the calculated equivalent noise level for reference period day–evening–night L_{den} and for night period L_{night} resulted in values only 0.4 dB(A) and 0.7 dB(A) lower than the measured data, respectively.

Table 5. SEL fits comparison for night U-SPB and SPB in T1. *a* and *b* units are dB(A).

Category	Lane	Parameter	U-SPB Monitoring Station	SPB-L Mic. 1.2 m	SPB-L Mic. 3 m
Light	near	$a \pm da$	75.1 ± 0.5	72.6 ± 0.6	75.2 ± 0.6
		$b \pm db$	20.0 ± 4.4	22.3 ± 4.5	21.6 ± 4.5
		χ^2	0.100	0.228	0.082
	far	$a \pm da$	75.7 ± 0.6	76.4 ± 0.6	76.1 ± 0.6
		$b \pm db$	23.8 ± 4.5	20.5 ± 5.4	20.2 ± 5.4
		χ^2	0.215	0.371	0.252
Two wheelers	near	$a \pm da$	76.6 ± 1.1	72.2 ± 0.8	73.9 ± 0.7
		$b \pm db$	20.2 ± 11.0	15.2 ± 5.4	15.7 ± 5.4
		χ^2	94	75	74
	far	$a \pm da$	76.4 ± 0.7	74.3 ± 1.0	73.1 ± 1.0
		$b \pm db$	20.4 ± 5.7	30.1 ± 12	30.9 ± 12
		χ^2	26	61	61
Heavy	near	$a \pm da$	78.2 ± 1.0	74.4 ± 0.5	76.9 ± 0.4
		$b \pm db$	22.4 ± 8.4	46.5 ± 8.4	44.2 ± 6.3
		χ^2	107	17	10
	far	$a \pm da$	80.8 ± 0.7	78.4 ± 1.2	78.2 ± 1.3
		$b \pm db$	17.0 ± 3.0	34.5 ± 16	32.1 ± 17
		χ^2	44	52	55

Table 6. Number of analyzed pass-bys in T1 (night dataset and SPB-L).

Category	Lane	No. of Events SPB-L	No. of Events U-SPB
Light	near	114	129
	far	102	98
Two wheelers	near	13	15
	far	11	12
Heavy	near	10	15
	far	11	20

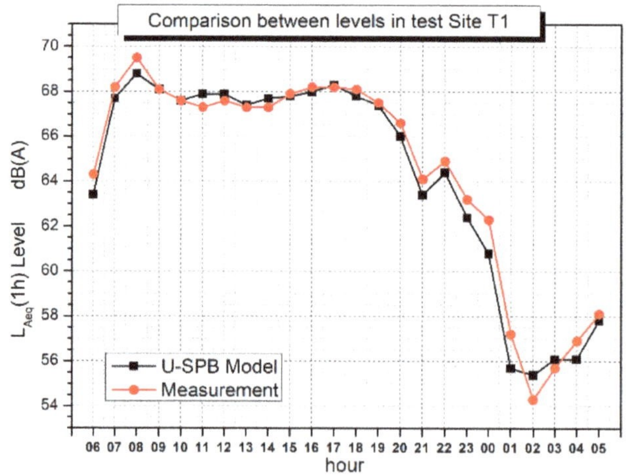

Figure 6. U-SPB $L_{Aeq,h}$ model and measured $L_{Aeq,h}$ comparison in T1.

5.2. Validation

The validation phase was performed in V (Sesto Fiorentino), where a long, straight road section with a free field condition was available to correctly perform SPB measurement with a suitable sample for a daytime, attended section. The road was paved with a low-noise surface. The number of acquired events is reported in Table 7. A single night acquisition was needed for U-SPB analysis of light vehicles and two nights for the heavy and two wheeler categories. A further differentiation of heavy vehicles into two-axis and multi-axis vehicles is reported since it came out that such a distinction improved the result on this road.

Table 7. Number of analyzed pass-bys in V.

Category	Lane	No. of Events SPB-L	No. of Events U-SPB
Light	near	65	120
	far	35	114
Two wheelers	near	13	31
	far	9	34
Heavy	near	8 two axes + 3 multiple axes	12 two axes + 10 multiple axes
	far	8 two axes	13 two axes + 12 multiple axes

The identification of truck classes at the monitoring station was performed based on vehicle length, a procedure that can give some unfortunate mismatching. This further differentiation led us to consider a further category and to compare SPB-L with only two-axis trucks of U-SPB since the number of multi-axis vehicles in the SPB-L sample was lower. Table 8 shows the fit results according to Equation (2).

Table 8. SEL fit comparison for night U-SPB and SPB-L in V. a and b units are dB(A).

Category	Lane	Parameter	Monitoring Station	Mic 1.2 m
Light	near	$a \pm da$	69.3 ± 0.5	69.5 ± 0.6
		$b \pm db$	17.7 ± 4.0	14.7 ± 9.3
		χ^2	1.2	0.5
	far	$a \pm da$	68.5 ± 0.4	69.0 ± 0.8
		$b \pm db$	17.3 ± 4.6	14.5 ± 10.8
		χ^2	3.6	0.2
Two wheelers	near	$a \pm da$	73.9 ± 1.2	73.4 ± 0.6
		$b \pm db$	8.7 ± 10.9	9.3 ± 4.6
		χ^2	0.2	40
	far	$a \pm da$	74.5 ± 0.8	73.5 ± 0.9
		$b \pm db$	21.7 ± 7.4	15.7 ± 9.5
		χ^2	2.4	44
Heavy	near	$a \pm da$	72.0 ± 0.9	74.3 ± 1.5
		$b \pm db$	11.5 ± 10.3	9.3 ± 4.6
		χ^2	108	183
	far	$a \pm da$	70.6 ± 1.0	73.5 ± 0.9
		$b \pm db$	15.8 ± 9.4	15.7 ± 9.5
		χ^2	113	39

The two methods' results were equivalent in terms of the A parameter, apart from the heavy vehicle category in the far lane. Some buses were included in the SPB sample that were not clearly identified in the U-SPB one.

With all the information, it was possible to estimate the SPBI with Equation (3), calculating for a low-speed road, i.e., with a reference speed equal to 50 km/h, as reported in Table 9.

Table 9. SPBI comparison for U-SPB and SPB standard in V. Values in dB(A).

Lane	SPBI (U-SPB)	SPBI (SPB-L)
near	71.1 ± 3.5	70.6 ± 12.5
far	70.4 ± 7.6	69.7± 5.4

SPBI values results were equivalent, with a less than 1 dB(A) difference, even if associated uncertainties were high due to the low number of heavy vehicle events. Thus, the procedure can also be considered as validated on a low-noise pavement.

Finally, the authors decided to also verify the coherence between pass-by and noise level measurements, including for the heavy vehicle category. An estimation based on flows was performed using the estimated SEL of each category; heavy vehicle categories were considered using the average SEL of two-axis and multi-axis trucks and the summed flow. The estimated level was compared with the measurement, as shown in Figure 7; while the U-SPB model underestimated the night levels due the influence of other sources (such as birds, roadside fluorescent lamps, etc.), on the measured ones, a substantial equivalence was achieved during the day between the estimated and measured experimental levels, as for the test in T1 (SP 224).

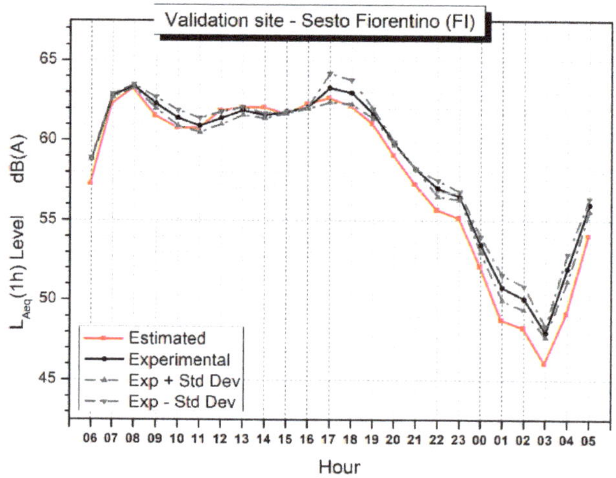

Figure 7. U-SPB $L_{Aeq,h}$ model and measured $L_{Aeq,h}$ comparison in V.

The comparison of noise equivalent level indicators estimated with measurement (hereafter L_{meas}) and the pass-by model according to Equation (1) (L_{model}) is reported in Table 10. Apart from night levels, when other sources might be relevant, the model was an efficient way to estimate noise levels, with prediction always being slightly lower than measured, as expected.

Table 10. Noise levels according to measurements and pass-by model.

Noise Metric	L_{meas} (dB(A))	L_{model} (dB(A))	$L_{meas} - L_{model}$ (dB(A))
L_D	62.0	61.8	0.2
L_E	59.1	58.3	0.8
L_N	54.0	52.4	1.6
L_{DEN}	63.0	62.1	0.9

5.3. Application

The measurement procedure was then applied on LIFE NEREiDE sites, where six different pavements (named stretches 1 to 6) were implemented, and their effects in terms of local fleet (ante- and post-operam values) were evaluated in the project. U-SPB was used to solve the problem while it also allowed the evaluation of the local influence on the overall measured noise brought by other noise sources. This was made possible by comparing the differences between the U-SPB-derived model for road traffic noise with the L_{den} and L_{night} measured values. The model derived from U-SPB parameters for each stretch was applied to Equation (1), taking into account measured traffic flows. The comparison is reported in the present chapter for four stretches among those laid in the project (note that ante-operam values for stretches 2 and 4 were unavailable, so their contribution was eliminated from the present work). Tables 11 and 12 show that the model was able to sufficiently predict the measured level with small differences both ante- and post-operam. In terms of noise mitigation action, its efficiency was also demonstrated since almost all the noise was due to the traffic on the main road, i.e., the one where the noise mitigation was implemented. At night, some differences were seen for stretches 3 and 6, and they might be explained by the presence of a T intersection and a parking area, respectively.

Table 11. L_{den} values ante- and post-operam in Massarosa. Levels in dB(A).

	Ante-Operam			Post-Operam		
Stretch	L_{meas}	L_{model}	$L_{meas} - L_{model}$	L_{meas}	L_{model}	$L_{meas} - L_{model}$
1	67.8	66.5	1.3	64.8	64.5	0.3
3	70.5	69.1	0.6	65.5	63.6	1.9
5	72.7	72.4	0.3	69.4	68.5	0.9
6	71.8	71.0	0.8	67.6	67.2	0.4

Table 12. L_n values ante- and post-operam in Massarosa. Levels in dB(A).

	Ante-Operam			Post-Operam		
Stretch	L_{meas}	L_{model}	$L_{meas} - L_{model}$	L_{meas}	L_{model}	$L_{meas} - L_{model}$
1	59.4	58.1	1.3	56.7	56.7	0.0
3	62.6	61.2	1.4	57.4	55.8	1.6
5	64.8	64.1	0.7	61.1	60.1	1.0
6	63.8	62.4	1.4	59.7	59.1	0.6

6. Discussion and Conclusions

Evaluating the efficacy of low-noise surfaces has started to be a highly demanded activity since the publication of the European minimum environmental criteria. As a consequence, the measurement procedure involved deserves the highest degree of confidence and should be applicable in all contexts. Among the different methods, the Statistical Pass-By (SPB-ISO) method evaluates road traffic noise by means of microphone and speed acquisitions at the roadside in real traffic flow conditions. The original measurement setup does not allow its application in urban areas, which is where most inhabitants live and, thus, where the noise mitigation would have a larger effect.

The present paper reported a method for on-field acquisitions, named the Urban Pass-By (U-SPB) method, which adapts the SPB-ISO to urban context. The event detection and their extraction are performed in lab, as well as the consequent analysis. The evolution the method represented allows the performance of measurements of vehicle pass-bys without the presence of operators, with a consequent saving of man hours. The method also uses the good achievements of previous attempts to adapt the method, such as:

- The use of higher acoustic sensor position and the use of an SEL indicator instead of L_{Amax}, as suggested in the HARMONOISE/IMAGINE projects;
- The use of statistical data binning before fit, as in the LEOPOLDO project.

The testing, validation and application of U-SPB were based on multiple measurement campaigns performed for different sites and road conditions. The results of the procedure were in good agreement with the standard ones, with the additional advantage of needing a reduced time to produce them. In fact, as an unattended methodology, U-SPB can use data acquired even during the night and then reduce the overall time required for monitoring. Furthermore, the availability of night noise and traffic data allows the estimation of long-term L_{den} from long-term traffic counts or statistics, without the need for long noise measurement campaigns, and the evaluation of the presence of other sources influencing local disturbance. Such an estimation is more than a noise model as it is calibrated on site.

It must be pointed out that the procedure can be used as before/after evaluation of mitigation measurement, allowing the comparison of levels measured in the same place in different traffic conditions. However, attention should be paid to possible factors influencing results, such as long queues, which might cause category mismatching, and temperature [25] and, more generally, seasons. Indeed, the predominance of specific vehicle categories such as mopeds can influence results when looking only at SPBI.

U-SPB can also be used to estimate the contribution of different sources to the overall measured sound level in cases where more than one road source is present. For this eventuality, traffic flows for each of the investigated roads are required. The same approach can be extended to the traffic composition evaluation, leading U-SPB to also be suitable for pointing out whether the noisiness is mainly due to light or heavy traffic or no traffic at all. In this regard, the method results are a valid support in driving the noise mitigation evaluation. Further applications of the method on different road types will allow the establishment of a sound methodology for better event detection and feature extraction; in particular, a possible improvement would be to use more precise methods such as the ones proposed in [24] to distinguish between vehicle categories and to find a specific minimum separation time interval for road type.

Finally, while extending the evaluation of the efficiency of low-noise laying in an urban context, U-SPB combines the possibility of estimating long-term noise levels and traffic information for other environmental noise purposes, such as noise mapping and action planning, proving to be a useful tool in environmental noise management.

Further developments of the method could also test the ability to tune conventional prediction models to the local fleet based on U-SPB measurements. Application can be crucial in specific contexts where the fleet is different from the one assumed in the conventional model, i.e., a predominance of old vehicles or of electric vehicles.

Author Contributions: Conceptualization, E.A. and M.C.; Data curation, E.A. and M.C.; Formal analysis, E.A. and M.C.; Funding acquisition, L.F. and G.L.; Investigation, E.A. and M.C.; Methodology, E.A., M.C., L.F. and G.L.; Project administration, G.L.; Resources, M.C. and G.L.; Software, M.C.; Supervision, G.L.; Validation, M.C. and L.F.; Visualization, L.F.; Writing—original draft, E.A. and L.F.; Writing–review and editing, E.A., M.C., L.F. and G.L. All authors have read and agreed to the published version of the manuscript.

Funding: This research was funded by the European Commission Executive Agency for Small and Medium-Sized Enterprises LIFE program (LIFE15 ENV/IT/000268).

Institutional Review Board Statement: Not applicable.

Informed Consent Statement: Not applicable.

Acknowledgments: Authors acknowledge Sonia Giannecchini, Duccio Simonetti, Erica Marchetti, Federica Rampino and Marco Nastasi for contributing to the analysis.

Conflicts of Interest: The authors declare no conflict of interest. The funders had no role in the design of the study; in the collection, analyses or interpretation of data; in the writing of the manuscript; or in the decision to publish the results.

References

1. Peris, E. *Environmental Noise in Europe—2020 EEA*; Report No 22/2019; European Environment Agency: København, Denmark, 2020; ISBN 978-92-9480-209-5.
2. Vaitkus, A.; Čygas, D.; Vorobjovas, V.; Andriejauskas, T. Traffic/road noise mitigation under modified asphalt pavements. *Transp. Res. Procedia* **2016**, *14*, 2698–2703. [CrossRef]
3. Li, B.; Cao, K.; Zhou, J.; Li, A.; Sun, M. Laboratory investigation on influence of mixture parameters on noise reduction characteristics of porous asphalt concrete. *Int. J. Pavement Eng.* **2022**, 1–15. [CrossRef]
4. Sandberg, U.; Ejsmont, J.A. *Tyre/Road Noise Reference Book*; INFORMEX Ejsmont & Sandberg Handelsbolag: Kisa, Sweden, 2002.
5. Fredianelli, L.; Del Pizzo, L.G.; Licitra, G. Recent developments in sonic crystals as barriers for road traffic noise mitigation. *Environments* **2019**, *6*, 14. [CrossRef]
6. Lagarrigue, C.; Groby, J.P.; Tournat, V. Sustainable sonic crystal made of resonating bamboo rods. *J. Acoust. Soc. Am.* **2013**, *133*, 247–254. [CrossRef] [PubMed]
7. Kim, S.H.; Lee, S.H. Air transparent soundproof window. *AIP Adv.* **2014**, *4*, 117123. [CrossRef]
8. Den Boer, L.C.; Schroten, A. Traffic noise reduction in Europe. *CE Delft* **2007**, *14*, 2057–2068.

9. *ISO 11819-2:2017*; Acoustics—Measurement of the Influence of Road Surfaces on Traffic Noise—Part 2: The Close-Proximity Method. International Organisation for Standardisation: Geneva, Switzerland, 2017.
10. *ISO/DIS 11819-1*; Acoustics—Measurement of the Influence of Road Surfaces on Traffic Noise Part 1: Statistical Pass-By method. International Organisation for Standardisation: Geneva, Switzerland, 1997.
11. *ISO 13472-1:2022*; Acoustics—Measurement of Sound Absorption Properties of Road Surfaces in Situ—Part 1: Extended Surface Method. International Organisation for Standardisation: Geneva, Switzerland, 2022.
12. *ISO 10534-1:1996*; Acoustics—Determination of Sound Absorption Coefficient and Impedance in Impedance Tubes—Part 1: Method Using Standing Wave Ratio. International Organisation for Standardisation: Geneva, Switzerland, 1996.
13. SILENCE. Quieter Surface Transport in Urban Areas, EU FP6 Project. Available online: https://trimis.ec.europa.eu/project/quieter-surface-transport-urban-areas (accessed on 5 August 2022).
14. Licitra, G.; Cerchiai, M.; Teti, L.; Ascari, E.; Fredianelli, L. Durability and variability of the acoustical performance of rubberized road surfaces. *Appl. Acoust.* **2015**, *94*, 20–28. [CrossRef]
15. Bühlmann, E. Improvements in the CPX method and its ability to predict traffic noise emissions. In *INTER-NOISE and NOISE-CON Congress and Conference Proceedings*; Institute of Noise Control Engineering: Washington, DC, USA, 2019; Volume 259, pp. 7064–7075.
16. De León, G.; Del Pizzo, L.G.; Teti, L.; Moro, A.; Bianco, F.; Fredianelli, L.; Licitra, G. Evaluation of tyre/road noise and texture interaction on rubberised and conventional pavements using CPX and profiling measurements. *Road Mater. Pavement Des.* **2020**, *21* (Suppl. S1), S91–S102. [CrossRef]
17. Knabben, R.M.; Triches, G.; Vergara, E.F.; Gerges, S.N.; van Keulen, W. Characterization of tire-road noise from Brazilian roads using the CPX trailer method. *Appl. Acoust.* **2019**, *151*, 206–214. [CrossRef]
18. Vieira, T.; Sandberg, U.; Eriksson, O. A round robin test on the close-proximity method: Comparison of results from four different CPX trailers measuring tyre/road noise properties of 10 Swedish road surfaces. *Noise Control. Eng. J.* **2020**, *68*, 421–440. [CrossRef]
19. Garbarino, E.; Rodriguez Quintero, R.; Donatello, S.; Wolf, O. *Revision of Green Public Procurement Criteria for Road Design, Construction and Maintenance. Procurement practice Guidance Document*; Publications Office of the European Union: Luxembourg, 2016. [CrossRef]
20. Kephalopoulos, S.; Paviotti, M.; Anfosso-Lédée, F. *Common Noise Assessment Methods in Europe (CNOSSOS-EU)*; Publications Office of the European Union: Luxembourg, 2012; 180p. [CrossRef]
21. *NF S31-119-2:2000*; Acoustics—In Situ Characterization of the Acoustic Qualities of Road Surfaces—Pass by Acoustic Measurement—Part 2: Controlled Pass-by Method. Afnor: Paris, France, 2000.
22. LIFE E-VIA (LIFE18 ENV/IT/000201): Electric Vehicle Noise Control by Assessment and Optimisation of Tyre/Road Interaction. Available online: https://life-evia.eu/ (accessed on 5 August 2022).
23. Praticò, F.G.; Fedele, R. Electric vehicles diffusion: Changing pavement acoustic design? *Noise Mapp.* **2021**, *8*, 281–294. [CrossRef]
24. Fredianelli, L.; Carpita, S.; Bernardini, M.; Del Pizzo, L.G.; Brocchi, F.; Bianco, F.; Licitra, G. Traffic flow detection using camera images and machine learning methods in ITS for noise map and action plan optimization. *Sensors* **2022**, *22*, 1929. [CrossRef] [PubMed]
25. Geluykens, M.; de Barros, A.G.; Goubert, L.; Vuye, C. Empirical Study on Temperature Influence on Noise Measurements with the Statistical Pass-By Method. *Sustainability* **2022**, *14*, 2099. [CrossRef]
26. Li, T. A state-of-the-art review of measurement techniques on tire–pavement interaction noise. *Measurement* **2018**, *128*, 325–351. [CrossRef]
27. Del Pizzo, L.G.; Bianco, F.; Moro, A.; Schiaffino, G.; Licitra, G. Relationship between tyre cavity noise and road surface characteristics on low-noise pavements. *Transp. Res. Part D Transp. Environ.* **2021**, *98*, 102971. [CrossRef]
28. *ISO/PAS 11819-4:2013*; Acoustics—Method for Measuring the Influence of Road Surfaces on Traffic Noise—Part 4: SPB Method USING backing Board. International Organization for Standardization: Geneva, Switzerland, 2013.
29. Life Nereide Project (LIFE15 ENV/IT/000268) Noise Efficiently Reduced by Recycled Pavements. Available online: https://www.nereideproject.eu/en/ (accessed on 5 August 2022).
30. Bianco, F.; Fredianelli, L.; Lo Castro, F.; Gagliardi, P.; Fidecaro, F.; Licitra, G. Stabilization of ap-u Sensor Mounted on a Vehicle for Measuring the Acoustic Impedance of Road Surfaces. *Sensors* **2020**, *20*, 1239. [CrossRef] [PubMed]
31. Licitra, G.; Cerchiai, M.; Teti, L.; Ascari, E.; Bianco, F.; Chetoni, M. Performance assessment of low-noise road surfaces in the Leopoldo project: Comparison and validation of different measurement methods. *Coatings* **2015**, *5*, 3–25. [CrossRef]
32. Licitra, G.; Teti, L.; Cerchiai, M. A modified Close Proximity method to evaluate the time trends of road pavements acoustical performances. *Appl. Acoust.* **2014**, *76*, 169–179. [CrossRef]
33. *ISO 1996-1:2016*; Acoustics—Description, Measurement and Assessment of Environmental Noise—Part 1: Basic Quantities and Assessment Procedures. International Organisation for Standardisation: Geneva, Switzerland, 2016.
34. SILVIA. Sustainable Road Surfaces for Traffic Noise Control, EU FP5 Project. 2005. Available online: https://trimis.ec.europa.eu/project/sustainable-road-surfaces-traffic-noise-control (accessed on 5 August 2022).
35. HARMONOISE. Harmonised, Accurate and Reliable Prediction Methods for the EU Directive on the Assessment and Management of Environmental Noise, EU FP5 Project 2004. Available online: https://cordis.europa.eu/project/id/IST-2000-28419 (accessed on 5 August 2022).

36. IMAGINE. Improved Methods for the Assessment of the Generic Impact of Noise in the Environment, EU FP 6 Project. 2006. Available online: http://cordis.europa.eu/result/rcn/47869_en.html. (accessed on 5 August 2022).
37. NT ACOU. 109 Approved 2001-11 Nordtest Method Vehicles: Determination of Immission Relevant Noise Emission; Nordtest Tekniikantie 12 FIN–02150 Espoo Finland. 2001. ISSN 0283-7145. Available online: http://www.nordtest.info/wp/2001/11/01/ (accessed on 10 November 2022).
38. ROSANNE. *Rolling Resistance, Skid Resistance, And Noise Emission Measurement Standards for Road Surfaces Deliverable D2.3 Report on the Analysis and Comparison of Existing Noise Measurement Methods for Noise Properties of Road Surfaces*; ROSANNE: Seattle, WA, USA, 2013.
39. Haider, M.; Conter, M.; Wehr, R.; Sandberg, U.; Anfosso-Lédée, F. *Project ROSANNE: Rolling Resistance, Skid Resistance, and Noise Emission Measurement Standards for Road Surfaces*; Internoise: Melbourne, Australia, 2014; 6p, Available online: https://hal.archives-ouvertes.fr/hal-01470451 (accessed on 5 August 2022).
40. Guarnaccia, C.; Quartieri, J. Analysis of road traffic noise propagation. *Int. J. Math. Model. Methods Appl. Sci.* **2012**, *6*, 926–933.
41. Licitra, G.; Cerchiai, M.; Ascari, E.; Simonetti, D.; Chiari, C. Giannecchini Ante operam noise monitoring within Life Nereide: Methods and Results. In Proceedings of the Euronoise 2018, Crete, Greece, 27–31 May 2018.
42. Zei, R.; Cerchiai, M.; Licitra, G. Studio dell'applicazione di un taglio dinamico al metodo SPB (in Italian). In Proceedings of the 41° Convegno Nazionale dell'Associazione Italiana di Acustica, Pisa 2014, Paris, France, 17–19 June 2014.
43. Lafferty, G.D.; Wyatt, T.R. Where to stick your data points: The treatment of measurements within wide bins. Nuclear Instruments and Methods in Physics Research Section A: Accelerators, Spectrometers, Detectors and Associated Equipment. *Sci. Direct* **1995**, *355*, 541–547. [CrossRef]

sensors

Article

Optimization Algorithm for Delay Estimation Based on Singular Value Decomposition and Improved *GCC-PHAT* Weighting

Shizhe Wang, Zongji Li †, Pingbo Wang † and Huadong Chen *

Academy of Weapony Engineering, Naval University of Engineering, Wuhan 430033, China
* Correspondence: 18971569979@189.cn
† These authors contributed equally to this work.

Abstract: The accuracy of time delay estimation seriously affects the accuracy of sound source localization. In order to improve the accuracy of time delay estimation under the condition of low SNR, a delay estimation optimization algorithm based on singular value decomposition and improved *GCC-PHAT* weighting (*GCC-PHAT*-$\rho\gamma$ weighting) is proposed. Firstly, the acoustic signal collected by the acoustic sensor array is subjected to singular value decomposition and noise reduction processing to improve the signal-to-noise ratio of the signal; then, the cross-correlation operation is performed, and the cross-correlation function is processed by the *GCC-PHAT*-$\rho\gamma$ weighting method to obtain the cross-power spectrum; finally, the inverse transformation is performed to obtain the generalized correlation time domain function, and the peak detection is performed to obtain the delay difference. The experiment was carried out in a large outdoor pool, and the experimental data were processed to compare the time delay estimation performance of three methods: *GCC-PHAT* weighting, *SVD-GCC-PHAT* weighting (meaning: *GCC-PHAT* weighting based on singular value decomposition) and *SVD-GCC-PHAT*-$\rho\gamma$ weighting (meaning: *GCC-PHAT*-$\rho\gamma$ weighting based on singular value decomposition). The results show that the delay estimation optimization algorithm based on *SVD-GCC-PHAT*-$\rho\gamma$ improves the delay estimation accuracy by at least 37.95% compared with the other two methods. The new optimization algorithm has good delay estimation performance.

Keywords: delay estimation; singular value decomposition; *GCC-PHAT*-$\rho\gamma$ weighting; generalized cross-correlation

Citation: Wang, S.; Li, Z.; Wang, P.; Chen, H. Optimization Algorithm for Delay Estimation Based on Singular Value Decomposition and Improved *GCC-PHAT* Weighting. *Sensors* **2022**, *22*, 7254. https://doi.org/10.3390/s22197254

Academic Editor: Yuxing Li and Luca Fredianelli

Received: 19 August 2022
Accepted: 13 September 2022
Published: 24 September 2022

1. Introduction

In recent years, time delay estimation technology [1–7] has been widely used in navigation and positioning, underwater AUV, radar detection and other fields. Especially in the research of sound source localization, the localization method based on time delay estimation has become the most commonly used and important method due to its low cost and reliable accuracy. The accuracy of time delay estimation directly determines the accuracy of sound source localization.

The generalized cross-correlation method [8–14] has the advantages of simple principle, high stability, and a small amount of computation when dealing with the delay estimation problem, so it has received extensive attention. However, this method has poor anti-noise ability. As the signal-to-noise ratio decreases, the peak value of the correlation function is no longer sharp, and the performance of the delay estimation drops sharply. However, there will be large noise interference in many practical scenarios, so the noise reduction processing is particularly important.

Based on the above research, it can be seen that there are two keys to improving the accuracy of delay estimation: one is how to effectively reduce noise to improve the signal-to-noise ratio; the other is how to improve the weighting function to make the correlation peak sharper. Reference [15] introduces an improved wavelet threshold function for signal denoising and reconstruction, and uses the GCC method for delay estimation. This method overcomes the shortcomings of traditional soft and hard threshold functions and improves the performance of delay estimation. However, based on the wavelet threshold, the denoising of the signal is restricted by factors such as wavelet base selection and decomposition scale. Reference [16] proposed a data processing method combining wavelet packet noise reduction algorithm and EMD decorrelation algorithm, which can effectively suppress modal aliasing and white noise interference at the same time, but lacks the means of suppressing impulse noise. Reference [17] used the cepstral method to separate the glottal impulse mechanism and the vocal tract response, and combined the spectral subtraction method to design an effective scheme to distinguish the useful frequency bands of noisy signals. Finally, using the spectral characteristics of the eigenmode function after empirical mode decomposition and signal reconstruction, this method significantly improves the anti-noise performance of the delay estimation algorithm, but the algorithm is complex, the application scenario is single, and performance testing in different scenarios is required.

The singular value decomposition method has the advantages of an obvious denoising effect and strong stability. Using this method for signal preprocessing can not only preserve the signal deformation information, but also better filter out the noise pollution. The collected sound source signals are mixed with various environmental noise interference signals. Therefore, the singular value decomposition is to construct a matrix that also contains noise, and the elements in the non-zero singular value matrix represent the distribution of signal and noise energy concentration. Large non-zero singular values represent useful signals, and small ones are interfering noise. The purpose of eliminating noise can be achieved by zeroing the smaller one. For underwater shock signals, the energy of the useful signal segment is relatively high. When the singular value decomposition method is used for noise reduction, the energy distribution of the signal represented by the elements in the non-zero singular value matrix and the noise are obviously different. Therefore, the singular value decomposition method can obtain a good noise reduction effect.

In sound source localization, the accuracy of time delay estimation is very important, which plays a decisive role in the final localization accuracy. Therefore, improving the accuracy of time delay estimation is of great significance to improve the accuracy of sound source localization. However, in some actual measurement environments, the collected acoustic signal contains a lot of noise information, which will have a great impact on the accuracy of time delay estimation. Therefore, it is very important to preprocess the signal with noise reduction first. Secondly, in the time-delay estimation algorithm, because there may be some correlation between noise and noise in the measured signal and between the target signal and noise, the traditional cross-correlation time-delay estimation method based on correlation analysis has difficulty obtaining a more accurate time-delay estimation value, and it is also of great significance to optimize the time-delay estimation method.

The papers [18,19] study the generalized cross-correlation time-delay estimation method based on singular value decomposition, and [20] studies the generalized cross-correlation time-delay estimation method based on improved Phat weighting, which can improve the time-delay estimation accuracy.

This paper combines the two and proposes a time delay estimation optimization algorithm based on singular value decomposition noise reduction. The method first uses singular value decomposition to preprocess the signal to improve the signal-to-noise ratio; secondly, it uses the *GCC-PHAT-*$\rho\gamma$ weighting function to process the cross power spectrum and sharpen the correlation peak to improve the accuracy of delay estimation.

2. Theoretical Introduction of Delay Estimation Optimization Algorithm

2.1. TDOA Signal Model

In the actual scenario of positioning based on TDOA delay estimation, the signal model [21–23] can be expressed as:

$$x_1(n) = s(n) + n_1(n) \tag{1}$$

$$x_2(n) = As(n - D) + n_2(n) \quad (2)$$

where $s(n)$ is the target signal, $x_1(n)$ and $x_2(n)$ are the two acquisition signals, and $n_1(n)$ and $n_2(n)$ are noise. Assuming that the noise and the target signal are uncorrelated, D is the time delay and A is the attenuation factor.

2.2. Singular Value Decomposition Noise Reduction Theory

The singular value decomposition [24–32] method is referred to as the SVD method, which is an important matrix decomposition method when dealing with mathematical problems. The noise reduction principle is to decompose the matrix on the basis of phase space reconstruction, set the singular value corresponding to the noise to zero, and then use the inverse operation to reconstruct the signal, so as to achieve the purpose of noise reduction.

Select the *Hankel* matrix as the trajectory matrix of singular value decomposition, set the noisy signal $X = [x(1), x(2), \cdots x(N)]$, construct the *Hankel* matrix $A_{m\times n}$:

$$A = \begin{bmatrix} x(1) & x(2) & \cdots & x(n) \\ x(2) & x(3) & \cdots & x(n+1) \\ \vdots & \vdots & & \vdots \\ x(m) & x(m+1) & \cdots & x(N) \end{bmatrix} \quad (3)$$

The value range of n is $N/20 - N/2$, $m + n - 1 = N$, and n is taken as $N/2$ in this paper. Matrix A is an $m \times n$ matrix with rank r. There are $m \times m$ orthogonal matrix U and $n \times n$ orthogonal matrix V, so that $A = U\Lambda V^T$.

$$\Lambda = \begin{bmatrix} \Sigma & 0 \\ 0 & 0 \end{bmatrix} = \begin{bmatrix} \delta_1 & 0 & 0 & \cdots & \cdots & \cdots & 0 \\ 0 & \delta_2 & & & & & 0 \\ 0 & & \ddots & & & & 0 \\ \vdots & & & \delta_r & & & \vdots \\ \vdots & & & & 0 & & \vdots \\ \vdots & & & & & \ddots & \vdots \\ 0 & 0 & 0 & \cdots & \cdots & \cdots & 0 \end{bmatrix} \quad (4)$$

The diagonal elements of matrix Λ are singular values of matrix A, and it satisfies $\delta_1 \gg \delta_2 \gg \cdots \gg \delta_r > 0$, where different singular values reflect different energy concentrations. The law of singular values corresponding to different signals is analyzed, the singular value corresponding to noise is set to zero, the appropriate singular value is selected to reconstruct the signal, and the effective signal can be restored to achieve the purpose of denoising. There are three ways to select singular values:

(1) Singular value difference spectrum method

After the *Hankel* matrix is decomposed, $\delta_1, \delta_2, \cdots, \delta_r$ are formed. If the first k singular values are significantly larger than the last $r - k$ singular values, that is, the singular value mutates at the k-th point, the first k singular values are the ideal signals to be extracted.

$$b_i = \delta_i - \delta_{i+1}, i = 1, 2, \cdots, r - 1 \quad (5)$$

The b_i sequence is called the difference spectrum of singular values. The difference spectrum can effectively and automatically determine the maximum mutation point b_k, set the following r_k singular values to zero, reconstruct the matrix, and realize the noise reduction of the signal.

(2) Feature mean method

The singular values $\delta_1, \delta_2, \cdots, \delta_r$ obtained by the decomposition of the *Hankel* matrix are the square roots of the eigenvalues $\lambda_1, \lambda_2, \cdots, \lambda_r$ of the square matrix AA^T. The eigenvalues smaller than the mean of all eigenvalues are set to zero, and the matrix is reconstructed to realize the denoising of the signal.

$$\delta_i = \sqrt{\lambda_i}, i = 1, 2, \cdots, r \tag{6}$$

(3) Singular value median method

Arrange the singular values $\delta_1, \delta_2, \cdots, \delta_r$ obtained by the decomposition of the *Hankel* matrix in ascending or descending order, obtain the median value of the sequence, set the eigenvalues less than the median value of all singular values to zero, reconstruct the matrix, and realize the noise reduction of the signal.

In the following, for the same noise-containing signal, the singular value difference spectrum method, the characteristic mean method and the median method are used for processing, to compare the noise reduction effect. The simulated signal source is: $x = sin3\pi t \cdot cos10\pi t + sin(20\pi t + sin30\pi t) + \alpha(n)$, where $\alpha(n)$ is Gaussian white noise with variance 0.3 and mean 0, and the sampling time is 0.005 s. The mean square error and the signal-to-noise ratio are used to measure the noise reduction effect, and the definitions are as shown in Equations (7) and (8). Among them, x is the actual signal, y is the signal after noise reduction, and n is the signal length. The difference in the noise reduction effect of the three methods is shown in Table 1.

$$\text{RMSE} = \sqrt{\frac{\sum(x_i - y_i)^2}{n}} \tag{7}$$

$$\text{SNR} = 10 \times lg\frac{\sum x_i^2}{\sum(x_i - y_i)^2} \tag{8}$$

Table 1. Comparison of noise reduction effects of three methods.

Method	RMSE	SNR
Singular value difference spectrum method	0.6722	2.8962
Feature mean method	0.2487	11.5337
Singular value median method	0.1429	16.3440

It can be seen from Table 1 that the singular value median method has the highest SNR of 16.3440 and the smallest mean square error of 0.1429. The singular value median method has a significant denoising effect, and can restore the original characteristics of the signal well, and the noise reduction effect is better than the other two methods. Figure 1 shows the effect of noise reduction processing using three methods, respectively.

2.3. *Improved PHAT Weighted Generalized Cross-Correlation Delay Estimation Algorithm*

The *GCC-PHAT* weighting method [12,33–36] can achieve better delay estimation results when the signal-to-noise ratio is high and the reverberation is weak. This leads to false peaks and increases the delay estimation error. In order to solve this problem, the parameter ρ is introduced, and the value of ρ is determined according to the signal-to-noise ratio in the actual environment. On the other hand, because the signal energy is small under the condition of low signal-to-noise ratio, using *GCC-PHAT* weighting will make the denominator of the weighting function approach 0, which will bring greater error, so consider adding a coherence factor $\gamma_{12}(\omega)$ to the denominator of the weighting function. The new weighting function is given below:

$$\psi_{12}(\omega) = \frac{1}{|G_{12}(\omega)|^{\rho} + |\gamma_{12}^2(\omega)|}, 0 \le \rho \le 1 \tag{9}$$

In the formula, $\gamma_{12}(\omega) = \frac{G_{12}(\omega)}{\sqrt{G_{11}(\omega)G_{22}(\omega)}}$ is the coherence factor, $G_{11}(\omega)$ is the self-power spectrum of the signal $x_1(t)$, $G_{22}(\omega)$ is the self-power spectrum of the signal $x_2(t)$, $G_{12}(\omega)$ is the cross-power spectrum of the signals $x_1(t)$ and $x_2(t)$.

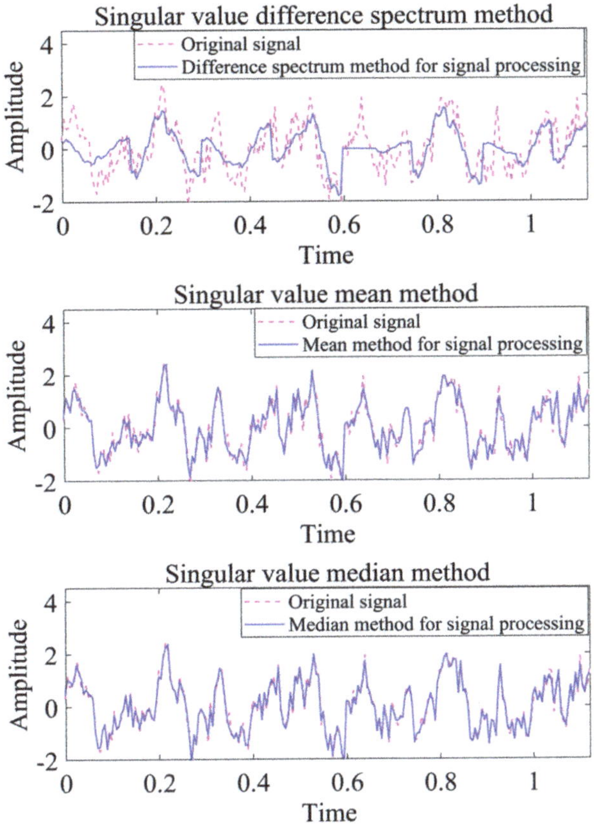

Figure 1. Effect diagram of three methods of noise reduction.

3. Principle of Delay Estimation Optimization Algorithm

The flowchart of the improved time delay estimation algorithm is shown in Figure 2. The calculation process of the improved time delay estimation algorithm is as follows:

(1) Performing singular value decomposition noise reduction processing on the original signals x_1 and x_2 to obtain denoised signals x_1' and x_2';
(2) Performing cross-correlation on the denoised signals x_1' and x_2' to obtain the cross-correlation function R_{12};
(3) Using the improved $PHAT$ weighting function to process the cross-correlation function, the power spectrum function is obtained;
(4) Performing inverse Fourier transform on the power spectrum function to obtain the generalized cross-correlation time-domain function;
(5) Performing peak detection on the generalized cross-correlation time-domain function to obtain the delay difference.

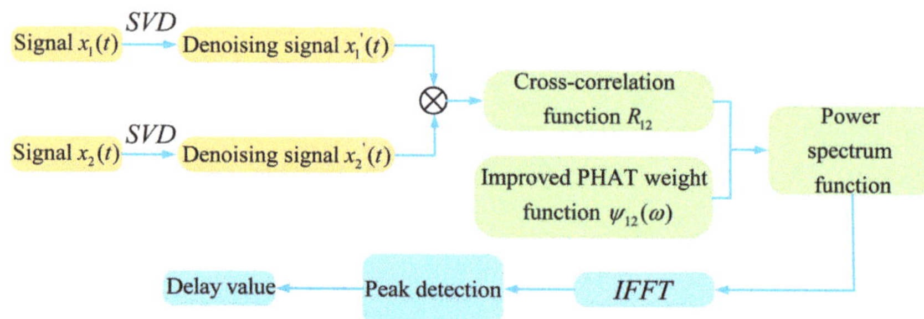

Figure 2. Flow chart of improved time delay estimation algorithm.

4. Simulation Analysis of Analog Signals

This chapter uses the test library speech signal numbered MSJS1-SI869.WAV in the TIMIT standard library as the sound source signal, the signal sampling frequency is 16,000 Hz, and Gaussian white noise is added manually to verify the delay estimation accuracy of the optimization algorithm in this paper. The time-domain diagram, spectrogram, time-frequency spectrum and Hilbert spectrum of the MSJS1-SI869 test library speech signal are shown in Figure 3. After manually adding Gaussian white noise, SNR = 0, and the time-domain diagram of the noisy signal is shown in Figure 4. The singular value median method is used to process the noisy signal, the eigenvalues of the Hankel matrix are shown in Figure 5, and the comparison between the noise reduction result and the original signal is shown in Figure 6.

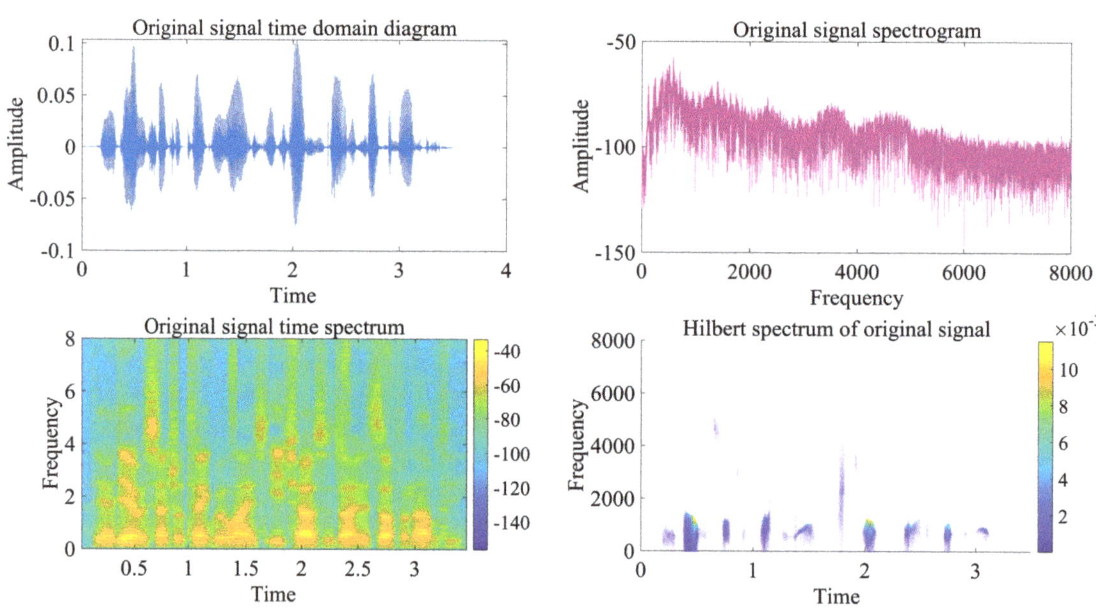

Figure 3. Original signal time-frequency analysis.

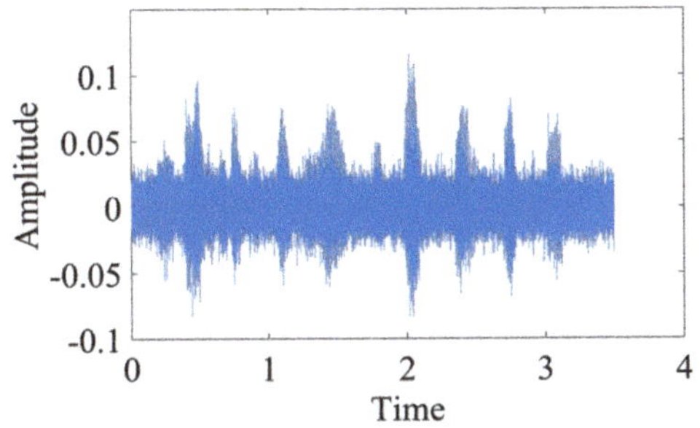

Figure 4. Time domain figure of noisy original signal.

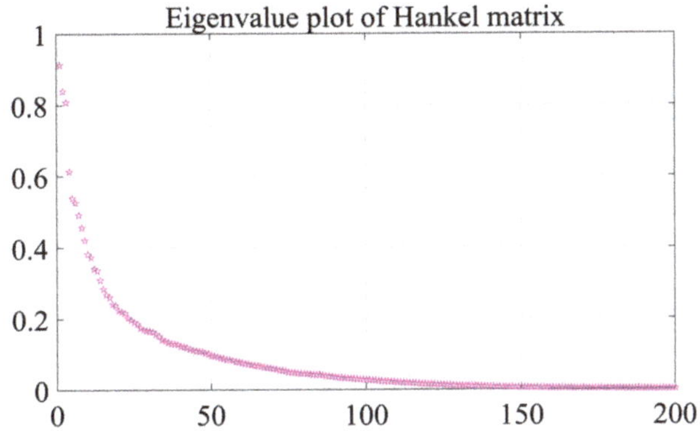

Figure 5. Eigenvalue figure of Hankel matrix.

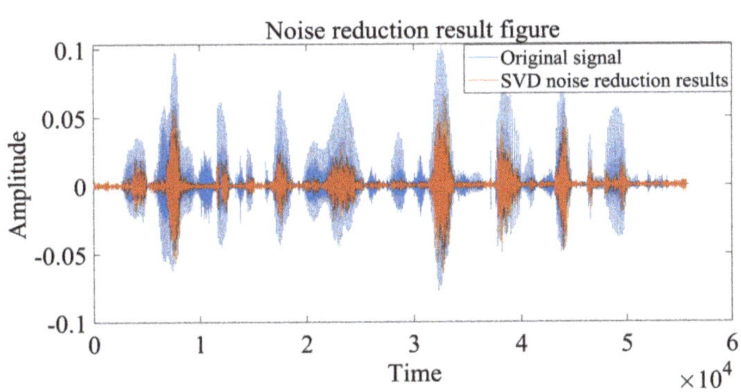

Figure 6. Comparison of SVD noise reduction results and original signals.

It can be seen that the noise reduction effect of the singular value median method is obvious, and the signal-to-noise ratio is greatly improved. In order to verify the time delay estimation performance of the three methods of *GCC-PHAT* weighting, *SVD-GCC-PHAT* weighting and *SVD-GCC-PHAT*-$\rho\gamma$ weighting, the original signal was manually delayed by 50 sampling points and then Gaussian white noise was added. The signal-to-noise ratio variation range is −20 dB–10 dB. The results are shown in Figures 7–10. The abscissa represents the number of delay points, and the ordinate represents the degree of correlation. Fifty random experiments were performed, and Figure 11 shows the root mean square error curves of the three methods under different signal-to-noise ratio conditions.

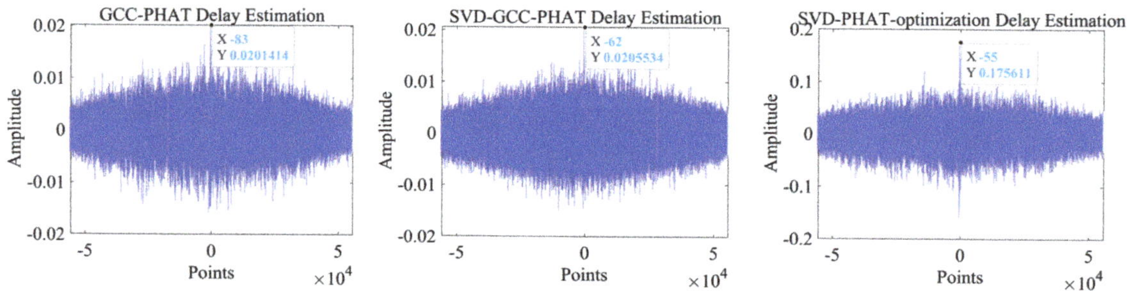

Figure 7. Comparison of delay estimation peak values of three methods when SNR = −20.

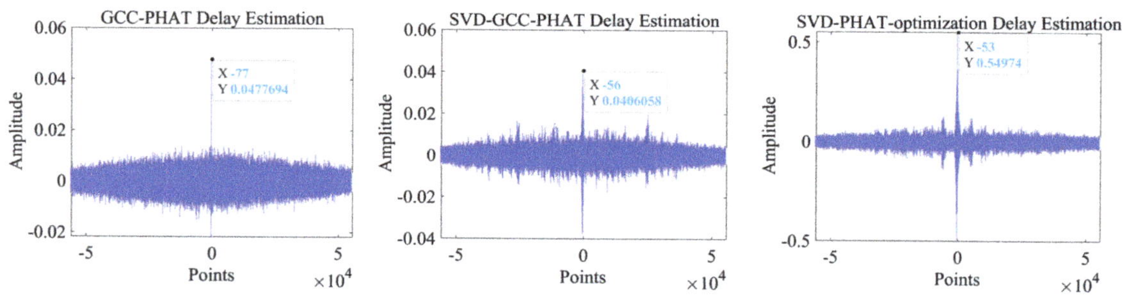

Figure 8. Comparison of delay estimation peak values of three methods when SNR = −10.

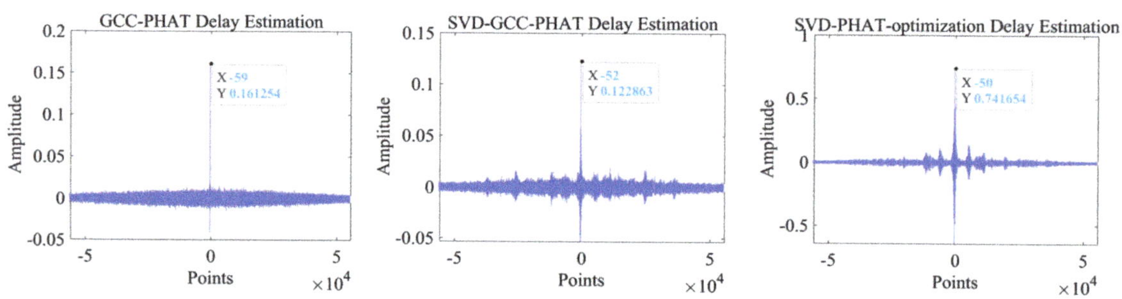

Figure 9. Comparison of delay estimation peak values of three methods when SNR = 0.

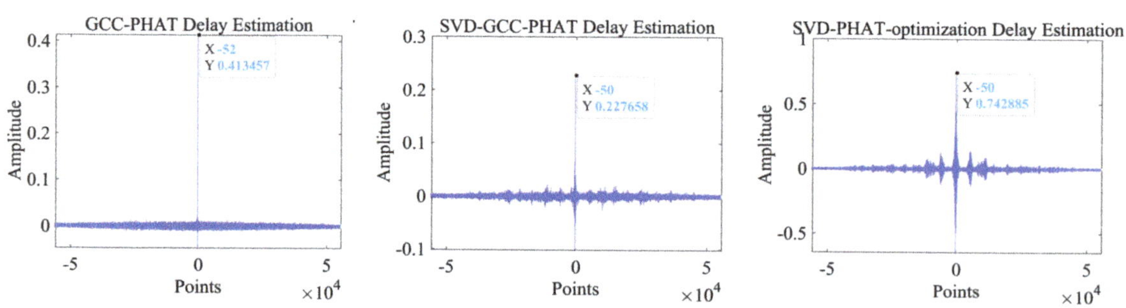

Figure 10. Comparison of delay estimation peak values of three methods when SNR = 10.

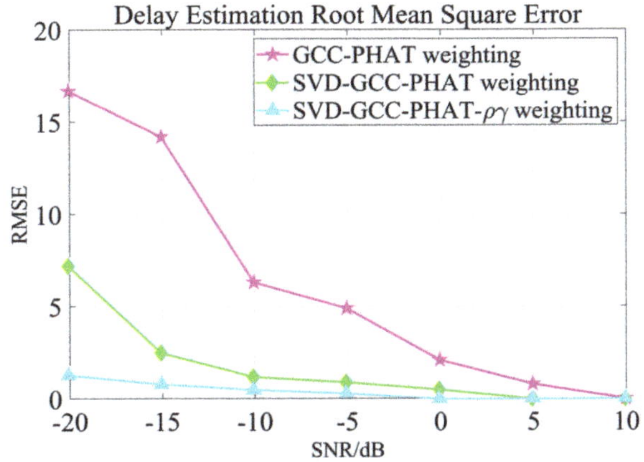

Figure 11. RMSE of three methods for delay estimation.

It can be seen from Figures 7–10 that under the condition of high signal-to-noise ratio, *GCC-PHAT* weighting, *SVD-GCC-PHAT* weighting and *SVD-GCC-PHAT-*$\rho\gamma$ weighting can all accurately estimate the delay difference. However, as the signal-to-noise ratio decreases, interference gradually appears around the weighted peak of *GCC-PHAT*, the correlation peak is no longer sharp, and even false peaks appear, resulting in a large delay estimation error. The weighted estimation accuracy of *SVD-GCC-PHAT* also gradually deviates, but compared with the *GCC-PHAT* weighting method, the singular value decomposition method is used to denoise the original signal, and the delay estimation error is lower than that of the *GCC-PHAT* weighting method. The *SVD-GCC-PHAT-*$\rho\gamma$ weighting not only uses *SVD* to denoise the signal, but also improves the weighting function, which strengthens the useful components in the signal and sharpens the correlation peak. A more accurate delay difference value is obtained, which shows certain advantages. It can also be seen from Figure 11 that as the SNR increases, the delay estimation accuracy of the three methods improves, but under the condition of low SNR, the delay estimation of the *SVD-GCC-PHAT-*$\rho\gamma$ weighting method has more superior performance.

5. Experiment and Performance Analysis

5.1. Experimental System Construction

In this chapter, the actual pool experiment was carried out, and the measured signal was analyzed and processed to verify the reliability of the method in this paper.

The experiment was carried out in a larger outdoor pool, which was about 120 m long, 70 m wide and 3 m high. A three-element linear acoustic sensor array with an array element spacing of 5 m is arranged on the wall of a broad side of the pool, and the depth of the three hydrophones is about 2 m. The underwater sound signal is generated by throwing heavy objects at different points in the pool, the schematic diagram of the experimental arrangement is shown in Figure 12, the experimental site is shown in Figure 13. The sensitivity of the acoustic sensor is less than −200 dB when detecting the sound source in the frequency range of 10–50,000 Hz, the sampling frequency of the acquisition instrument is 128 kHz, and each acquisition time is set to 10 s. The day of the experiment was sunny, breezy, and the water surface of the pool was calm.

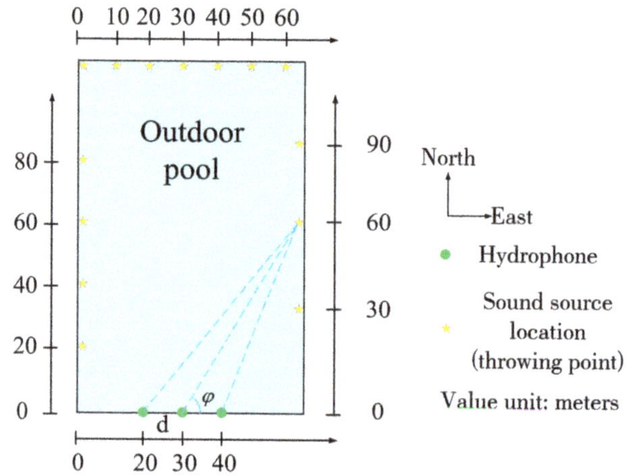

Figure 12. Schematic diagram of experimental layout.

Figure 13. Experimental site diagram.

5.2. Experimental Signal Processing

The collected acoustic signals are analyzed according to the following steps:

(1) Firstly, time-frequency domain analysis is performed to analyze the characteristics of the acquired signal;
(2) Then, the singular value decomposition process is performed to obtain the denoised signal, and the difference between the acquired signal and the denoised signal is compared, and the reliability and correctness of the singular value decomposition noise reduction are verified by the frequency domain analysis of the denoised signal;

(3) Finally, based on the *GCC-PHAT-ργ*-weighted delay estimation algorithm, the three-way signal is estimated for pairwise delays, and the delay difference is obtained, which is compared with the theoretical delay difference.

Taking the acoustic signal emitted at a position near (120 m, 90°) as an example, analyze and process the signal according to the above steps.

As can be seen from Figures 14–16, the impact signal generated by throwing heavy objects into the pool in the three-way signal is very obvious, occurring between 4–5 s. The singular value median method is used for noise reduction processing. Figure 17 is a comparison effect diagram of the original value of the three-way signal and the noise reduction processing value. Figure 18 is a comparison effect diagram of the original spectrum of the three-channel signal and the spectrum after noise reduction.

It can be seen from Figure 17 that the denoising effect of the singular value median method is obvious, which not only removes the interference of noise, but also fully retains the characteristics of the shock signal, which is beneficial to the subsequent delay estimation. It can be seen from Figure 18 that the signal after noise reduction by the singular value median method almost completely retains the characteristics of the effective signal, and there is no distortion phenomenon. Figures 17 and 18 both verify the reliability and correctness of singular value decomposition noise reduction processing of the shock signal.

5.3. Algorithm Verification and Analysis

In order to further verify the important role of the research in this paper in improving the accuracy of delay estimation, the acoustic signal in the previous section is analyzed, and the delay estimation performance of the three methods is compared: direct generalized cross-correlation (*GCC-PHAT*) weighting, generalized cross-correlation (*SVD-GCC-PHAT*) weighting after *SVD* processing, and improved generalized cross-correlation (*SVD-GCC-PHAT-ργ*) weighting after *SVD* processing. For the improved generalized cross-correlation weighting method after SVD processing (*SVD-GCC-PHAT-ργ*), the value range of ρ is 0 to 1, and its value is related to the signal-to-noise ratio of the measured environment. In this paper, the selection of a value of ρ is as follows: set a certain step, traverse ρ from 0 to 1, and select the value of ρ corresponding to the smallest delay estimation error as the final value.

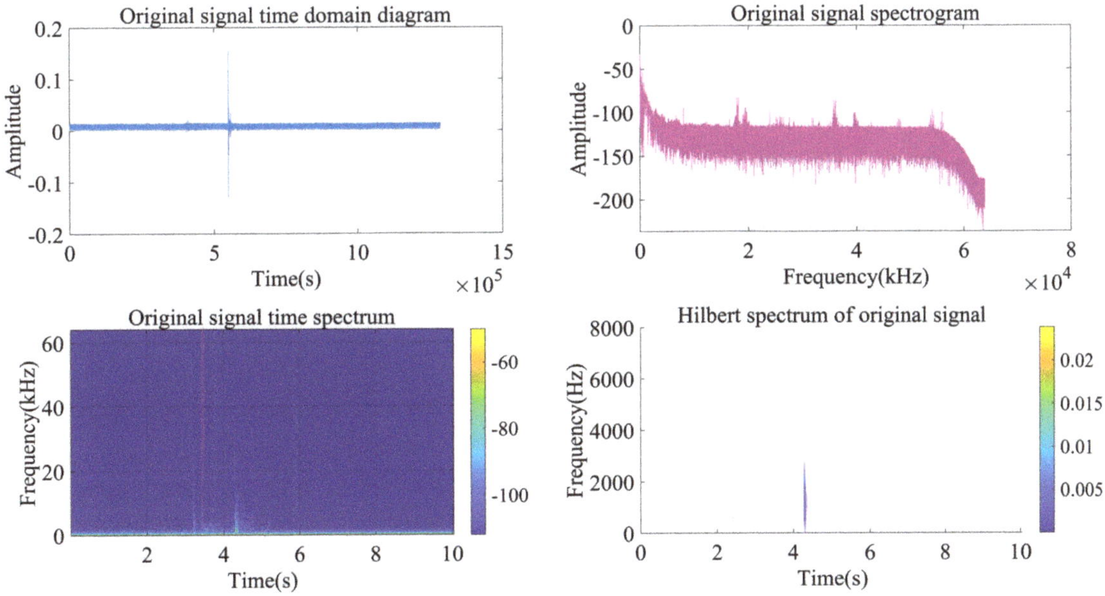

Figure 14. First-channel acquisition signal processing and analysis.

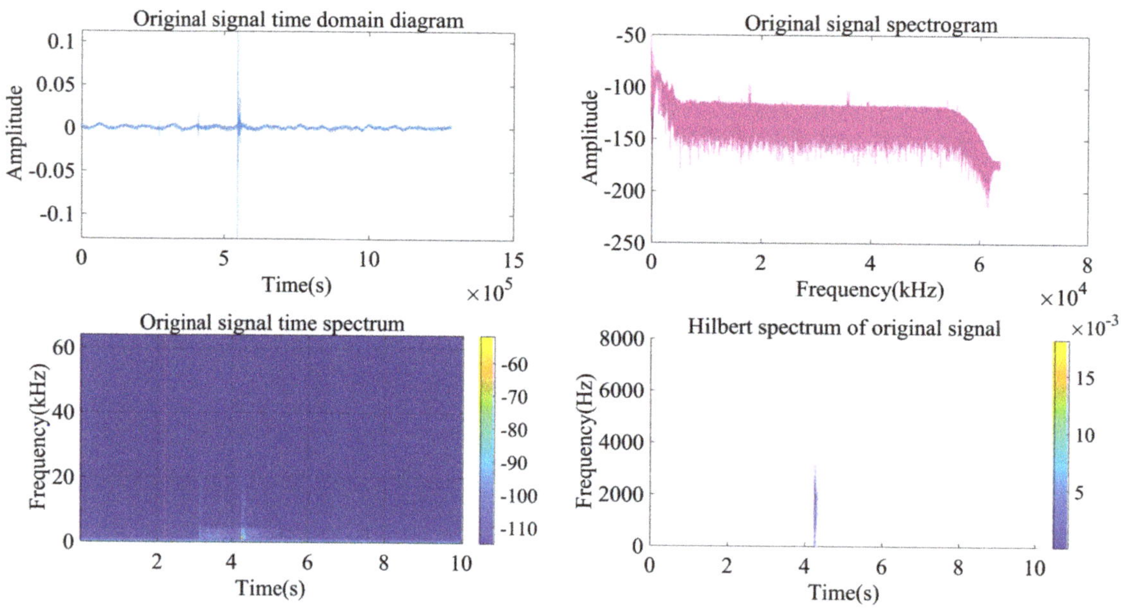

Figure 15. Second-channel acquisition signal processing and analysis.

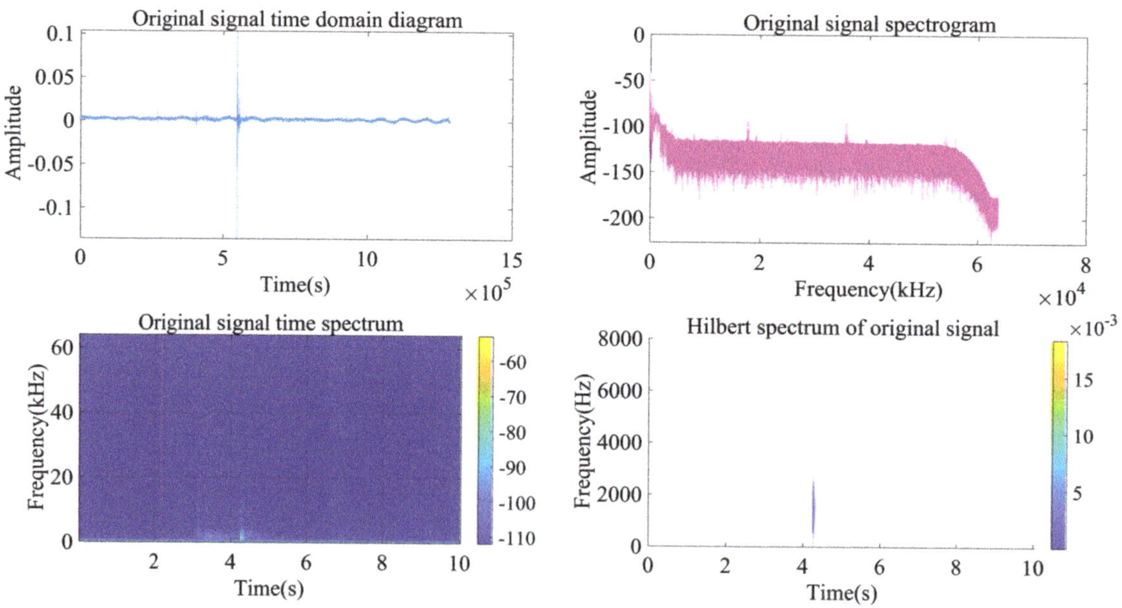

Figure 16. Third-channel acquisition signal processing and analysis.

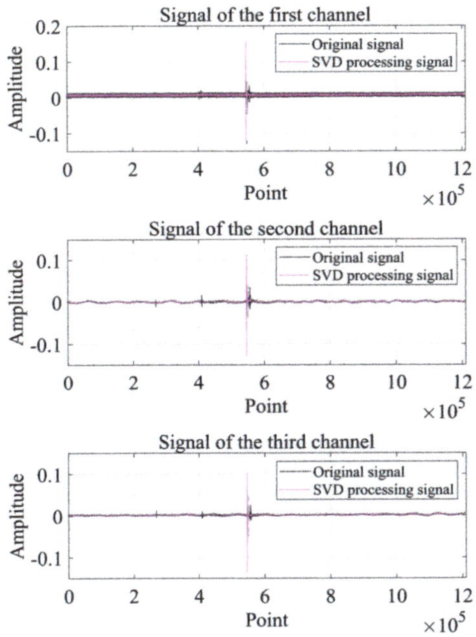

Figure 17. Comparison of the original value of the three-channel signal and the noise reduction value.

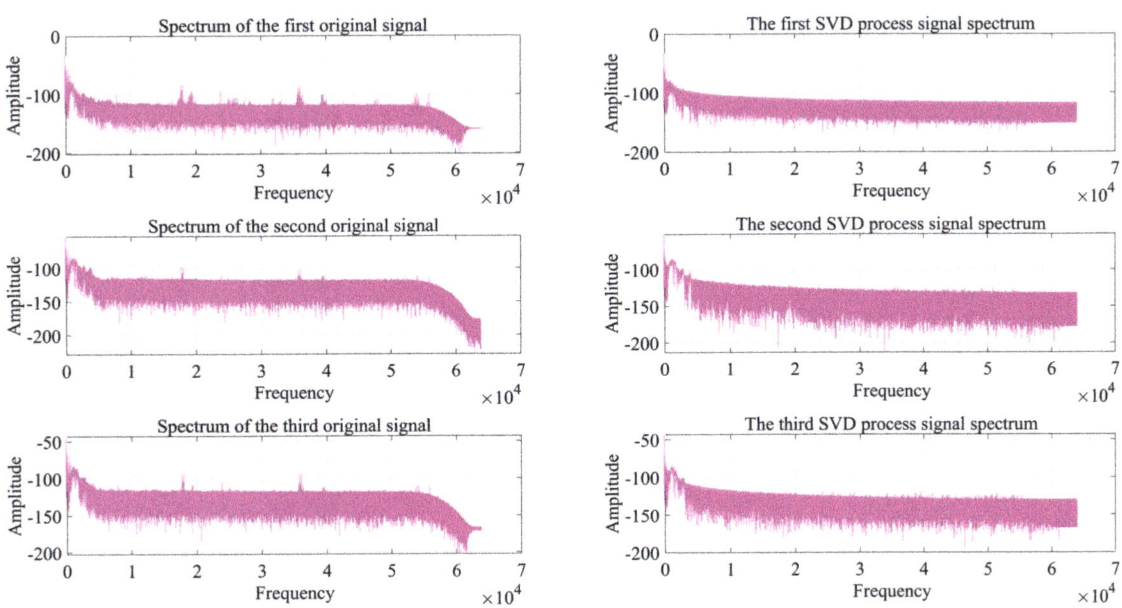

Figure 18. Comparison of the original spectrum of the three-channel signal and the spectrum after noise reduction.

The delay estimation results are shown in Table 2. The performance evaluation index is the number of delay points.

Table 2. Comparison of results of different time delay estimation methods.

Method	τ_{12} $(Points)$	τ_{13} $(Points)$	τ_{23} $(Points)$
Theoretical delay value	−9.27	−37.03	−27.76
GCC-PHAT weighting	−3	−21	−22
SVD-GCC-PHAT weighting	−4	−24	−22
SVD-GCC-PHAT-$\rho\gamma$ weighting	−6	−31	−26

It can be seen from Figure 19 that as the signal undergoes singular value noise reduction processing and weighting function improvement, the peak sharpness of the correlation function increases significantly, the resolution becomes more and more accurate, and the accuracy of delay estimation is greatly improved.

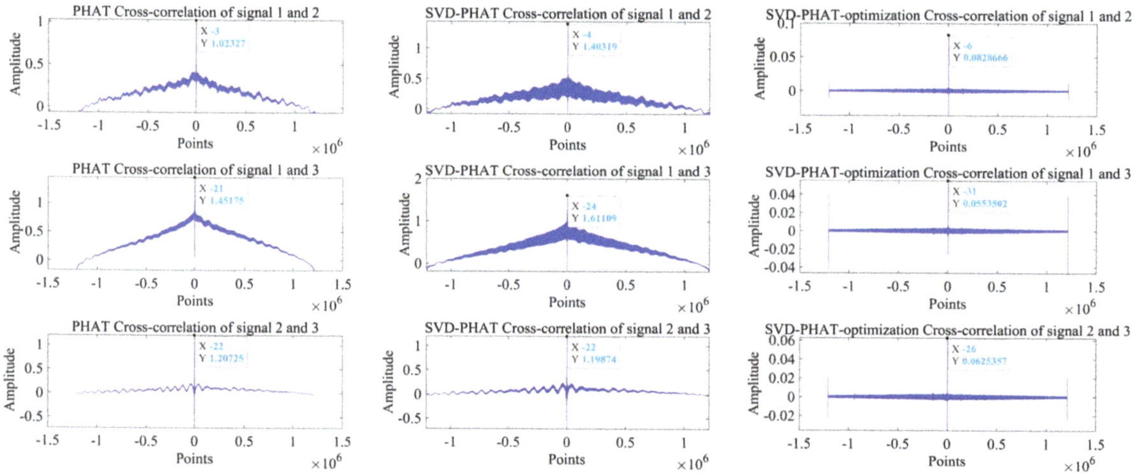

Figure 19. Delay estimation performance of *GCC-PHAT* weighting, *SVD-GCC-PHAT* weighting and *SVD-GCC-PHAT-*$\rho\gamma$ weighting.

Compared with the direct PHAT weighted delay estimation method (*GCC-PHAT*), the delay estimation method based on singular value decomposition noise reduction and improved PHAT weighting function (*SVD-GCC-PHAT-*$\rho\gamma$) improves the estimation accuracy of τ_{12}, τ_{13} and τ_{23}. They were increased by 47.85%, 62.38%, and 69.44%, respectively. Compared with the time delay estimation method based on singular value decomposition noise reduction and PHAT weighting (*SVD-GCC-PHAT*), τ_{12}, τ_{13} and τ_{23} estimation accuracy increased by 37.95%, 53.72%, and 69.44%, respectively. The improvement of time delay estimation accuracy fully demonstrates the significance of singular value noise reduction and improved *PHAT* weighting function.

6. Conclusions

In this paper, an optimization algorithm for delay estimation based on singular value decomposition and *GCC-PHAT-*$\rho\gamma$ weighting is proposed. The proposed delay estimation method improves anti-noise performance and sharpens correlation peaks. The time delay estimation performance and anti-noise performance of the three methods are verified through the analysis and processing of the simulated signal. By analyzing and processing the measured outdoor data, and comparing the three methods based on *GCC-PHAT* weighting, *SVD-GCC-PHAT* weighting and *SVD-GCC-PHAT-*$\rho\gamma$ weighting, the accuracy of delay estimation of the new algorithm in this paper is verified. Compared with the time-delay estimation method based on basic correlation analysis, this method does not greatly increase the calculation amount, but the accuracy of time-delay estimation is improved, the research in this paper can be applied to the actual scene of TDOA sound source localization, which is of great significance for target detection, localization and tracking based on time delay estimation, and has certain application prospects.

Author Contributions: S.W.: conceptualization, methodology, software, writing—original draft preparation. Z.L.: software, methodology, data curation. P.W.: conceptualization, methodology, writing—reviewing and editing. H.C.: conceptualization, methodology, software. All authors have read and agreed to the published version of the manuscript.

Funding: This research was funded by the Naval University of Engineering, PLA Scientific Research Development Fund Self-establishment Program (Grant No. 425317S091).

Institutional Review Board Statement: Not applicable.

Informed Consent Statement: Not applicable.

Data Availability Statement: Not applicable.

Conflicts of Interest: The authors declare no conflict of interest.

References

1. Zhang, X.; Liu, J.; Gao, Q.; Ju, Z. Adaptive robust decoupling control of multi-arm space robots using time-delay estimation technique. *Nonlinear Dyn.* **2020**, *100*, 2449–2467. [CrossRef]
2. Kali, Y.; Ayala, M.; Rodas, J.; Saad, M.; Doval-Gandoy, J.; Gregor, R.; Benjelloun, K. Current control of a six-phase induction machine drive based on discrete-time sliding mode with time delay estimation. *Energies* **2019**, *12*, 170. [CrossRef]
3. Mustafa, G.I.Y.; Wang, H.P.; Tian, Y. Vibration control of an active vehicle suspension systems using optimized model-free fuzzy logic controller based on time delay estimation. *Adv. Eng. Softw.* **2019**, *127*, 141–149. [CrossRef]
4. Han, S.; Wang, H.; Tian, Y. Model-free based adaptive nonsingular fast terminal sliding mode control with time-delay estimation for a 12 DOF multi-functional lower limb exoskeleton. *Adv. Eng. Softw.* **2018**, *119*, 38–47. [CrossRef]
5. Zhang, X.; Wang, H.; Tian, Y.; Peyrodie, L.; Wang, X. Model-free based neural network control with time-delay estimation for lower extremity exoskeleton. *Neurocomputing* **2018**, *272*, 178–188. [CrossRef]
6. Ahmed, S.; Wang, H.; Tian, Y. Adaptive high-order terminal sliding mode control based on time delay estimation for the robotic manipulators with backlash hysteresis. *IEEE Trans. Syst. Man Cybern. Syst.* **2019**, *51*, 1128–1137. [CrossRef]
7. Han, S.; Wang, H.; Tian, Y.; Christov, N. Time-delay estimation based computed torque control with robust adaptive RBF neural network compensator for a rehabilitation exoskeleton. *ISA Trans.* **2020**, *97*, 171–181. [CrossRef]
8. Cobos, M.; Antonacci, F.; Comanducci, L.; Sarti, A. Frequency-sliding generalized cross-correlation: A sub-band time delay estimation approach. *IEEE/ACM Trans. Audio Speech Lang. Process.* **2020**, *28*, 1270–1281. [CrossRef]
9. Wang, Y.; Li, P.; Zhang, J.; Liu, X.; Bai, Q.; Wang, D.; Zhang, M.; Jin, B. Distributed optical fiber vibration sensor using generalized cross-correlation algorithm. *Measurement* **2019**, *144*, 58–66. [CrossRef]
10. Catur, H.A.H.B.B.; Saputra, H.M. Azimuth estimation based on generalized cross correlation phase transform (GCC-PHAT) using Equilateral triangle microphone array. In Proceedings of the 2019 International Conference on Radar, Antenna, Microwave, Electronics, and Telecommunications (ICRAMET), Tangerang, Indonesia, 23–24 October 2019; pp. 89–93.
11. Lim, J.S.; Cheong, M.J.; Kim, S. Improved generalized cross correlation-phase transform based time delay estimation by frequency domain autocorrelation. *J. Acoust. Soc. Korea* **2018**, *37*, 271–275.
12. Padois, T.; Doutres, O.; Sgard, F. On the use of modified phase transform weighting functions for acoustic imaging with the generalized cross correlation. *J. Acoust. Soc. Am.* **2019**, *145*, 1546–1555. [CrossRef] [PubMed]
13. Glentis, G.O.; Angelopoulos, K. Using Generalized Cross-Correlation estimators for leak signal velocity estimation and spectral region of operation selection. In Proceedings of the 2022 IEEE International Instrumentation and Measurement Technology Conference (I2MTC), Ottawa, ON, Canada, 16–19 May 2022; pp. 1–6.
14. Lim, J.S.; Kim, S. Study on the pre-processors to improve the generalized-cross-correlation based time delay estimation under the narrow band single tone signal environments. *J. Acoust. Soc. Korea* **2020**, *39*, 207–215.
15. Ye, D.; Lu, J.Y.; Zhu, X.J.; Lin, H. Generalized cross corre-lation time delay estimation based on improved waveletthreshold function. In Proceedings of the 2016 Sixth International Confer-ence on Instrumentation & Measurement, Computer, Communication and Control, Harbin, China, 21–23 July 2016; pp. 629–633.
16. Zhang, X.; Lei, X.; Liu, Q. Wheel/Rail Force Signal Denoising Based on Wavelet Packet and Improved EMD. *Noise Vib. Control* **2016**, *36*, 104–107.
17. Zhou, Z.; Sun, S.; Li, Y.; Chen, P.; Shen, J. Second correlation time delay estimation based on empirical mode de-composition reconstruction. *Telecommun. Eng.* **2016**, *56*, 562–567.
18. Wei, W.; Mao, Y. Cross-correlation delay estimation optimization algorithm based on singular value decomposition. *Electron. Meas. Technol.* **2020**, *43*, 52–56.

19. Zhang, Y.; Yan, T.; Yang, Z. Generalized Cross Correlation Time Delay Estimation Based on Singular Value Decomposition. *J. Lanzhou Jiaotong Univ.* **2017**, *36*, 47–51.
20. Zhang, C.; Liu, L.; Wei, J.; Tang, Q.; Wang, N.; Ji, H. Time delay estimation simulation of robot fish obstacle avoidance based on PHAT-β generalized cross-correlation in the transformer. *Sci. Technol. Eng.* **2021**, *21*, 6354–6360.
21. Fokin, G.; Kireev, A.; Al-odhari, A.H.A. TDOA positioning accuracy performance evaluation for arc sensor configuration. In Proceedings of the 2018 Systems of Signals Generating and Processing in the Field of on Board Communications, Moscow, Russia, 14–15 March 2018; pp. 1–5.
22. Xiong, W.; Schindelhauer, C.; So, H.C.; Bordoy, J.; Gabbrielli, A.; Liang, J. TDOA-based localization with NLOS mitigation via robust model transformation and neurodynamic optimization. *Signal Process.* **2021**, *178*, 107774. [CrossRef]
23. Wu, R.; Zhang, Y.; Huang, Y.; Xiong, J.; Deng, Z. A novel long-time accumulation method for double-satellite TDOA/FDOA interference localization. *Radio Sci.* **2018**, *53*, 129–142. [CrossRef]
24. Schanze, T. Compression and noise reduction of biomedical signals by singular value decomposition. *IFAC-PapersOnLine* **2018**, *51*, 361–366. [CrossRef]
25. Govindarajan, S.; Subbaiah, J.; Cavallini, A.; Krithivasan, K.; Jayakumar, J. Partial discharge random noise removal using Hankel matrix-based fast singular value decomposition. *IEEE Trans. Instrum. Meas.* **2019**, *69*, 4093–4102. [CrossRef]
26. Li, H.; Liu, T.; Wu, X.; Chen, Q. A bearing fault diagnosis method based on enhanced singular value decomposition. *IEEE Trans. Ind. Inform.* **2020**, *17*, 3220–3230. [CrossRef]
27. Liang, X.; Li, Y.; Zhang, C. Noise suppression for microseismic data by non-subsampled shearlet transform based on singular value decomposition. *Geophys. Prospect.* **2018**, *66*, 894–903. [CrossRef]
28. Liu, H.; Dong, H.; Ge, J.; Liu, Z.; Yuan, Z.; Zhu, J.; Zhang, H. A fusion of principal component analysis and singular value decomposition based multivariate denoising algorithm for free induction decay transversal data. *Rev. Sci. Instrum.* **2019**, *90*, 035116. [CrossRef] [PubMed]
29. Li, H.; Liu, T.; Wu, X.; Chen, Q. Research on bearing fault feature extraction based on singular value decomposition and optimized frequency band entropy. *Mech. Syst. Signal Process.* **2019**, *118*, 477–502. [CrossRef]
30. Xu, X.; Luo, M.; Tan, Z.; Pei, R. Echo signal extraction method of laser radar based on improved singular value decomposition and wavelet threshold denoising. *Infrared Phys. Technol.* **2018**, *92*, 327–335. [CrossRef]
31. Zhong, J.; Bi, X.; Shu, Q.; Chen, M.; Zhou, D.; Zhang, D. Partial discharge signal denoising based on singular value decomposition and empirical wavelet transform. *IEEE Trans. Instrum. Meas.* **2020**, *69*, 8866–8873. [CrossRef]
32. Devi, H.S.; Singh, K.M. Red-cyan anaglyph image watermarking using DWT, Hadamard transform and singular value decomposition for copyright protection. *J. Inf. Secur. Appl.* **2020**, *50*, 102424. [CrossRef]
33. Villadangos, J.M.; Ureña, J.; García-Domínguez, J.J.; Jiménez-Martín, A.; Hernández, Á.; Pérez-Rubio, M.C. Dynamic adjustment of weighted gcc-phat for position estimation in an ultrasonic local positioning system. *Sensors* **2021**, *21*, 7051. [CrossRef]
34. Padois, T.; Doutres, O.; Nelisse, H.; Sgard, F. Acoustic imaging using different weighting functions with the generalized cross correlation based on the generalized mean. In Proceedings of the 26th International Congress on Sound and Vibration, Montreal, QC, Canada, 7–11 July 2019.
35. Yang, X.; Bao, C.; Cui, Z. Weighting function modification used for phase transform-based time delay estimation. *China Commun.* **2022**. [CrossRef]
36. Tan, C.S.; Mohd-Mokhtar, R.; Arshad, M.R. Improved Generalized Cross Correlation Phase Transform Algorithm for Time Difference of Arrival Estimation. In *Proceedings of the 10th National Technical Seminar on Underwater System Technology 2018*; Springer: Singapore, 2019; pp. 315–322.

sensors

MDPI

Article

A Time-of-Flight Estimation Method for Acoustic Ranging and Thermometry Based on Digital Lock-In Filtering

Qi Liu [1], Bin Zhou [1,*], Jianyong Zhang [2], Ruixue Cheng [2], Xuhao Zhao [1], Rong Zhao [1], Minglu Dai [1], Bubin Wang [1] and Yihong Wang [1]

[1] School of Energy and Environment, Southeast University, Nanjing 210096, China; liuqi667@seu.edu.cn (Q.L.); seuzxh@seu.edu.cn (X.Z.); zhaorong@seu.edu.cn (R.Z.); danilo@seu.edu.cn (M.D.); wbb@seu.edu.cn (B.W.); wyh@seu.edu.cn (Y.W.)

[2] School of Computing, Engineering and Digital Technologies, Teesside University, Middlesbrough TS1 3BX, UK; j.zhang@tees.ac.uk (J.Z.); r.cheng@tees.ac.uk (R.C.)

* Correspondence: zhoubinde@seu.edu.cn

Abstract: Accurate ranging and real-time temperature monitoring are essential for metrology and safety in electrical conduit applications. This paper proposes an acoustic time-of-flight (TOF) estimation method based on the digital lock-in filtering (DLF) technique for conduit ranging and thermometry. The method establishes the relationship between the frequency and the time domain by applying a linear frequency modulated Chirp signal as the sound source and using the DLF technique to extract the first harmonic of the characteristic frequencies of the transmitted and received signals. Acoustic TOF estimation in the conduit is then achieved by calculating the mathematical expectation of the time difference between each characteristic frequency in the time-frequency relationship of the two signals. The experimental results with enhanced noise interference on different conduit lengths and various temperature conditions, proved that the proposed DLF method can establish a robust linear time-frequency relationship according to the characteristics of the Chirp signal, and the measurement accuracy of TOF has also been confirmed. Compared to the conventional method, the DLF method provides the lowest absolute error and standard deviation for both distance and temperature measurements with an enhanced robustness.

Keywords: acoustic ranging; acoustic thermometry; digital lock-in filtering; electrical conduit; time-of-flight estimation

Citation: Liu, Q.; Zhou, B.; Zhang, J.; Cheng, R.; Zhao, X.; Zhao, R.; Dai, M.; Wang, B.; Wang, Y. A Time-of-Flight Estimation Method for Acoustic Ranging and Thermometry Based on Digital Lock-In Filtering. *Sensors* **2022**, *22*, 5519. https://doi.org/10.3390/s22155519

Academic Editors: Yuxing Li and Luca Fredianelli

Received: 27 June 2022
Accepted: 21 July 2022
Published: 24 July 2022

1. Introduction

In conduit construction and the maintenance process in the power industry, the indirect calculation of the conduit length is required. Generally, the conduits buried underground or in complex systems are long and mostly bent. Accurate length measurement and real-time temperature monitoring are challenging and indispensable technical safeguards in conduit laying and maintenance processes.

Traditional contact measurement methods such as scale and thermocouple cannot be applied to lengthy underground conduits. Ranging and thermometry methods based on laser [1,2] and imaging [3,4] technologies are constrained because light can only travel in straight line, so it is unsuitable for lengthy, curved underground pipeline applications. Ultrasonic and microwave ranging [5,6] and thermometry [7,8] methods require considerable power for long-distance measurements due to their shorter wavelengths and faster attenuation, while increased power means an increased size of the sensor, and hence they are inconvenient for such application, particularly for small diameter conduits. The acoustic method has its advantages over the mentioned techniques above.

For convenience, the audio range with an operating frequency falling between 20 Hz and 20 kHz is referred to as acoustic hereinafter in distinction from ultrasound. Different from the laser and camera imaging, when a sound wave propagates in a curved conduit encountering local obstructions such as sludge, pipe bulge, and cable obstacles, the propagation can continue through diffraction. The advantages of the acoustic method also lay in its low cost and high anti-interference natures.

According to the measurement mode, acoustic-based ranging and thermometry are mainly divided into two methods, namely the resonance method [9–11] and the pulsed time difference method [12,13]. In the pulsed time difference method [14,15] the length and temperature are measured by estimating the acoustic time-of-flight (TOF) in the conduit, which is the focus of this paper.

In the acoustic TOF estimation, the cross-correlation algorithm is widely used, in which the time delay is computed by identifying the similarity of transmitted and received signals. Based on the direct cross-correlation (DCC) algorithm [16], there have been various transformations and developments. The generalized cross-correlation (GCC), weighted cross-correlation [17], and phase-corrected cross-correlation [18] are such examples to mention. However, these algorithms have a common requirement of wide signal bandwidth, because it is inversely related to the measurement error for TOF. Another well-known method is the least mean-square time delay estimation (LMSTDE) [19,20], in which an adaptive FIR filter is used to model the time difference and interpolate the filter weights to obtain the time delay. Many adjustments and deformations [21,22] have been made to LMSTDE in order to reduce the effect of noisy input caused by the limited filter length, however an accurate TOF could still not be achievable under a low signal-to-noise ratio (SNR) if a small number of filter taps are used. The higher-order statistical time-delay estimation algorithm [23,24] has an advantage to suppress the smoothly correlated Gaussian white noise and extract the signal amplitude and phase information by extending the higher-order spectrum of the multidimensional Fourier transform [25]. However, to achieve a reliable accuracy, strict requirements on the sampling length and resolution of the A/D converter must be satisfied. For other techniques such as the over-zero detection method [26,27], the amplitude squared coherence function method [28], the phase spectrum estimation method [29,30], and the adaptive TOF estimation method [31], although they have been found various applications, none of them are suitable for conduit length measurement and temperature monitoring in relatively harsh field conditions due to inferior and unreliable measurement results at low SNR, limitations on sampling period and accuracy, and high computational load.

Conduits for field applications are generally buried underground or in a complex system close to a production or auxiliary equipment, transportation line or building site. Hence a variety of noise and vibration interferences in a conduit can be expected. Moreover, the bending of a conduit, the obstruction of internal cables, and the sand and dirt mixture left inside during construction pose severe challenges to the acoustic measurement method. It is critical for the acoustic method [32] to be able to suppress noise and overcome physical obstacle interference while ensuring the accuracy and precision of acoustic TOF estimation.

This paper proposes a novel acoustic TOF estimation method based on a digital lock-in filter (DLF) to improve the accuracy of electrical conduit length measurement and temperature monitoring. A linear frequency modulated Chirp signal as the acoustic source is used and the relationship between the first harmonic of the multi-frequency component and the Chirp signal moment is established through digital phase-locking and low-pass filtering of the signal. The acoustic TOF estimation in the conduit is achieved by calculating the mathematical expectation of the time difference between the received and transmitted signals at each characteristic frequency. The accuracy of the DLF method for conduit length and temperature monitoring is verified experimentally by comparison with the classical GCC and GCC-SCOT algorithms. The robustness of the method is tested at various levels of noise intensity with different time spans. The results confirm that both the accuracy and robustness are improved and satisfactory.

2. Principle of Acoustic Ranging and Thermometry

In an electrical conduit system shown in Figure 1, although there are some stones, sand, dents, and joints in the conduit, the sound wave will be transmitted from the loudspeaker (the transmitter) at one end to the microphone (the receiver) fixed at the other end of the conduit through the gas medium due to continuity of the gas in the conduit. The length and temperature of the conduit can be found indirectly by estimating the TOF between the transmitted and received signals at the source and at the receiver by calculation of the time delay algorithm.

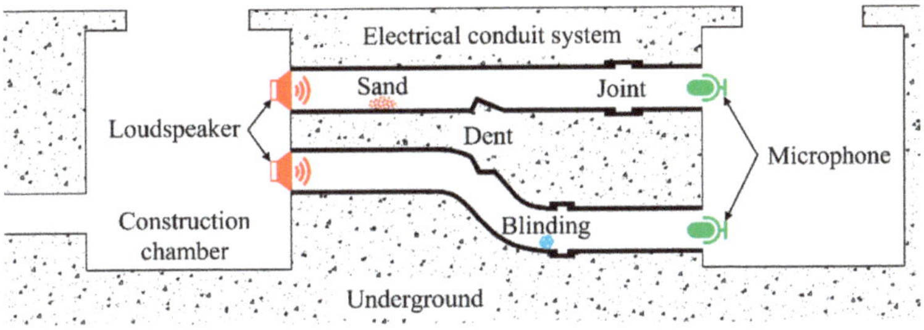

Figure 1. Schematic of the acoustic ranging and thermometry method for electric conduits.

Since there is little variation in the enclosed conduit's electromagnetic, gas state, and composition, the acoustic wave can be assumed to be a linear propagation process. As expressed in Equation (1), in the pre-construction stage, the conduit length L is calculated at ambient temperature through the product of the sound speed c and the TOF τ. Once the construction of the conduit is completed, online measurement of conduit temperature [33,34] can be achieved based on the calibrated conduit length.

$$L = \tau c = \tau \sqrt{\frac{\gamma R}{m} T} \tag{1}$$

where γ is the isentropic exponent of the medium, R is the universal gas constant, m is the molar mass, and T is the gas temperature inside the conduit.

It can be found from Equation (1) that the accurate estimation of the acoustic TOF is the priority either for both distance calculating and temperature monitoring. Since the proposed DLF and other time delay estimation methods have already been elaborated elsewhere with the acoustic source parameters, only the sound source signal model will be introduced in Section 2.1.

2.1. Acoustic Source Signal Model

The signal model with linear variation in frequency [33,34] has been proven to differentiate signal better from noise hence leading to improved SNR. In this research, the sound source signal is modulated with the widely used linear frequency Chirp. Assuming the presence of additive noise in both the transmitted and received signals of the loudspeaker:

$$\begin{cases} x_1(t) = s(t) + \mu_1(t) \\ x_2(t) = \xi s(t - \tau) + \mu_2(t) \end{cases} \tag{2}$$

where $s(t)$ is the acoustic source signal. $\mu_1(t)$ and $\mu_2(t)$ are the random noises contained in the signals. ξ is the attenuation coefficient of the acoustic signal. τ denotes the TOF between the two signals. $s(t)$, $\mu_1(t)$, and $\mu_2(t)$ are assumed to be uncorrelated.

The transmitted signal $x_1(t)$ is:

$$x_1(t) = A\cos(2\pi f_{\text{Linear}} t + \Delta\varphi) + \mu_1(t) \quad t \in [0, \tau_s] \tag{3}$$

where A is the signal amplitude, τ_s is the pulse width, $\Delta\varphi$ is the signal phase. f_{Linear} denotes the sweeping frequency of the signal, which is expressed as:

$$f_{\text{Linear}} = f_0 + \frac{B_w}{\tau_s} t \quad t \in [0, \tau_s] \tag{4}$$

where f_0 is the starting frequency at $t = 0$, B_w is the bandwidth in hertz.

In order to avoid overlap of the source signal with the low-frequency noise in the field and in line with the best frequency response of the loudspeaker, in this research, the starting frequency and bandwidth of the source signal are set to 4 kHz. The frequency change with time is shown in Figure 2, where the amplitude A of the sound source signal is 1 V, and the waveform is a continuously varying cosine from 0 s to 0.2 s. The instantaneous frequency of the signal increases linearly with the time. The frequency varies linearly from 4 kHz to 8 kHz in 0.2 s.

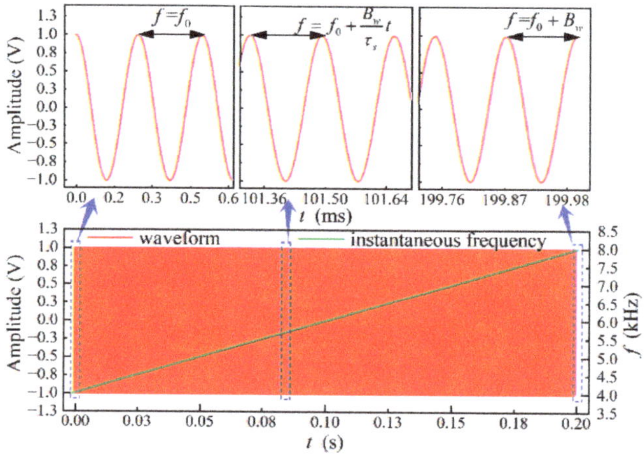

Figure 2. The source signal and its instantaneous frequency.

2.2. *The DLF-Based Acoustic Time-of-Flight Estimation Method*

The digital lock-in filtering technique is a versatile signal processing method with a wide range of applications in spectroscopy [2], electricity [35], and magnetism [36] to extract signals at specific frequencies from harsh interference environments. Based on the DLF technique, this paper proposes a novel method that can be used to accurately estimate the TOF.

Once the acoustic signal propagates through the conduit, its waveform may be distorted and attenuated. The intrinsic mode decomposition [37,38] of the acoustic signal reveals that its frequency characteristics generally remain unchanged. By analyzing the instantaneous frequency of the received signal, a new time-frequency relationship can be derived, from which, it can be seen that the offset between the two time-frequency relationships is the acoustic TOF. Traditional frequency domain analysis methods are constrained by the Fourier transform characteristics so an accurate relationship between the time and frequency domains cannot be established. In this paper, digital lock-in filtering techniques are adopted to extract the instantaneous frequencies of the transmitted and received signals, respectively, which can better characterize the local features of the two acoustic signals at different moments.

The DLF method can be used to demodulate the amplitude and phase of each component of the same frequency in the received and transmitted signals by means of the phase-sensitive detection principle. The proposed TOF estimation method uses digital phase-locking and low-pass filtering to extract the first harmonics of the different characteristic frequencies of the acoustic signal. It demodulates the moment of occurrence at that frequency by detecting the peak of the first harmonic, thus establishing the time-frequency relationship of multiple characteristic frequencies. The procedure for calculating the time difference at each characteristic frequency is shown in Figure 3.

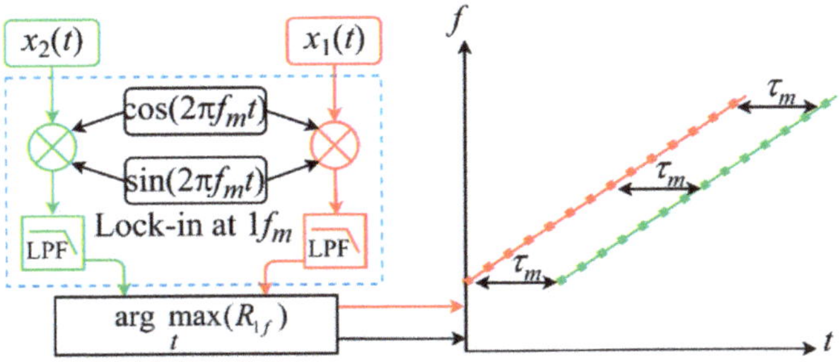

Figure 3. Flow chart of the DLF method for establishing time-frequency relationships.

In the DLF method, the reference signal $Z_1(t)$ and its orthogonal signal $Z_2(t)$ with 90° phase shift are assumed to be:

$$\begin{cases} Z_1(t) = B\sin(2\pi f_m t) \\ Z_2(t) = B\cos(2\pi f_m t) \end{cases} \tag{5}$$

where B is the amplitude of the reference and orthogonal signals with characteristic frequency f_m ranging from f_0 to $B_w + f_0$. Since the first harmonic detection range is determined from the transmitted signal, the expressions for the reference and orthogonal signals are known.

In the DLF method, first the acoustic signal with the reference phase is multiplied by the orthogonal signals to realize the function of phase-sensitive detection. For illustrative purposes, the following is an example of the transmitted signal $x_1(t)$, which goes through the multiplier as follows:

$$X_{1f}(t) = x_1(t) * Z_2(t) = \frac{AB}{2}[\cos(2\pi f_{\text{Linear}}t + 2\pi f_m t + \Delta\varphi) + \cos(2\pi f_{\text{Linear}}t - 2\pi f_m t + \Delta\varphi)] + B\cos(2\pi f_m t)\mu_1(t) \tag{6}$$

$$Y_{1f}(t) = x_1(t) * Z_1(t) = \frac{AB}{2}[\sin(2\pi f_{\text{Linear}}t + 2\pi f_m t + \Delta\varphi) - \sin(2\pi f_{\text{Linear}}t - 2\pi f_m t + \Delta\varphi)] + B\sin(2\pi f_m t)\mu_1(t) \tag{7}$$

When the characteristic frequencies of the reference and orthogonal signals are the same as the signal to be detected, i.e., $f_{\text{Linear}} = f_m$, Equations (6) and (7) can be simplified as:

$$X_{1f}(t) = \frac{AB}{2}[\cos(2\pi f_m t + \Delta\varphi) + \cos(\Delta\varphi)] + B\cos(2\pi f_m t)\mu_1(t) \tag{8}$$

$$Y_{1f}(t) = \frac{AB}{2}[\sin(2\pi f_m t + \Delta\varphi) - \sin(\Delta\varphi)] + B\sin(2\pi f_m t)\mu_1(t) \tag{9}$$

After being low-pass filtered, the output of the two signals is obtained as:

$$X_{1f}(t) = \frac{AB}{2}\cos(\Delta\varphi) \tag{10}$$

$$Y_{1f}(t) = \frac{AB}{2}\sin(\Delta\varphi) \tag{11}$$

Therefore, the magnitude of the first harmonic R_{1f} can be calculated as:

$$R_{1f} = \sqrt{X^2{}_{1f} + Y^2{}_{1f}} \tag{12}$$

The moment t corresponding to the characteristic frequency f_m is obtained from the peak detection of the first harmonic:

$$t_{fm} = \underset{t}{\text{argmax}}(R_{1f}) \tag{13}$$

Therefore, the time delay for each frequency f_m is expressed as:

$$\tau_m = t_{\text{received\_}f_m} - t_{\text{transmitted\_}f_m} \tag{14}$$

where $t_{\text{received\_}fm}$ and $t_{\text{transmitted\_}fm}$ are the moment of the received and transmitted signals at the mth frequency f_m, respectively.

The m time-frequency relationships t_{fm} are obtained by detecting the time differences at all m number of characteristic frequencies. TOF can then be obtained by calculating the mathematical expectation of the time difference between the received and transmitted signals over the passband B_w.

$$\tau = E(\tau_m) = \frac{1}{B_w} \sum_{f_m = f_0}^{f_0 + B_w} (\tau_m) \tag{15}$$

The Chirp signal frequency varies linearly over the pulse width. The analysis of the dynamics of the received acoustic signals by means of nonlinear dynamics metric [39,40] shows that if the loudspeaker's response is inadequate, this can result in intermodulation distortion of the output acoustic signal. The signal conditioner also causes harmonic distortion of the signal after it has been received by the microphone, leading to a non-linear [41] relationship between frequency and time of the acoustic signal. Due to the known time-frequency characteristics of this pair of acoustic signals, the relationship is fitted linearly by means of a least-squared-based regression method in order to minimize the effects of non-linearity. The linear nature of the Chirp signal is combined to improve the fitting accuracy of the time-frequency relationship by minimizing the sum of squares of the errors. Assume that the linear equation of the time-frequency relationship is:

$$f_i = b + kt_i + \varepsilon \tag{16}$$

where b and k are the intercept and slope, respectively, and ε is the residual between the actual and the fitting frequency values. The residual sum of squared error $\Delta f(b,k)$ can be determined as follows:

$$\Delta f(b,k) = \sum_{i=1}^{n} \varepsilon^2 = \sum_{i=1}^{n} (f_i - b - kt_i)^2 \tag{17}$$

where f_i is the actual value. $b + kt_i$ is the targeting value. By making the partial derivatives of the residual sum $\Delta f(b,k)$ relative to b and k zeros, the optimized b and k can be identified to achieve the minimum $\Delta f(b,k)$.

In order to present the process of estimating TOF with the DLF method, an example of an acoustic signal sampled at a conduit length of 3 m is shown in Figure 4. The signal sampling rate is 200 kS/s, the sampling period is 0.22 s, and the starting frequency and bandwidth of the acoustic source signal are 4 kHz.

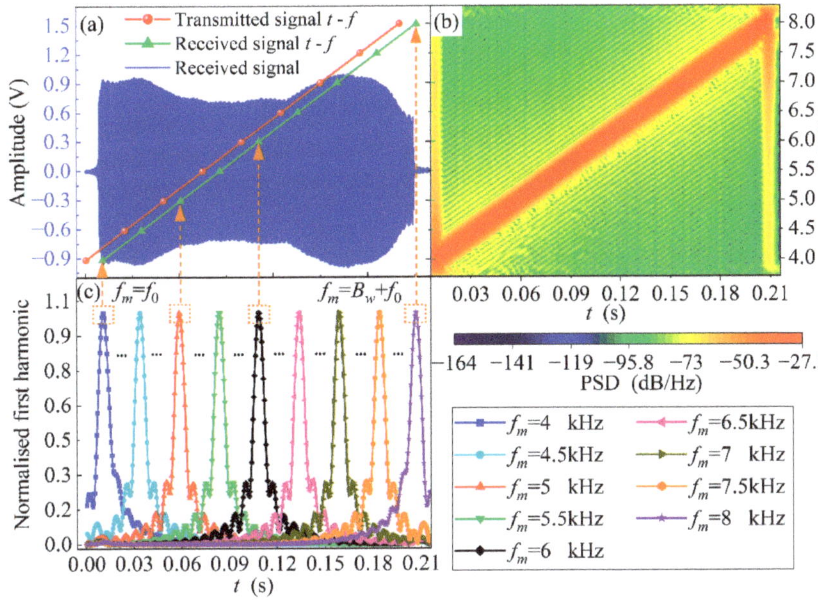

Figure 4. Example of TOF estimation process with DLF method. (**a**) The time-domain waveform of the received signal and time-frequency relationships of transmitted and received signals. (**b**) The acoustic spectrogram of the received signal. (**c**) Normalized first harmonic shown at 0.5 kHz intervals within the signal bandwidth from the starting frequency.

Figure 4a shows that the received acoustic signal is distorted in waveform amplitude compared to the source signal in Figure 2 after propagating 3 m inside the conduit. The effective pulse width is 0.2 s, which is consistent with the acoustic source. The time lag at the front of the waveform is the acoustic TOF. Corresponding to Figure 4a, the acoustic spectrum of the received signal is depicted in Figure 4b. The acoustic spectrogram is essentially a short-time Fourier transform, which does not provide for both time and frequency resolution of the analyzed signal, but the overall trend of the signal time-frequency correspondence can be clearly seen from this figure. Shown in Figure 4a, the frequency of the received signal varies linearly from 4 kHz to 8 kHz with time over the pulse width, which is the same as the frequency-time relationship of the source signal, except that there is a significant hysteresis in between. The multiple characteristic frequencies of the received signal are digitally phase-locked and filtered. Figure 4c shows the normalized first harmonics at 0.5 kHz intervals from the starting frequency. The peak of the first harmonic coincides with the moment at which that frequency is located. Hence, it can be concluded that a discrete sequence of time-frequency relationships can be derived by detecting the peaks of multiple first harmonics, and the linear interpolation can be assumed between discrete points. The time-frequency relationship of the received signal in Figure 4a is determined with the least-squares-based regression fitting method detailed with Equations (16) and (17).

Similarly, the time-frequency relationship of the transmitted signal is also shown in Figure 4a. Thus, each eigenfrequency corresponds to a time delay as described in Equation (14). The TOF of the acoustic wave can be estimated based on the mathematical expectation of the time delays corresponding to the multiple eigenfrequencies.

2.3. Cross-Correlation Time-of-Flight Estimation Method

The cross-correlation algorithm is widely used in the study of acoustic TOF estimation. The operation result reflects the similarity strength of the two signals, and the time delay of the two signals is detected when the similarity is most potent, where the peak position of the cross-correlation function occurs.

Compared to the traditional DCC algorithm, the generalized cross-correlation (GCC) is more flexible in its operation [17]. The cross-power spectrum is obtained by performing a Fourier transform on the two time-domain acoustic signals and, if necessary, pre-processing the cross-power spectrum in the frequency domain by weighting. The inverse Fourier transform is then applied to the cross-power spectrum. The process is as follows:

$$G_{x_1x_2}(\omega) = F[x_1(t)]F^*[x_2(t)] \tag{18}$$

$$R_{x_1x_2}(\tau) = F^{-1}[G_{x_1x_2}(\omega)] \tag{19}$$

where $G_{x1x2}(\omega)$ is the cross-power spectrum of the two signals; $F[\cdot]$ stands for Fourier transform; * denotes the complex conjugate; $R_{x1x2}(\tau)$ is the GCC function; $F^{-1}[\cdot]$ represents inverse Fourier transform function.

The GCC algorithm is whitened and pre-weighted in the frequency domain to enhance the suppression of noise interference. The weighting process of the Smoothed Coherent Transform (SCOT) [42] has a significant suppression effect on noise and is widely used for accurate time delay estimation. The SCOT weighting function is essentially an improvement on the ROTH weighting function [17], which is expressed as follows:

$$\psi_{x_1x_2}(\omega)_{\text{SCOT}} = [G_{x_1x_1}(\omega)G_{x_2x_2}(\omega)]^{-1/2} \tag{20}$$

where $G_{x1x1}(\omega)$ and $G_{x2x2}(\omega)$ are the self-power spectra of transmitted and received signals, respectively. Therefore, the GCC-SCOT function is defined as:

$$R^{\text{SCOT}}_{x_1x_2}(\tau) = F^{-1}[\psi_{x_1x_2}(\omega)_{\text{SCOT}}G_{x_1x_2}(\omega)] \tag{21}$$

Described with Equation (22), in GCC, the accurate estimation of the acoustic TOF is determined by detecting the peak position of the cross-correlation function, and so is in the GCC-SCOT.

$$\tau = \underset{\tau}{\text{argmax}}[R_{x_1x_2}(\tau)] \tag{22}$$

3. Results Analysis and Discussion

Following the sequence of electrical conduit construction and testing requirements, the calibration measurements of geometric distances were first carried out for different electrical conduits. A laser distance meter (LDM) was used to measure the length of the electrical conduit chosen for the experiments in its straightened state. The measurement was repeated several times to determine the reference standard. The same conduit was measured with the acoustic waveform when the conduit was bent to verify the applicability and feasibility of the proposed DLF method. The experimental study of continuous online monitoring of the conduit temperature was also carried out under disturbing conditions of different noise energy levels, and with mud and gravel inside at different temperatures, by doing so, to validate the method in practical application environments. As a comparison, the GCC, GCC-SCOT, and DLF methods are employed for calculation simultaneously.

3.1. Experiment Setup

The experiment measurement system shown in Figure 5 was set up to better imitate the field conditions of the conduit in the laboratory. A linear frequency modulated Chirp signal with a starting frequency and bandwidth of 4 kHz was used as the sound source, sent from the data acquisition card to the loudspeaker via a power amplifier with adjustable gain. At one end of the conduit, the acoustic sound and noise were emitted by Loudspeakers S1 and S2 respectively and transmitted through the conduit. The microphone at the other end was used to receive the signals. The loudspeakers and the microphone were installed utilizing fixed-size acoustic waveguides. The acoustic waveguide installed with S1 and S2 was an internal tee structure. The microphone was 1/2-inch in diameter, with a sensitivity of 50.4 mV/Pa, and a dynamic range of 20 dB to 136 dB, its frequency response range was from 20 Hz to 20 kHz.

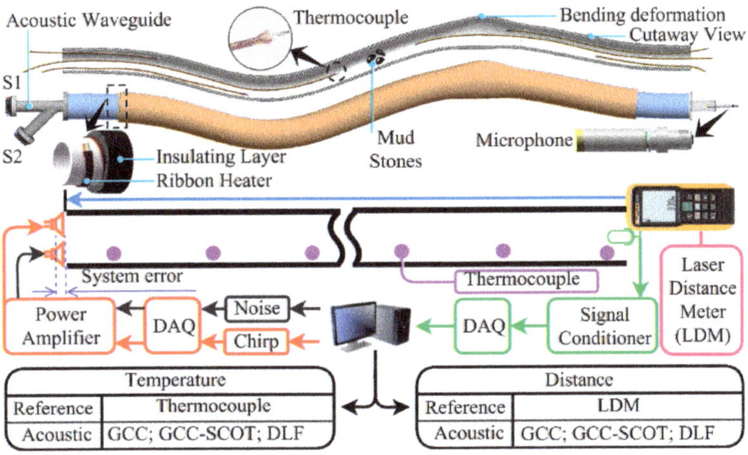

Figure 5. Schematic of the measurement experiment platform.

The inner diameter of the conduit was 45 mm. A ribbon heater was uniformly wound around the pipe, which was wrapped with insulation material to create different temperature experiment environments. Six type K thermocouples (GG-K-30-SLE made in Shanghai Yaogeng Automation Instrument Co., Shanghai, China) with a diameter of 0.255 mm were evenly spaced 250 mm apart inside the conduit. As can be seen in the cross-sectional view in Figure 5, the thermocouples were wrapped in a breathable glass fiber cloth with a wire diameter of 1 mm in order to avoid direct contact with inner pipe wall to prevent mismeasurement. The gas pressure inside the conduit was kept almost constant when the conduit was closed, and the average value obtained from several thermocouples was used as a reference or standard temperature. The model of the laser distance meter (LDM) was Fluke 404E (Made in Fluke Co., Beijing, China), with a range of 0.2 m to 40 m and measurement precision of ± 2.0 mm + 5×10^{-5} *L. It was used to calibrate the length of the conduit when it was in the straightened condition. The noise transmitted by S2 was to simulate the noise disturbance from construction, equipment, and traffic in field applications. Also included in the system were: a data acquisition card NI USB6356 (Made in National Instruments Co., Austin, TX, USA) with a sampling rate of 1.25 MS/s and a resolution of 16 bits; a signal conditioner containing phantom power unit and signal amplification circuit.

3.2. System Error Calibration

As shown in Figure 5, the distance between the sound output of the loudspeaker and the coil-wrapped sound diaphragm results in a systematic error in the acoustic ranging and thermometry. Since both the measurement system and the loudspeaker structure remain unchanged, this systematic error can be calibrated employing the least-squares method.

$$L_r = \varsigma L_e + \Delta L \tag{23}$$

where L_r is the reference length measured by the LDM, L_e is the estimated length calculated through the acoustic method, ΔL is the system error, and ς is a scale factor by reason that the acoustic signal distortion loss due to conduit bends and obstacles in the pipe. In the experiment, the correction factor and system error were 1.0024 and 8.16 mm, respectively, so the correct distance could be calculated.

3.3. Analysis of the Measurement Results

3.3.1. Acoustic Ranging and TOF Estimation in Different Conduit Length

Following the construction of the electrical conduits, the different conduit lengths were first measured using three different time-delay estimation algorithms to investigate the accuracy of the proposed DLF method for different TOF. The internal temperatures of the closed conduits with different lengths were kept the same and maintained during the measurements and the six thermocouples arranged at the ends of the conduits were used to indicate the actual temperature. The electrical conduits were cut into 8 different pieces with the lengths of 0.503 m, 1.021 m, 2.032 m, 3.022 m, 4.017 m, 5.023 m, 6.993 m, and 10.022 m, respectively, based on LDM measurements at their straight states. Once the experimental bench and sensors were all installed and arranged, the temperature in the conduit was monitored using thermocouples, and as the experiments for length measurement using acoustic methods were carried out the temperature became stable. The average temperature of the six thermocouples was 26.31 °C when the thermal equilibrium was reached, and the theoretical speed of sound in the conduit was 345.82 m/s.

The length for each of the mentioned conduit above was measured in the pre-construction phase, which was free from noise and vibration interference. The Chirp acoustic waves were transmitted from S1, and received by the microphone to estimate the acoustic TOF across each of eight different conduits over their full lengths. The number of measurements for each conduit was 200.

In the example of the propagated acoustic signal travelling a conduit of 7 m, it can be seen from Figure 6a that a slight waveform distortion occurred due to the conduit bending, gravel, obstruction of the thermocouple wire, and gas attenuation. A significant time delay of the received signal to the transmitted signal is clearly shown, this time is the TOF to be estimated. From the acoustic spectrum of the two signals in Figure 6b,c, it is evident that the attenuation of the waveforms did not affect the time-frequency relationship of the received signals. This time delay is also obviously reflected in the corresponding acoustic spectrum of the received signal in Figure 6c. This phenomenon provides an explanation of the proposed DLF method.

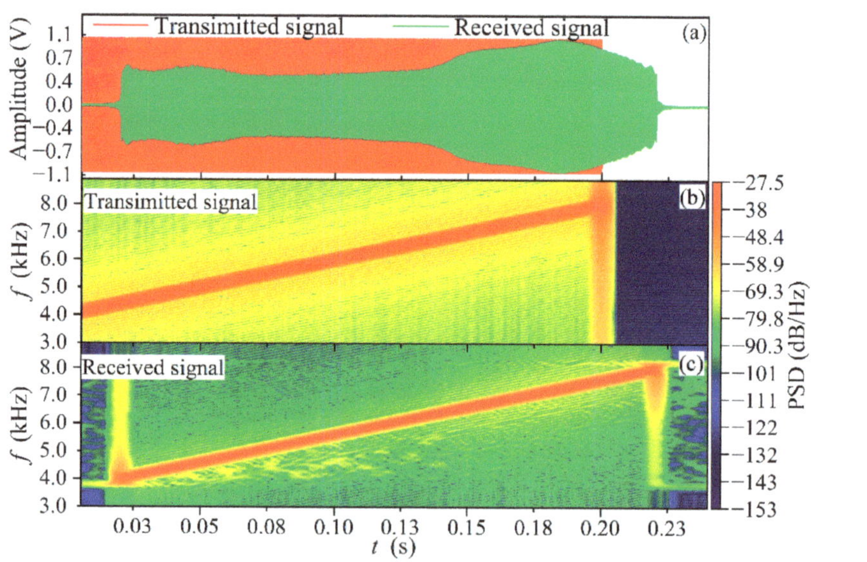

Figure 6. Signal waveform and acoustic spectrum at conduit length of 7 m. (**a**) Transmitted and received acoustic signal waveforms for a conduit length of 7 m. (**b**) Acoustic spectrum of the transmitted signal. (**c**) Acoustic spectrum of the received signal.

The DLF, GCC, and GCC-SCOT methods were used to process the acoustic signals obtained from the eight conduits with different lengths mentioned above. Figure 7 shows the time-frequency relationship for each conduit at the given length calculated using the DLF method. The time frequency graphs of TOF are parallel for different conduit lengths, indicating the distinct time taken for the acoustic wave travelling a given journey.

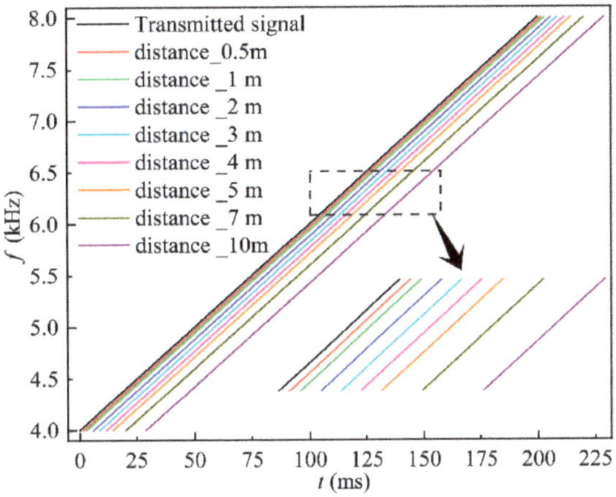

Figure 7. The time-frequency relationships of different lengths of conduits are extracted by the DLF method.

Figure 8 provides the details of the analysis, where the measured TOFs were converted to length for the three acoustic methods. Taking the lengths determined with LDM measurements as references, the relative error (RE) and standard deviation (SD) of the three acoustic methods were compared. The results are depicted in Figure 8. Apparently, the graphs in this figure confirm that both RE and SD for all three different methods increase with distance, and both errors with the GCC-SCOT are slightly smaller than with the GCC over the entire range. The DLF method has the best performance in this regard, the maximum RE under DLF is 6.26%, and 16.99% smaller than that derived with GCC and GCC-SCOT, respectively. The DLF method therefore offers the best measurement accuracy, and the lowest SD also demonstrates the best stability over multiple repeated measurements.

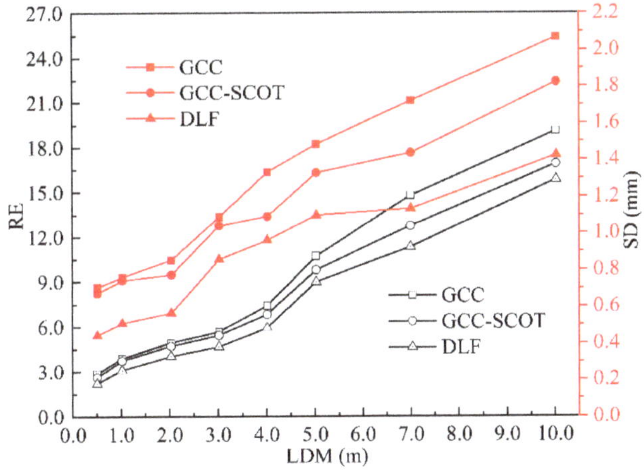

Figure 8. RE and SD results for acoustic methods at different conduit lengths.

3.3.2. Experiments on Temperature Monitoring with Noise Interference at Different Energy Levels

The production or auxiliary equipment, transport, and building construction close to an electrical conduit can generate various noises and vibrations inside, which can interfere with the temperature monitoring by acoustic methods. The vibration effect is not considered in this paper as underground vibrations are mainly low frequency, well below the bandwidth of the signal used for measurement. The previous studies have shown that the noise in underground power conduits is mainly Gaussian white noise [9]. The noise immunity of the DLF method was verified by adding Gaussian white noise to the conduit via S2 and the noise energy level was set by adjusting the power amplifier's gain. In this section, the thermocouples were evenly aligned in a conduit of 3 m length. The experiments were carried out without heating at seven interfering SNR levels, and the results are shown in Figure 9. The number of measurements at each SNR level was 200.

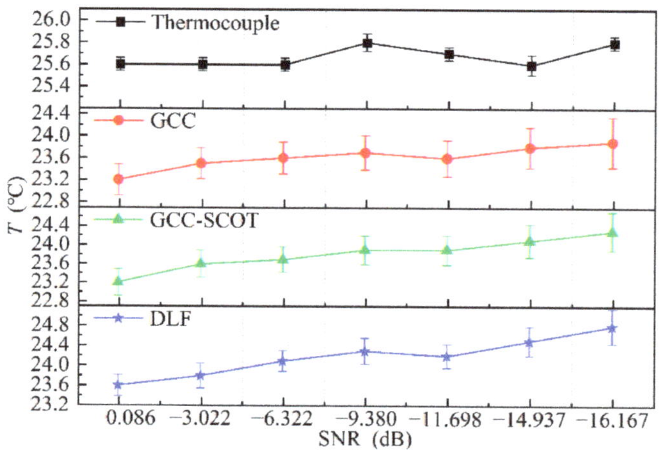

Figure 9. Temperature measurement results in different SNR conditions.

Due to the closeness of the measured results from the three acoustic methods and the tendency for the error bars to overlap, Figure 9 displays them separately for clarity. The graphs show that the results based on all of the three acoustic thermometry are lower than that given by the thermocouple. The temperature measured by the thermocouple increases by approximately 0.3 °C as the measurement time increases, and the trend is followed by the acoustic measurement results. But the temperatures measured using the DLF method is closest to that obtained using the thermocouple. The maximum absolute error of the DLF is 16.7% and 16.3% lower than that of GCC and GCC-SCOT methods. Thus, it can be concluded that DLF provided the best measurement accuracy.

3.3.3. Experiments on Temperature Monitoring at Different Temperatures

The use of cables in electrical conduits is accompanied by heat generation, which can be dangerous when too much heat builds up. In the experiments, the temperature of the ribbon heater was adjusted utilizing a temperature controller to produce different temperature conditions. The temperature conditions are shown in Figure 10, each of which was measured after the temperature controller had been adjusted and a steady temperature reading was reached. The noise was added in the experiments to simulate interference signal. The SNR of the experiment environment was approximately equal to −9.380 dB, and the number of measurements per temperature condition was 200.

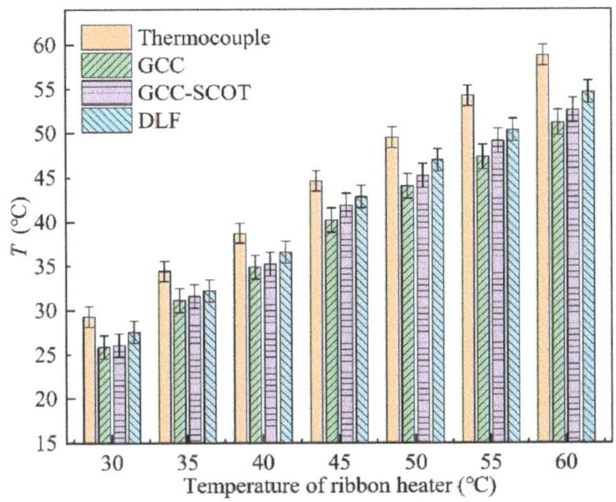

Figure 10. Temperature results of the four methods in different temperature conditions.

Since the upper limit of the temperature controller is 65 °C, seven different temperature conditions were set up ranging from 30 °C up to 60 °C. Figure 10 shows that the thermocouple readings are higher than that of acoustic methods and have small spread (indicated by shorter error bars), probably due to the fact that the glass fibers wrapped around the thermocouple were in contact with the inner conduit wall for a long time, resulting in the temperature being closer to the wall by heat transfer while the acoustic method measured the gas temperature inside the conduit and therefore gave lower results than the thermocouple. The DLF temperature results were higher than the GCC and GCC-SCOT methods with maximum temperature differences of 6.8% and 5.7% times higher and were closest to the thermocouple measurements. The GCC-SCOT results were better than the GCC method in terms of temperature and error. The DLF method had the lowest error, with maximum errors of 9.5% and 7.2% lower than the GCC and GCC-SCOT methods, respectively, and had the best measurement stability.

4. Conclusions

This paper proposes a new acoustic time-of-flight estimation method to achieve high accuracy in ranging and thermometry systems. The proposed DLF method is more interference immune which can extract the first harmonics of time-domain signals of low SNR from various noise and vibration disturbing environments. The relationship between time and frequency signals is established by harmonic peak detection. As both the transmitted and received signals have highly similar time-frequency characteristics, an accurate estimation of the acoustic TOF is required that can be attained by solving for the mathematical expectation of the time difference between the two signals so that the conduit length and the temperature can be calculated.

In order to simulate more closely to the on-site conduit conditions, an experiment platform and system for length and temperature measurement were constructed in the laboratory. The signal interference encountered in the field measurement due to physical disturbances from sand, gravel, and mud in the conduit, as well as caused by nearby construction and traffic, and different temperature environments was assumed to be Gaussian white noise. Compared with the conventional TOF estimation methods GCC and GCC-SCOT, the proposed DLF method can give more accurate measurement for conduit length and temperature where the temperature may fluctuate. The DLF method has the advantages of an extensive measurement range, higher accuracy, and robustness so it is more suitable for field applications.

Author Contributions: Conceptualization, Q.L. and B.Z.; methodology, Q.L. and Y.W.; software, Q.L.; validation, J.Z., B.Z. and R.C.; formal analysis, Q.L.; investigation, X.Z.; resources, R.Z.; data curation, M.D.; writing—original draft preparation, Q.L.; writing—review and editing, J.Z.; visualization, B.W.; supervision, Y.W.; project administration, B.Z.; funding acquisition, B.Z. All authors have read and agreed to the published version of the manuscript.

Funding: This research was funded in part by the National Natural Science Foundation of China (Grant No. 50976024 and No. 50906013), and the National Key Research and Development Program of China (Grant No. 2017YFB0603204).

Institutional Review Board Statement: Not applicable.

Informed Consent Statement: Informed consent was obtained from all subjects involved in the study.

Data Availability Statement: Not applicable.

Conflicts of Interest: The authors declare no conflict of interest.

References

1. Godbout, M.; Deschênes, J.-D.; Genest, J. Spectrally Resolved Laser Ranging with Frequency Combs. *Opt. Express OE* **2010**, *18*, 15981–15989. [CrossRef] [PubMed]
2. Wang, Y.; Zhou, B.; Liu, C. Calibration-Free Wavelength Modulation Spectroscopy Based on Even-Order Harmonics. *Opt. Express* **2021**, *29*, 26618–26633. [CrossRef]
3. Wang, P.; Xiao, X.; Zhang, Z.; Sun, C. Study on the Position and Orientation Measurement Method with Monocular Vision System. *Chin. Opt. Lett. COL* **2010**, *8*, 55–58. [CrossRef]
4. Dai, M.; Zhou, B.; Zhang, J.; Zuo, B.; Wang, Y. Experimental and Simulation Investigation of 3-D Soot Temperature and Volume Fraction Fields of Afterburner Flame. *Case Stud. Therm. Eng.* **2022**, *33*, 101932. [CrossRef]
5. Wu, J.; Zhu, J.; Yang, L.; Shen, M.; Xue, B.; Liu, Z. A Highly Accurate Ultrasonic Ranging Method Based on Onset Extraction and Phase Shift Detection. *Measurement* **2014**, *47*, 433–441. [CrossRef]
6. Megej, A.; Beilenhoff, K.; Schussler, M.; Ziroff, A.; Mottet, B.; Yilmazoglu, O.; Mutamba, K.; Hamann, C.D.; Baican, R.; Hartnagel, H.L. Integrated Microwave Sensors for Cavity-Length Measurement in Machine Engineering. *IEEE Trans. Microw. Theory Tech.* **2002**, *50*, 3070–3076. [CrossRef]
7. Hwang, O.; Lee, M.C.; Weng, W.; Zhang, Y.; Li, Z. Development of Novel Ultrasonic Temperature Measurement Technology for Combustion Gas as a Potential Indicator of Combustion Instability Diagnostics. *Appl. Therm. Eng.* **2019**, *159*, 113905. [CrossRef]
8. Moran, D.S.; Eliyahu, U.; Heled, Y.; Rabinovitz, S.; Hoffman, J.; Margaliot, M. Core Temperature Measurement by Microwave Radiometry. *J. Therm. Biol.* **2004**, *29*, 539–542. [CrossRef]
9. Tanaka, S.; Okamoto, M. Dynamic Model Based Length Measurement Using Stationary Waves. *Measurement* **2000**, *28*, 249–259. [CrossRef]
10. Okamoto, M.; Nakayama, M.; Tanaka, S. Accurate Pipe Length Measurement Under Noisy Environment Using Stationary Waves. In Proceedings of the ISCIE International Symposium on Stochastic Systems Theory and Its Applications, Saitama, Japan, 3–4 November 2004; Volume 2004, pp. 81–84. [CrossRef]
11. Tanaka, S.; Okamoto, M.; Nakayama, M. Length Measurement of Pipes with Open Ends Under Noisy Environment Using Stationary Wave. In Proceedings of the ISCIE International Symposium on Stochastic Systems Theory and its Applications, Ube, Japan, 30–31 October 2003; Volume 2003, pp. 114–118. [CrossRef]
12. Eriksson, H.; Borjesson, P.O.; Odling, P.; Holmer, N.-G. A Robust Correlation Receiver for Distance Estimation. *IEEE Trans. Ultrason. Ferroelectr. Freq. Control* **1994**, *41*, 596–603. [CrossRef]
13. Manabe, T.; Tsujimura, T.; Tomita, S.; Tabuta, T. Acoustic Inspection System for Underground Telecommunication Conduit. In Proceedings of the 10th Anniversary IMTC/94: Advanced Technologies in I & M. 1994 IEEE Instrumentation and Measurement Technolgy Conference (Cat. No.94CH3424-9), Hamamatsu, Japan, 10–12 May 1994; Volume 2, pp. 761–764.
14. Peng, C.; Shen, G.; Zhang, Y.; Li, Y.; Tan, K. BeepBeep: A High Accuracy Acoustic Ranging System Using COTS Mobile Devices. In Proceedings of the 5th International Conference on Embedded Networked Sensor Systems, Sydney, NSW, Australia, 6–9 November 2007; Association for Computing Machinery: New York, NY, USA, 2007; pp. 1–14.
15. Zhiganov, I.Y.; Skvortsov, B.V.; Sinnikov, S.G. Electronic-Acoustic Instruments for Measuring Pipe Lengths. *Meas. Tech.* **2002**, *45*, 769–772. [CrossRef]
16. Liu, Q.; Zhou, B.; Zhang, J.; Cheng, R. Development of Flue Gas Audio-Range Velocimeter Using Quadratic-Convex Frequency Sweeping. *IEEE Sens. J.* **2021**, *21*, 9777–9787. [CrossRef]
17. Knapp, C.; Carter, G. The Generalized Correlation Method for Estimation of Time Delay. *IEEE Trans. Acoust. Speech Signal Process.* **1976**, *24*, 320–327. [CrossRef]
18. Tu, Y.Q.; Shen, Y.L. Phase Correction Autocorrelation-Based Frequency Estimation Method for Sinusoidal Signal. *Signal Process.* **2017**, *130*, 183–189. [CrossRef]

19. Reed, F.; Feintuch, P.; Bershad, N. Time Delay Estimation Using the LMS Adaptive Filter–Static Behavior. *IEEE Trans. Acoust. Speech Signal Process.* **1981**, *29*, 561–571. [CrossRef]
20. So, H.C. Noisy Input–Output System Identification Approach for Time Delay Estimation. *Signal Process.* **2002**, *82*, 1471–1475. [CrossRef]
21. So, H.C.; Ching, P.C.; Chan, Y.T. A New Algorithm for Explicit Adaptation of Time Delay. *IEEE Trans. Signal Process.* **1994**, *42*, 1816–1820. [CrossRef]
22. So, H.C.; Ching, P.C. Comparative Study of Five LMS-Based Adaptive Time Delay Estimators. *IEE Proc. Radar Sonar Navig.* **2001**, *148*, 9–15. [CrossRef]
23. Porat, B.; Friedlander, B. Direction Finding Algorithms Based on High-Order Statistics. In Proceedings of the International Conference on Acoustics, Speech, and Signal Processing, Albuquerque, NM, USA, 3–6 April 1990; Volume 5, pp. 2675–2678.
24. Gonen, E.; Mendel, J.M. Applications of Cumulants to Array Processing. Part VI. Polarization and Direction of Arrival Estimation with Minimally Constrained Arrays. *IEEE Trans. Signal Process.* **1999**, *47*, 2589–2592. [CrossRef]
25. Hou, H.; Sheng, G.; Zhu, W.; Wang, H.; Jiang, X. Time-Delay Estimation Algorithm of Partial Discharge Ultra High Frequency Signals Based on Higher-Order Cumulant. *High Volt. Eng.* **2013**, *39*, 342–347. [CrossRef]
26. Kedem, B. Spectral Analysis and Discrimination by Zero-Crossings. *Proc. IEEE* **1986**, *74*, 1477–1493. [CrossRef]
27. Friedman, V. A Zero Crossing Algorithm for the Estimation of the Frequency of a Single Sinusoid in White Noise. *IEEE Trans. Signal Process.* **1994**, *42*, 1565–1569. [CrossRef]
28. Wang, S.; Tang, M. Exact Confidence Interval for Magnitude-Squared Coherence Estimates. *IEEE Signal Process. Lett.* **2004**, *11*, 326–329. [CrossRef]
29. Piersol, A. Time Delay Estimation Using Phase Data. *IEEE Trans. Acoust. Speech Signal Process.* **1981**, *29*, 471–477. [CrossRef]
30. Viola, F.; Walker, W.F. A Comparison between Spline-Based and Phase-Domain Time-Delay Estimators. *IEEE Trans. Ultrason. Ferroelectr. Freq. Control* **2006**, *53*, 515–517. [CrossRef] [PubMed]
31. Xia, W.; Jiang, W.; Zhu, L. An Adaptive Time Delay Estimator Based on ETDE Algorithm with Noisy Measurements. *Chin. J. Electron.* **2017**, *26*, 760–767. [CrossRef]
32. Liu, Q.; Zhou, B.; Zhao, R.; Dai, M.; Wang, B.; Wang, Y. Development of Acoustic Thermometer and Velocimeter with High Temporal Resolution and Noise Suppression Capability. In Proceedings of the 2022 11th International Conference on Communications, Circuits and Systems (ICCCAS), Singapore, 13–15 May 2022; pp. 71–74.
33. Zhang, S.; Shen, G.; An, L. Online Monitoring of Furnace Exit Gas Temperature in Power Plants. *Appl. Therm. Eng.* **2019**, *147*, 917–926. [CrossRef]
34. Zhang, K.; Feng, X.J.; Zhang, J.T.; Duan, Y.Y.; Lin, H.; Duan, Y.N. Determination of T–T90 from 234 K to 303 K by Acoustic Thermometry with a Cylindrical Resonator. *Metrologia* **2020**, *57*, 024004. [CrossRef]
35. Sun, S.J.; Xu, L.J.; Cao, Z.; Sun, J.T.; Yang, W.Q. Signal Demodulation Methods for Electrical Tomography: A Review. *IEEE Sens. J.* **2019**, *19*, 9026–9035. [CrossRef]
36. Tsukada, K.; Kiwa, T. Magnetic Property Mapping System for Analyzing Three-Dimensional Magnetic Components. *Rev. Sci. Instrum.* **2006**, *77*, 063703. [CrossRef]
37. Li, Y.; Tang, B.; Yi, Y. A Novel Complexity-Based Mode Feature Representation for Feature Extraction of Ship-Radiated Noise Using VMD and Slope Entropy. *Appl. Acoust.* **2022**, *196*, 108899. [CrossRef]
38. Li, Y.; Tang, B.; Jiang, X.; Yi, Y. Bearing Fault Feature Extraction Method Based on GA-VMD and Center Frequency. *Math. Probl. Eng.* **2022**, *2022*, e2058258. [CrossRef]
39. Li, Y.; Geng, B.; Jiao, S. Dispersion Entropy-Based Lempel-Ziv Complexity: A New Metric for Signal Analysis. *Chaos Solitons Fractals* **2022**, *161*, 112400. [CrossRef]
40. Li, Y.; Gao, P.; Tang, B.; Yi, Y.; Zhang, J. Double Feature Extraction Method of Ship-Radiated Noise Signal Based on Slope Entropy and Permutation Entropy. *Entropy* **2022**, *24*, 22. [CrossRef] [PubMed]
41. Zeqiri, B.; Cook, A.; Rétat, L.; Civale, J.; ter Haar, G. On Measurement of the Acoustic Nonlinearity Parameter Using the Finite Amplitude Insertion Substitution (FAIS) Technique. *Metrologia* **2015**, *52*, 406–422. [CrossRef]
42. Zhang, C.; Shen, S.; Huang, H.; Wang, L. Estimation of the Vehicle Speed Using Cross-Correlation Algorithms and MEMS Wireless Sensors. *Sensors* **2021**, *21*, 1721. [CrossRef]

sensors

MDPI

Article

A Novel Deep-Learning Method with Channel Attention Mechanism for Underwater Target Recognition

Lingzhi Xue, Xiangyang Zeng * and Anqi Jin

School of Marine Science and Technology, Northwestern Polytechnical University, Xi'an 710072, China; 2018100384@mail.nwpu.edu.cn (L.X.); jinaq@mail.nwpu.edu.cn (A.J.)
* Correspondence: zenggxy@nwpu.edu.cn

Abstract: The core of underwater acoustic recognition is to extract the spectral features of targets. The running speed and track of the targets usually result in a Doppler shift, which poses significant challenges for recognizing targets with different Doppler frequencies. This paper proposes deep learning with a channel attention mechanism approach for underwater acoustic recognition. It is based on three crucial designs. Feature structures can obtain high-dimensional underwater acoustic data. The feature extraction model is the most important. First, we develop a ResNet to extract the deep abstraction spectral features of the targets. Then, the channel attention mechanism is introduced in the camResNet to enhance the energy of stable spectral features of residual convolution. This is conducive to subtly represent the inherent characteristics of the targets. Moreover, a feature classification approach based on one-dimensional convolution is applied to recognize targets. We evaluate our approach on challenging data containing four kinds of underwater acoustic targets with different working conditions. Our experiments show that the proposed approach achieves the best recognition accuracy (98.2%) compared with the other approaches. Moreover, the proposed approach is better than the ResNet with a widely used channel attention mechanism for data with different working conditions.

Keywords: feature extraction; target recognition; neural networks; underwater acoustic signals

Citation: Xue, L.; Zeng, X.; Jin, A. A Novel Deep-Learning Method with Channel Attention Mechanism for Underwater Target Recognition. *Sensors* **2022**, *22*, 5492. https://doi.org/10.3390/s22155492

Academic Editors: Yuxing Li and Luca Fredianelli

Received: 6 July 2022
Accepted: 21 July 2022
Published: 23 July 2022

1. Introduction

The traditional methods of target recognition include feature extraction techniques based on mathematical modeling [1]. Using the entropy theory [2,3] as a feature to extract the radiation noise of a ship is one of the most common mathematical modeling methods. Additionally, a critical approach to recognition is to analyze the peaks of the spectrum to obtain the physical features, such as the propeller speed cavitation noise of the engine [4,5]. The spectrum will be distorted because of the Doppler effect when the ship moves toward the hydrophone receivers [6]. Wang proposes the multi-method spectra based on auditory feature extraction from the human ear and effectively extracts stable feature points under the Doppler effect [7]. Modeling the Doppler power spectrum of non-stationary underwater acoustic channels is another method to reduce the impact of the Doppler effect in underwater acoustic target recognition [8]. The information extracted by traditional methods is limited when the spectrum of signal changes with the Doppler effect. Li [9] uses the square root unscented Kalman filter to attenuate the Doppler phenomena in underwater acoustic signals.

Deep learning has an advantage in extracting the spectrum feature compared with the traditional method. However, it is often difficult to collect enough underwater acoustic signal data for training, which significantly limits the performance of deep neural networks in underwater target recognition. Nevertheless, researchers are still exploring the application of deep learning in underwater target recognition with the constraints of the available underwater acoustic data. Yang [10] et al. use deep auto-encoder networks combined with long short-term networks to extract target features and set different gates according to the temporal characteristics of underwater acoustic to extract feature information effectively. Auto-encoder networks can downscale high-dimensional data to low-dimensional data while retaining sufficient feature information, but the number of parameters is enormous. The convolution neural network (CNN) method can significantly reduce the number of parameters compared with DNN. Hence, CNN is better for underwater acoustic signals with limited samples. Hu [11] uses CNN to reduce the number of parameters and obtain better experimental results. Wang [12] investigates the intrinsic mechanism of convolution networks for underwater acoustic signals and displays the relationship between the waveform of the original data and the convolution kernel. Hu builds an underwater acoustic recognition model based on separable convolution operations according to the information collection mechanism of the auditory system, which is the first time grouped convolution models are applied in underwater acoustic recognition. Tian [13] applies a deep convolution stack to optimize CNN networks, which solves the lack of depth and structural imbalance of the networks. However, the above CNN model extracts single-scale features with the fixed size of the convolution kernel, which lose a lot of feature information. Hong [14] proposes a deep convolution stack network with a multi-scale residual unit (MSRU) to extract multi-scale features while exploring using generative adversarial networks (GAN) to synthesize underwater acoustic waveforms. The method modifies two advanced GAN models and improves their performance. GAN network with generators model and adversaries model uses the idea of the game to optimize the network. The generators can generate underwater acoustic samples when reaching the Nash equilibrium between the generators and adversaries models. We propose [15] an underwater acoustic target recognition model based on GAN, optimizing the recognition model with two model adversaries. The experiment verifies the better recognition ability of GAN than the other networks with small samples. The number of neural network layers has increased due to the urgent need to identify underwater acoustic data under different spatial and temporal conditions. Doan [16] applies the dense convolutional neural network to identify the target class, which addresses the over-fitting problem in a deep convolutional neural network with a limited number of samples. Gao [17] increases the number of samples using the GAN model, extracting underwater acoustic features with small samples in deeper network layers. To solve the recognition problem with a limited number of samples in deep networks, He [18,19] first proposes a ResNet model in image recognition, which uses the residual function to eliminate the gradient disappearance effectively. Wu [20] conducts deeper research in terms of the depth and width of ResNet models. Liu [21] applies the ResNet model to the study of underwater acoustic signals and acquires good experimental results.

Hu [22] first proposes the SE (squeeze and excitation) network, which uses channel weighting to discriminate the importance of information in different channels of ResNet. This network is a channel attention mechanism approach that can assign weights to different channel information according to their effectiveness and effectively remove channels with similar features. The channel attention mechanism can again adaptively optimize the neural network models, and different channel attention mechanisms for different research objects are required. Because of the low-pass filtering properties of the underwater acoustic channel, the high-frequency spectrum of the signal is decreased when increasing the distance between the target signal and the hydrophone. So, the underwater acoustic signal contains the spectrum of low frequency and the continuous spectrum. The continuous spectrum contains the ocean background noise signal, and the spectrum of low frequency contains the ship's radiation noise, propeller noise, machine noise, and other hull self-noise. Yang [23,24] uses an auditory inspired for ship type classification. The core of the underwater acoustic target recognition method is to extract the low-frequency spectrum [25,26]. However, the distance changes between the target and the hydrophone lead to a Doppler shift, which makes the information in the low-frequency spectrum disappear. This paper designs a camResNet (ResNet with channel attention mechanism) model to extract the low-frequency spectrum of underwater acoustic signals when the Doppler shift occurs. The channel attention mechanism of camResNet is divided into two parts. First, the signal channels are weighted by analysis of channel information. Second, the valid information points in each channel are extracted, and the complete information is weighted.

This paper is organized as follows. Section 2 introduces the structure of the SE_ResNet network. Section 3 describes the details of the underwater acoustic target recognition method based on camResNet. Section 4 describes the experimental data and shows the experimental results. Section 5 concludes the advantages and disadvantages of the proposed method.

2. Structure of ResNet

The ResNet model deals with network degradation caused by network layer deepening using residual learning methods. Hong [27] studied the characteristics of underwater acoustic signals and increased the recognition rate with an 18-layer residual network (ResNet18), which contains an embedding layer.

The ResNet model consists of many residual modules; the input of the modules is x, and the output of the convolutional structure of multi-layer stacking is $H(x)$, called the learned features. The learned features are difficult to optimize by backward gradient propagation with a network having too many layers, even if the nonlinear activation function performs very well. He finds that function $F(x) = H(x) - x$, called the residual function, is easier to optimize $H(x)$. The output of residual modules is the complex feature function $F(x) + x$, which is the residual function learned by the network summed with the original signal, and the output of residual modules is the input of the following residual modules. Figure 1 shows the architecture of the ResNet model, in which $H(x)$ is the residual function, and the mathematical expression is defined as

$$H(x) = x + w_N\delta(w_{N-1}(\delta(\ldots\delta(w_1x)))) \tag{1}$$

The $w_1 \cdots w_N$ in this equation denotes the weight of each module in the residual network. The function for x mathematical expression is defined as

$$\frac{\partial H(x)}{\partial x} = 1 + \frac{\partial(w_N\delta(w_{N-1}(\delta(\cdots\delta(w_1x)))))}{\partial x} \tag{2}$$

The first term of Equation (2) equals 1, and the second term is the gradient value of the weight function to x. Since it contains 1, the function $\frac{\partial H(x)}{\partial x}$ will not equal 0, even if the second term is small.

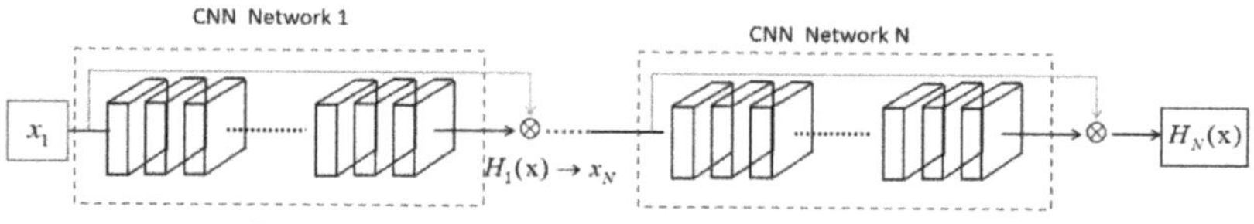

Figure 1. The architecture of the ResNet model.

3. Architecture of camResNet in Underwater Acoustic Target Recognition Method

3.1. Architecture of camResNet

The camResNet model is excellent for extracting classification-related feature information because it adds the channel attention mechanism based on the ResNet model. The process of the camResNet model includes three steps: feature structure building, feature extraction, and feature classification, as shown in Figure 2.

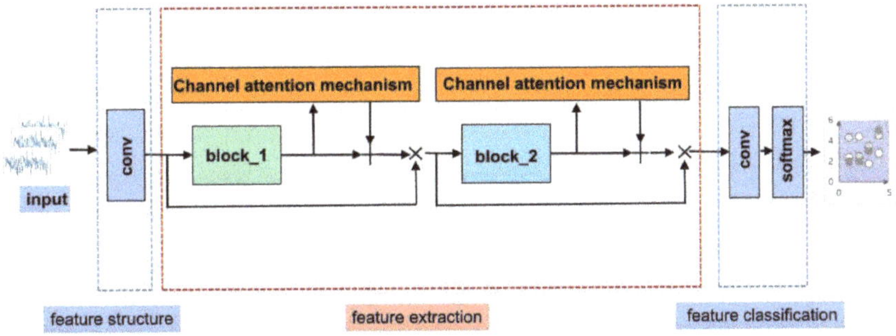

Figure 2. The architecture of the camResNet model.

The low-dimensional underwater acoustic signal limits the ability of convolution networks to extract high-dimensional abstract features. So, the feature structure building module decomposes the input acoustic signal into base signals using a set of one-dimensional convolutions as deep convolution filters, which can obtain high-dimensional input data. Different convolution kernels of N are set in the deep convolution filters $F(F_1, F_2 \cdots F_N)$, and each convolution layer contains a two-dimensional convolution kernel. The output of the feature module contains 16 groups of signals, so 16 one-dimensional convolution layers are needed. The specific formula is as follows:

$$y_i^m = f(x^m \times \omega_i^m + b_i^m) \tag{3}$$

where x^m is the m-th input sample, ω_i^m denotes the convolution kernel of the i-th output channel of the m-th sample, b_i^m denotes the bias function of the i-th output channel of the m-th sample, and y_i^m is the i-th channel output value of the m-th sample. The symbol $\times$ means dot product. Finally, the output feature group of the i-th layer is y_i^m, formed through the ReLU function $f(\cdot)$.

The number and frequency of the spectrum are the primary basis for underwater acoustic signal target recognition. The spectrum energy that will shift with the change of distance between the target and the hydrophone is called unstable spectra. The spectrum energy that will not shift with the change of distance between the target and the hydrophone is called stable spectra. The camResNet model can extract the stable spectrum of the underwater acoustic target as the feature to recognize the target category accurately when the spectra of the target are shifted due to the Doppler effect. The stable spectra contain many harmonic signals. The fundamental frequency is the shaft frequency signal of the propeller, and the relationship of the harmonic groups is the multiplier. For a B-bladed propeller, each B is a set of pulses with a period T, and the repetition period of the pulses is T/B. The $2N$ + 1st set of pulses in the time domain signal is selected, and its k-th Fourier transform is denoted as $F_N^k(\omega)$. The specific formula of power spectral density by this random process is as follows [28]:

$$\begin{aligned} s(\omega) &= E\{s_k(\omega)\} = E\left\{\lim_{N\to\infty}\frac{1}{(2N+1)T}\left|F_N^{(k)}(\omega)\right|^2\right\} \\ &= |g(\omega)|^2\left\{(2N+1)(U_1-U_2)+\sum_{p=-2N}^{2N}(2N+1-|p|)\cos\omega pT[U_2+U_3\cdot 2\cos\omega(T/2)+U_4\cdot\cos\omega(T/4)]\right\} \end{aligned} \tag{4}$$

where $E\{\cdot\}$ is the expected value, ω denotes angular frequency $g(\omega)$ Fourier spectrum, representing the time domain waveform. The specific formula of U is as follows:

$$\begin{cases} U_1 = \overline{a_0^2} + \overline{a_1^2} + \overline{a_2^2} + \overline{a_3^2} \\ U_2 = \overline{a}_0^2 + \overline{a}_1^2 + \overline{a}_2^2 + \overline{a}_3^2 \\ U_3 = \overline{a}_0 \cdot \overline{a}_1 + \overline{a}_1 \cdot \overline{a}_3 \\ U_4 = \overline{a}_0 \cdot \overline{a}_1 + \overline{a}_1 \cdot \overline{a}_2 + \overline{a}_2 \cdot \overline{a}_3 + \overline{a}_3 \cdot \overline{a}_0 \end{cases} \quad (5)$$

where a_i denotes the amplitude of the pulse number i in a set of signals. $\overline{a}_i$ denotes the average value of a_i. The fundamental frequency and the first group of harmonic signals can be used as stable signal characteristics because the modulation spectrum of the actual vessel radiation noise decays rapidly with the increasing number of groups of spectra. The obtained multidimensional information with the feature structure building module is called the original information, which is the input of the feature extraction module. The feature extraction module contains two ResNet models with the channel attention mechanism. A convolution kernel size of 1×64 is a good trade-off between the quality of the recognition and the computational cost of the model for underwater acoustic. The first layer of the residual network contains two convolutions. Each convolution operation maps 16 sets of base signals to another 16 sets of base signals to extract the deep features of the signal. The convolution operation consists of 16 convolution layers, each containing 16 different filters $F(F_1, F_2 \cdots F_N)$. So, 16×16 one-dimensional convolution layers are needed. The specific formula is as follows:

$$y_i^m = \sum_{k=1}^{N} f(x_k^m \times \omega_{ik}^m + b_{ik}^m) \quad (6)$$

where x_{ik}^m denotes the input value of the k-th channel in the m-th sample, ω_{ik}^m denotes the k-th convolution kernel of the i-th layer convolution of the m-th sample, b_{ik}^m denotes the k-th bias function of the i-th layer convolution of the m-th sample, and y_{ik}^m is the output of the i-th layer convolution of the m-th sample. The symbol $\times$ means dot product. The output feature group of the k-th convolution of the i-th convolution layer is formed through the activation function $f(\cdot)$, which uses the ReLU function.

Finally, all the convolution outputs in the i-th layer are summed up as the convolution output value of the i-th layer. The second convolution is the same as the first convolution operation in order to obtain deeper underwater acoustic features. A channel attention mechanism is added to each one-residual network to enhance the stable spectrum features and further enhance the network's performance in extracting underwater acoustic signals.

Section 3.2 describes the channel attention mechanism of the feature structure building module in detail.

The feature classification uses a fully convolutional network to map the high-dimensional features from the output of the feature extraction module to a lower dimension with the size of the classification class. The details are listed as follows.

Stage 1: In feature structure, the data shape of the input layer is a four-dimensional matrix $64 \times 1 \times 1 \times 800$. The shape changes from $64 \times 16 \times 1 \times 800$ to $64 \times 16 \times 1 \times 800$ by convolutional layer. The batch normalization layer is applied, followed by a ReLU activation function and max pooling with the stride of 2×1.

Stage 2: The feature extraction module contains two residual modules, called block-1 and block-2. The input shape of block-1 is $64 \times 1 \times 1 \times 400$. The shape changes from $64 \times 16 \times 1 \times 400$ to $64 \times 16 \times 1 \times 400$ by two convolutions with a convolution kernel of 64×1 and a stride of 1×1. Batch normalization is applied after each convolution and connected between the two convolutions using the activation function ReLU. Finally, add the channel attention mechanism, marked with the dashed yellow box in Figure 2, which will be described in detail in Section 3.2 of the paper. The obtained data are summed with the original data as the output of block-1.

Stage 3: The input shape of block-2 is $64 \times 1 \times 1 \times 400$. The shape changes from $64 \times 16 \times 1 \times 400$ to $64 \times 16 \times 1 \times 200$ by convolution with a convolution kernel 64×1 and a step of 2×1. Batch normalization and a ReLU activation function are applied. The second convolution does not change the shape of the data and adds the channel attention mechanism. The obtained data are summed with the original data as the output of block-2.

Stage 4: This paper uses a fully convolutional networks model, in which a cubic convolutional network is used to map high-dimensional features to low-dimensional features in the decision module.

3.2. Structure of Channel Attention Mechanism Based on Underwater Acoustic of camResNet

The changes in the distance between the target and the hydrophone lead to a Doppler effect, which is the frequency move. The Doppler frequency compensation is challenging, as the underwater acoustic channel is low-frequency filtering. The method in this paper can extract the stable spectral features under the Doppler frequency shift by the channel attention mechanism, which can automatically acquire the critical information in each feature channel by learning to enhance the valuable features and suppress the less useful features for the current task.

The amount of information on the channels is different, and the channel attention mechanism increases the weight to that of the channel with high information. It can improve the model's capability. First, squeeze the information out of each channel and then add a lightweight gating system to optimize the channel information and output the channel weights. The channel attention mechanism of this paper is divided into two parts. Figure 3 shows the channel attention mechanism model. The first part is the primary part, which weighs each channel, and the second part is the auxiliary part of formation extraction, which is another channel information after transposing the information.

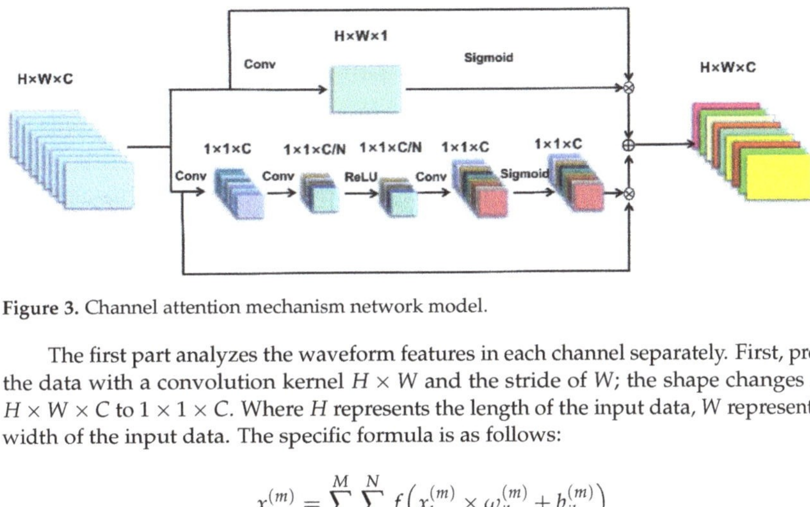

Figure 3. Channel attention mechanism network model.

The first part analyzes the waveform features in each channel separately. First, process the data with a convolution kernel $H \times W$ and the stride of W; the shape changes from $H \times W \times C$ to $1 \times 1 \times C$. Where H represents the length of the input data, W represents the width of the input data. The specific formula is as follows:

$$x^{(m)} = \sum_{i=1}^{M} \sum_{k=1}^{N} f\left(x_k^{(m)} \times \omega_{ik}^{(m)} + b_{ik}^{(m)}\right) \quad (7)$$

where $x_k^{(m)}$ denotes the k-th channel data in the input information of the channel attention mechanism module, $\omega_{ik}^{(m)}$ denotes the weight of the k-th channel of the i-th layer of convolution, $b_{ik}^{(m)}$ denotes the bias of the k-th channel of the i-th layer of convolution, and $x^{(m+1)}$ denotes the output value of $x^{(m)}$ after one convolution.

The data of each channel characterize the global features of each channel. In order to be able to learn the nonlinear characteristics between the channels independently, this paper uses a gating system with an activation function. The specific formula is as follows.

$$x^{(m+3)} = \sum_{i=1}^{M} \sum_{k=1}^{N} \sigma\left(\delta\left(x_k^{(m+1)} \times \omega_{ik}^{(m+1)} + b_{ik}^{(m+1)}\right) \times \omega_{ik}^{(m+2)} + b_{ik}^{m+2}\right) \quad (8)$$

where $x_k^{(m+1)}$ is the global feature of size $1 \times 1 \times C$. $\omega_{ik}^{(m+1)}$ and $\omega_{ik}^{(m+2)}$ are the weights of the network mapping. In order to obtain the features of the network channel, convolutional mapping is used, and the feature points before mapping are r times after mapping, so $\omega_{ik}^{(m+1)} \in R^{\frac{c}{r}\times c}$, $\omega_{ik}^{(m+2)} \in R^{c\times\frac{c}{r}}$. δ is the ReLU activation function, and σ is the sigmoid activation function.

The second part synthesizes the signal characteristics in all channels. Process the data with a convolution kernel 1×64 and the stride of 1; the shape changes from $H \times W \times C$ to $H \times W \times 1$. The multi-layer convolutional network has a solid ability to extract sufficient recognition information, and the output of the network contains a large number of stable signals with a small number of unstable signals. One-dimensional data of the same size are extracted from the network's output as the channel weights of the original signal, which can effectively enhance the spectrum energy contained in the channel.

The two parts of the channel attention mechanism weigh the signal features from different perspectives. Finally, the two weighted pieces of information are fused as the output of the channel attention mechanism.

4. Model Evaluation

4.1. Dataset

The eight hydrophones are fixed at the same level in eight different places at the same interval. This paper randomly selects four sets of hydrophones at equal intervals as input data. The data used in the experiments contain four classes of vessels, and the third of the four types of signals is the radiated noise of the iron vessel, while the first, second and fourth types are vessels of the same material and similar hull size.

To study the recognition effect of camResNett under different Doppler frequency shifts, four different working conditions were intercepted in each class of experimental data. Each class of the data obtained has four modes of operation: straight ahead at a constant speed, straight forward acceleration, straight-ahead deceleration, and turning. Figure 4 shows the spectrogram of different working conditions by the fourth type of vessel.

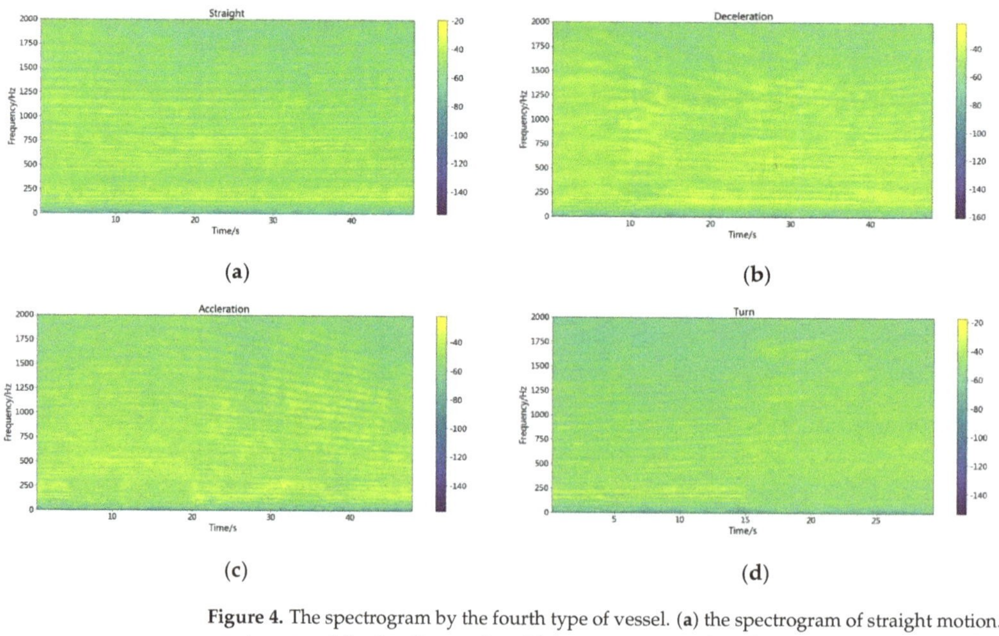

Figure 4. The spectrogram by the fourth type of vessel. (**a**) the spectrogram of straight motion. (**b**) the spectrogram of decelerating motion. (**c**) the spectrogram of accelerating motion. (**d**) the spectrogram of turning movement.

Figure 4a is the time-frequency relationship of the signal by the vessel of straight motion. It shows that there is acceleration when the vessel is just starting, and the frequency shifts to high frequency. The speed reaches stability within a brief period, and a stable spectrum characteristic appears, which contains line and continuous spectra. The formula with the Doppler shift is as follows:

$$f = f_0 \cdot \frac{1}{1 - \frac{u \cos \theta}{v}} \tag{9}$$

where f_0 is the original frequency of the vessel, v is the speed of the underwater acoustic signal propagating in the channel, u is the speed of the vessel motion, and f is the frequency after the Doppler shift. θ is the angle between the line of the vertical distance connecting the ship and the hydrophone and the line connecting the ship and the hydrophone. The signal will have a stable frequency shift when the vessel movement speed is constant. In the passive recognition process, the stable spectrum feature after the frequency shift is the primary information for recognizing the target. However, when the target accelerates, the u keeps changing, and the f varies with the change of u. Figure 4b,c are time-frequency diagrams of the ship in the motion state of acceleration and deceleration. The low-frequency spectra are the stable spectra, and the spectrum above 400 Hz will change with time. Figure 4d is the time-frequency diagram by the vessel of turning, and a large number of unstable spectra appear in the time-frequency diagram because the θ keeps changing.

To further observe the energy distribution of the frequencies with the vessel for different operating conditions, Figure 5 shows the power spectral density for the different operating conditions by the fourth type of vessel, which is the Fourier transform of the correlation function with the 0.5 s window length.

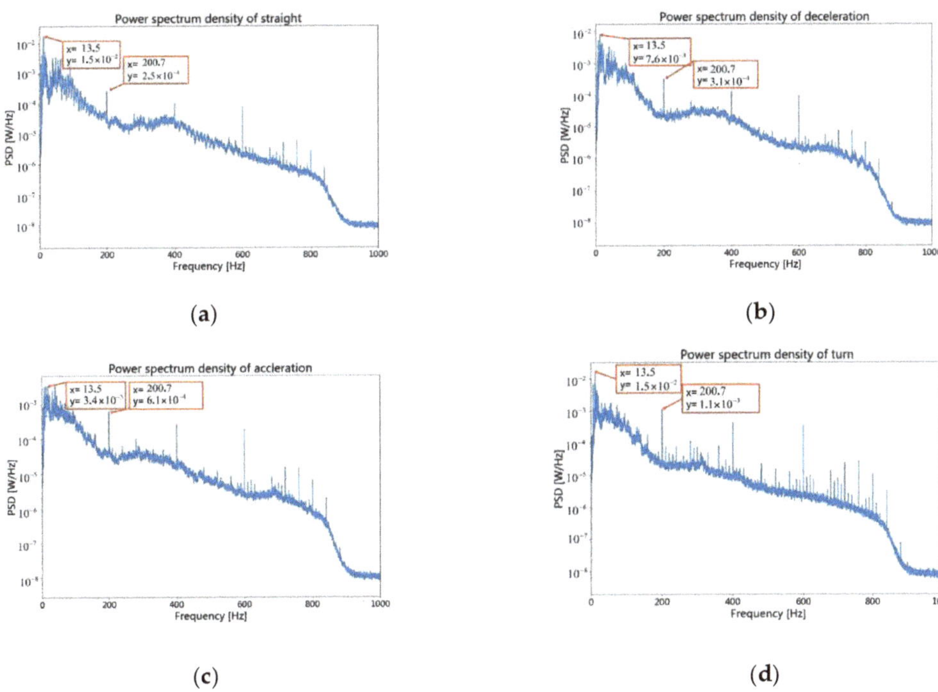

Figure 5. The power spectrum density by the fourth type of vessel. (**a**) The power spectrum density of straight motion. (**b**) The power spectrum density of decelerating motion. (**c**) The power spectrum density of accelerating motion. (**d**) The power spectrum density of turning movement.

Figure 5 shows the power spectrum density by the fourth type of vessel. A set of resonant waves at a fundamental frequency of 200 Hz occur stably under four different operating conditions. High-frequency points are shifted when the vessel is in an accelerated motion. The high-frequency spectral density varies significantly, and the low-frequency spectral density is more stable than the high frequency under different working conditions. Figure 5b,c show the acceleration and deceleration. Compared with Figure 5a, the power spectral density in high frequency is higher than in the straight motion, and some frequency points in the high frequency are changed. Figure 5d shows turning, and many spectral density power spikes appear in the high frequency compared with Figure 5a.

The same class of targets contains different Doppler shift signals, which will increase the difficulty of recognition, with the original signal compressed or broadened. This method extracts the stable features of the same class of vessels under different working conditions.

To study the difference between the categories with four types, the straight motion working condition of each type of vessel is chosen to exhibit a time-frequency relationship. Figure 6 shows the pictures and time-frequency diagrams of the four types of vessels, containing class I, class II, and class III and IV vessels. The background noise of the four vessels has relatively apparent differences, but there are similar low-frequency spectra.

As can be observed in Figure 6b,d, a clear line spectrum in the low-frequency band is very similar. Figure 6h has two precise line spectra, respectively, similar to the line spectra in Figure 6b,d. No clear line spectrum is observed in Figure 6f, but the energy distribution at low frequencies is similar to that in Figure 6h. Figure 6 shows that the spectrum is very similar to the different vessel types, in which the spectrum energy is concentrated in the low frequency and continuous. So, it is difficult to distinguish the vessel category with the traditional method.

4.2. Data Pre-Processing

There are 800 feature points (0.1 s) for a frame and no overlap between frames. If the maximum feature point of the sample is less than 0.1, eliminate the small value frame sample, ensuring that the recognition results are not affected by the particular sample points. After eliminating the small samples, the samples contain 7097 samples. Use 1/4 of the data as the test set and 3/4 of the data as the training set after normalizing the samples. The prepared data have 5322 samples as the training set and 1774 samples as the test set. In total, 200 samples are randomly selected as the validation set in each class, and the validation set contains 800 samples in total. The training method is a batch method, in which 64 samples are randomly selected in each batch, and the selected samples will not be used as alternative samples in the next batch.

4.3. Experimental Results

4.3.1. Discussion of Model Structure

This reports the experimental results of the model with Doppler shifts signals. The straight condition is considered a signal without a Doppler shift. The other conditions are considered a Doppler shift. The experiment chose four conditions as input data.

The first experiment illustrates the relationship between the recognition rate and the number of residual layers, where the size of the number of residual layers changes in the set of {1,2,3,4}. According to the results in Table 1, two residual layers have the best recognition effect, and the recognition rate will decrease by increasing the number of residual layers.

Table 1. Recognition rate for different numbers of residual layers.

Residual Layers.	**1**	**2**	**3**	**4**
Recognition rate of test set	96.3%	98.2%	96.5%	91.4%

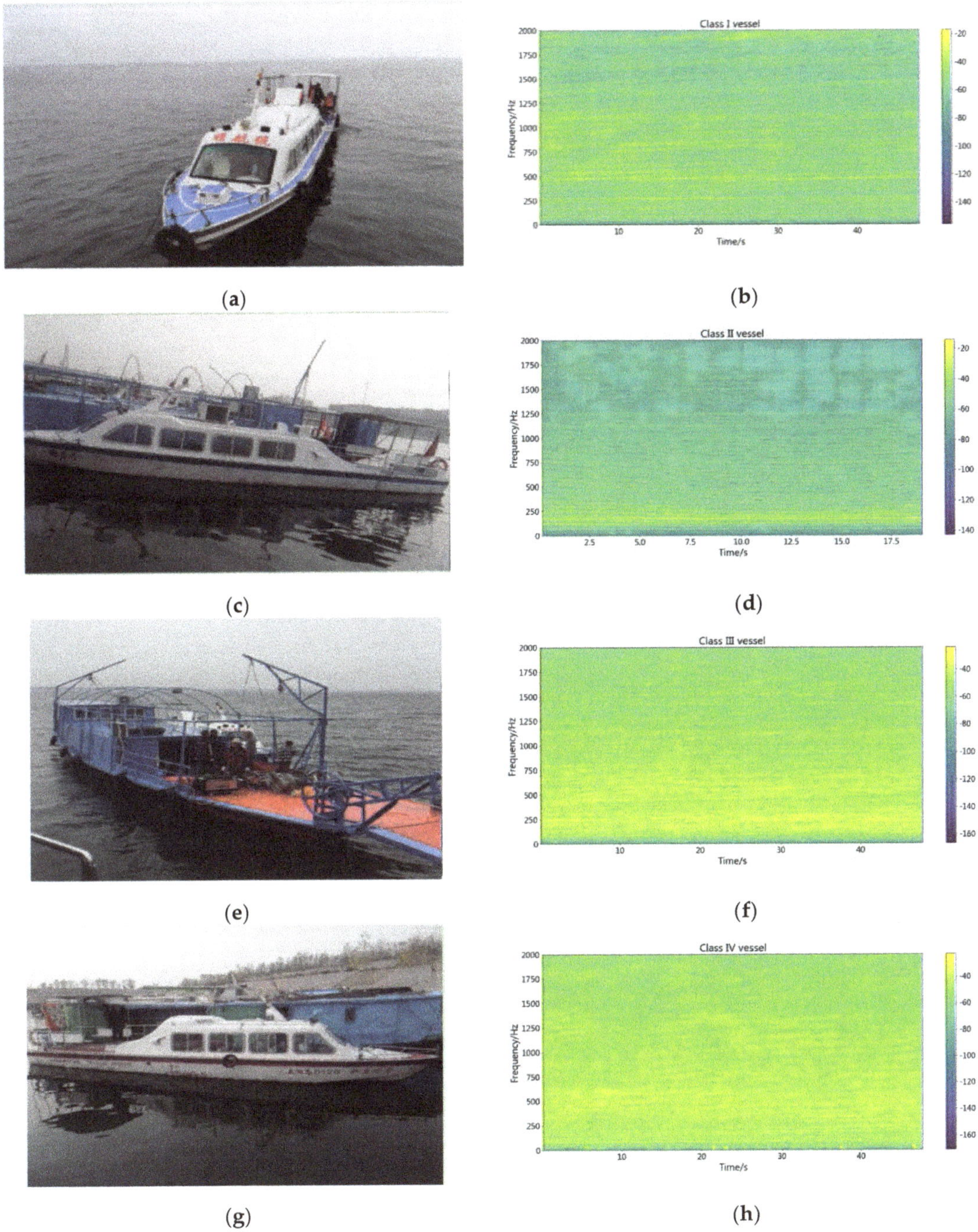

Figure 6. The pictures and spectrograms of the different vessels. (**a**) the pictures of class I vessel. (**b**) the spectrogram of class I vessel. (**c**) the pictures of class II vessel. (**d**) the spectrogram of class II vessel. (**e**) the pictures of class III vessel. (**f**) the spectrogram of class III vessel. (**g**) the pictures of class IV vessel. (**h**) the spectrogram of class IV vessel.

Two residual layers are appropriate for the number of samples in the experiment, and the different number of samples matches the different number of layers. If the ResNet network is not over-fitted or under-fitted, the over-fitting phenomenon will occur and decrease the recognition accuracy when adding the channel attention mechanism. If the ResNet network is under-fitted, adding the channel attention mechanism will compensate for this under-fitting phenomenon. The number of model parameters needs to match the number of samples, and the number of parameters increases after adding the channel attention mechanism.

The second experiment illustrates the relationship between the recognition rate and the size of the convolutional kernel. The size of the 1D convolutional kernel varies in the set of {3, 5, 7, 9, 11, 15, 17, 21, 25, 33, 41, 49, 57, 64, 75, 85, 95}. Table 2 shows that a kernel size of 64 is best for the recognition rate. The scale size of the target needs to match the actual perceptual field after the addition of convolutional kernels because the underwater acoustic target is submerged in background noise, and a large amount of ocean background noise is extracted if there is no match.

Table 2. Recognition rate for convolution kernels of different sizes.

Convolution Kernels	Recognition Rate of Validated Set	Recognition Rate of Test Set
1×3	91.6%	91.1%
1×5	92.0%	91.5%
1×7	94.3%	92.1%
1×11	97.1%	92.3%
1×15	98.2%	93.1%
1×17	97.9%	93.5%
1×21	97.8%	95.3%
1×25	98.1%	95.3%
1×33	98.1%	95.9%
1×41	98.2%	96.1%
1×49	98.3%	96.3%
1×57	99.5%	97.4%
1×64	99.9%	98.2%
1×75	98.9%	97.1%
1×85	98.3%	96.4%
1×95	99.5%	95.9%

4.3.2. Classification Experiment Results

In the experimental data, four-vessel classes are used to train different deep-learning network models, and the information of each network model is described below.

(1) The DBN model has an input layer, three hidden layers, and one output layer. The number of nodes in the input layer is 199, the number of nodes in the three hidden layers is 100, 50, and 20, and the number of nodes in the output layer is the number of sample categories. Each pair of adjacent layers constitutes an RBM network, and the three RBM networks are trained separately first, followed by the whole network. A batch method with a batch size of 64 is used for training. A gradient descent algorithm with a learning rate of 0.01 is used to optimize the training process.

(2) The GAN network model consists of two modules: generation and discrimination. The generation module consists of three convolutional layers, and the discrimination module consists of convolutional layers. The generative module comprises three convolutional layers, with 64, 128, and 800 filters with a filter size of 1×4 and a step size of 4. The discriminative model is a single-layer convolutional neural network with 16 filters with a filter size of 1×4 and a step size of 4. Batch training with a batch size of 64 is used, and the learning rate is 0.001.

(3) The DenseNet model is made up of three modules, each of which has three layers of a convolutional neural network. The data are normalized before each convolutional operation, and after convolution, the data are nonlinearly mapped using the elu

activation function. The convolutional operation with a convolutional kernel size of 1 × 64 and a step size of 1 is chosen. The batch method with a batch size of 64 is used for training. The optimization method is chosen during training using a gradient descent method, and the learning rate is 0.001. For optimization, the gradient descent algorithm is used.

(4) The U_Net model is made up of three down-sampling modules and three up-sampling modules. Each down-sampling module contains two convolutional layers and a pooling layer of the specified size of 1 × 2. There is a splicing layer, a deconvolution layer, and a pooling layer with a pooling size of 1 × 1 in each up-sampling module. The batch method is used for training, with a batch size of 64 and an optimization method of gradient descent with a learning rate of 0.001.

(5) The SE ResNet network is set up and trained in the same way as the camResNet UAS model network, with the exception that the channel attention mechanism is a three-layer auto-encoder network model.

A test set was used to evaluate the model's recognition ability. Table 3 shows the recognition rate with straight motion and four different working conditions. The recognition rate of amRestNet and SE_ResNet are similar when the data contain straight data. The recognition rate of amRestNet is higher than SE_ResNet when the data contain four different working conditions. Both amRestNet and SE_ResNet can extract valid feature information when the data contain a single working condition. However, the SE_ResNet is not as effective as amRestNet in extracting stabilization features when different working conditions are included and have different Doppler frequencies.

Table 3. Recognition rate of proposed model and compared models with straight motion data and four different working conditions.

The Input	Models	Recognition Rate	
		Straight Motion Data	Four Different Working Conditions
Time domain signal	camResNet	98.9%	98.2%
Frequency domain signal	DBN	85.6%	82.4%
Time domain signal	GAN	96.6%	96.3%
Frequency domain signal	DenseNet	97.3%	96.1%
Time domain signal	U_Net	93.9%	93.6%
Time domain signal	SE_ResNet	98.8%	97.1%

Table 3 shows that the camResNet model has a recognition rate of 98.2%, which is 1.1–15.8% higher than the other networks. The DBN model is a basic neural network model based on probabilistic statistics, and its input signal is a frequency domain signal. The GAN model is the adversarial model, which mainly contends with small-sample data, and its input signal is the time domain signal. The DenseNet model can simplify the network complexity and reduce network parameters by designing the dense block, and its input signal is the frequency domain signal. The ResNet model uses residual learning to update the network parameters, and its input signal is the time domain signal. The U_Net model uses up-sampling and down-sampling to extract multi-scale features, which can improve the recognition effect, and its input signal is the time domain signal.

The DBN model has different optimization methods compared to other models, which use probabilistic models to optimize the parameters, so the recognition rate of the DBN model is lower than other networks. The recognition rate of U_Net is lower than the GAN model and the DenseNet model because the up-sample and down-sample can lose some feature information. The SE_ResNet model has an excellent performance in recognition rate because the ResNet model has the balance between network depth and recognition rate of small samples. The camResNett model is better than the other models in terms of the recognition rate because the channel attention mechanism deals with underwater signals' sparsity and multi-scale characteristics.

In the display of recognition experiment results, we use recognition accuracy, recall rate, precision, and F1-score to evaluate the recognition performance of the networks. The formulae for each indicator are as follows.

$$Precision = \frac{TP}{TP + FP} \tag{10}$$

$$Recall = \frac{TP}{TP + FN} \tag{11}$$

$$Accuracy = \frac{TP + TN}{TP + TN + FP + FN} \tag{12}$$

$$F1\text{-}score = \frac{2TP}{2TP + FP + FN} \tag{13}$$

TP, *TN*, *FP*, and *FN* are true positive, true negative, false positive, and false negative. Table 4 shows the precision, recall rate, F1-score, and accuracy of the test sample, while Table 5 shows the confusion matrix.

Table 4. Recognition results of camResNet.

	Precision	Recall	F1-Score	Accuracy
Class I vessel	0.996	0.991	0.993	0.996
Class II vessel	0.984	0.981	0.982	0.991
Class III vessel	0.978	0.982	0.980	0.990
Class IV vessel	0.993	0.998	0.995	0.997

Table 5. Confusion matrix of camResNet.

	Class I Vessel	Class II Vessel	Class III Vessel	Class IV Vessel
Class I vessel	1783	1	7	9
Class II vessel	2	1764	32	2
Class III vessel	3	24	1772	1
Class IV vessel	2	0	1	1769

Class I of the vessel includes three acceleration signals, three deceleration signals, five straight-ahead signals, and seven turn signals. Class II of the vessel includes three acceleration signals, three deceleration signals, three straight-ahead signals, and six turn signals. Class III consists of three acceleration signals, deceleration signals, straight-ahead signals, and five turn signals. Class IV consists of three acceleration signals, deceleration signals, straight-ahead signals, and turn signals. The vessels of the different categories have similar sizes but different materials, and the third category material is significantly different to the materials from the other three. In Table 5, the probability of incorrectly recognizing Class II of the vessel as Class III of the vessel is the highest. This is followed by the probability of incorrectly recognizing Class III of the vessel as Class II of the vessel. This indicates that camResNet extracts shallow physical features and deep category features, which is related to the Doppler effect. Class II of the vessel and Class III of the vessel contain the most similar samples in the composition structure of working conditions, resulting in many samples with similar Doppler shifts. Table 4 shows that the recognition effects of Class I of the vessel and Class IV of the vessel are better than Class II of the vessel and Class III of the vessel, which may appear confusing.

The precision of Class I of the vessel is the highest, and the probability of incorrectly recognizing Class I of the vessel as Class I of the vessel is the highest because Class I of the vessel contains many straight samples and has a prominent stable spectrum without a Doppler shift. Class IV of the vessel has the highest recall, which indicates that the samples of different working conditions in Class IV are more balanced than the others and have more stable Doppler shift characteristics than the others.

4.3.3. Visualization of Energy Distribution by the Architecture of camResNet

Power Spectral Density

To further assess the feature extraction capability of the camResNet model, the trained camResNet model was fed by Class IV of the vessel because the spectrogram and the power spectral density are displayed in Figures 4 and 5. Figure 7 shows the time-frequency diagram and the power spectral density of the output.

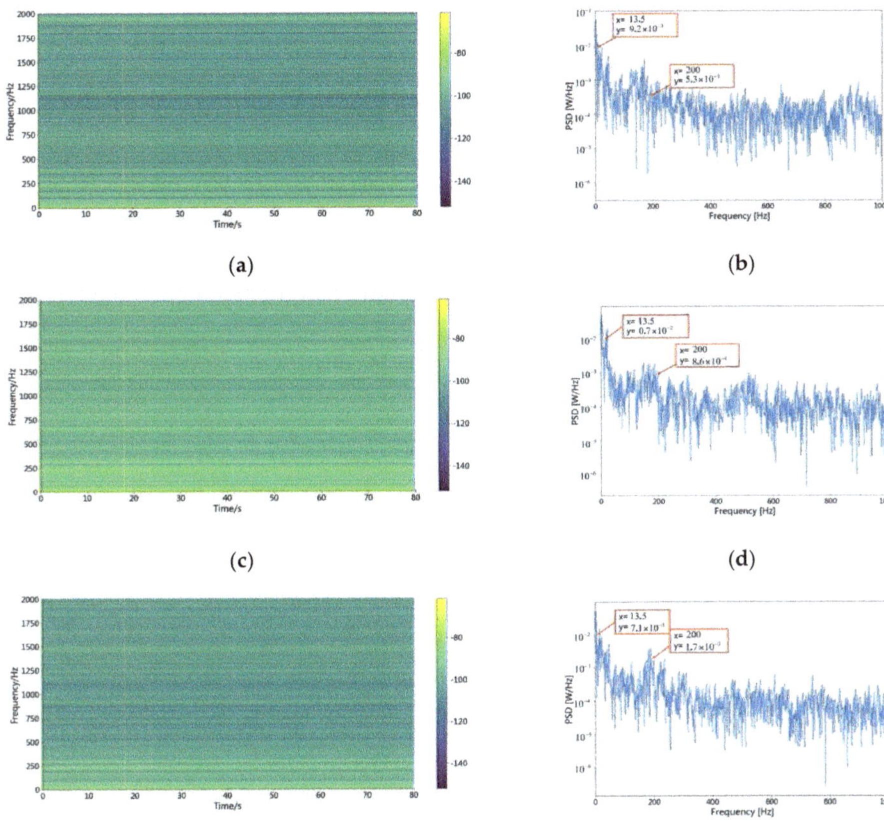

Figure 7. The spectrogram and power spectrum density of Class IV with camResNet. (**a**,**c**,**e**) The spectrogram of extracted features for Class IV vessel by camResNet; (**b**,**d**,**f**) The power spectrum density of extracted features for Class IV vessel by camResNet.

Figure 7a,c,e show the signal after the camResNet model by Class IV of the vessel, and Figure 4 shows the spectrogram of the original signal from Class IV vessel. The comparison indicates that the energy of the feature is still concentrated in the low frequency after the camResNet model. Figure 7b,d,f show the power spectrum density of Class IV of the vessel after processing the camResNet model, and Figure 5 shows the power spectrum density of the original signal for Class IV of the vessel. The comparison indicates that the apparent fundamental frequency signal in the original signal still exists after processing the camResNet model. In Figure 7, the camResNet model's output contains not only stable signals but also some high-frequency signals, which indicates the camResNet model can avoid extracting unstable signals that are quickly Doppler shifted and recovers stable signals that are submerged in high frequencies.

t-SNE Feature Visualization Graphs

The above experiment shows that the camResNet model can extract signals of stable frequencies in underwater acoustic signals. To further analyze the ability to extract features by camResNet, the distance of the original features and camResNet output features is visualized using the t-SNE method. Figure 8 shows the distance characteristics of the original signal and the output of the camResNet model when different working conditions are used as the input data. Figure 8a shows the t-SNE of the original underwater acoustic signal, which indicates that the original underwater acoustic signal has weak separability. Figure 8b shows the t-SNE of the output signals with the input of four different working conditions in the camResNet model. Figure 8c–f show the t-SNE of the output signals after putting straight motion, acceleration, deceleration, and turning conditions into the camResNet model, respectively.

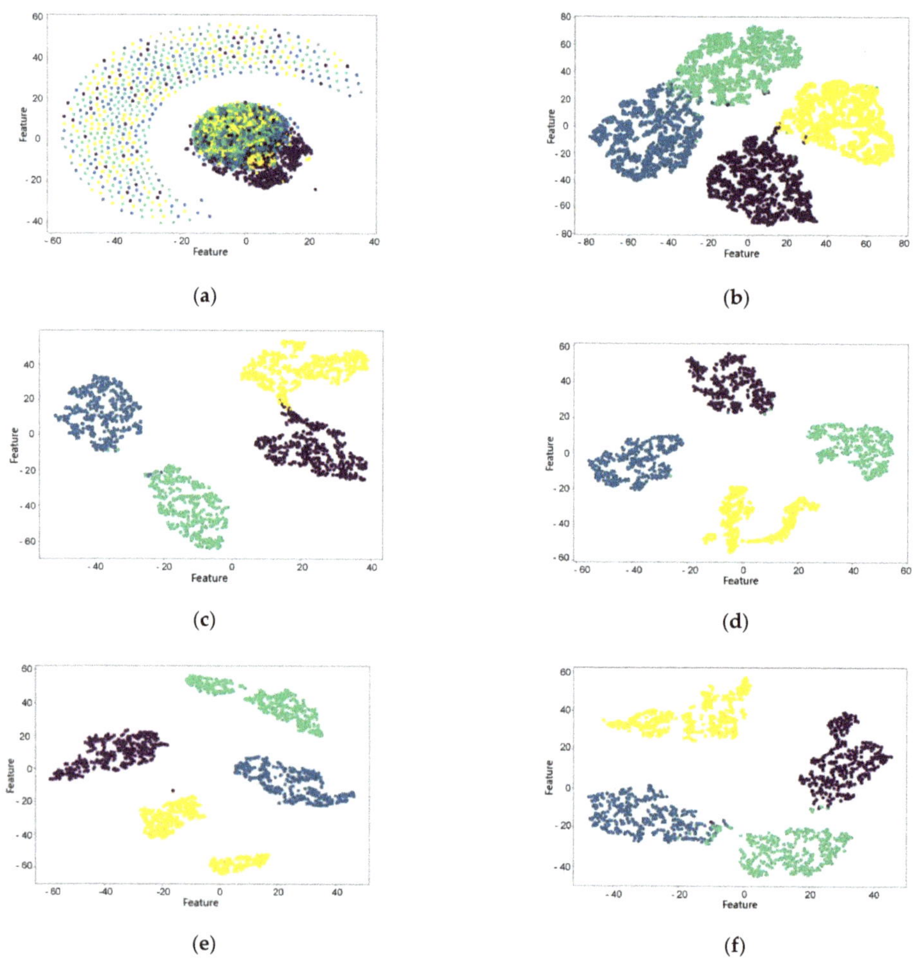

Figure 8. The t-SNE visualized graphs. (**a**) The t-SNE visualized graphs of original hydroacoustic signal; (**b**) The t-SNE visualized graphs' output by camResNet; (**c**) The t-SNE visualized graphs of straight motion by camResNet; (**d**) The t-SNE visualized graphs of accelerating motion by camResNet; (**e**) The t-SNE visualized graphs of decelerating motion by camResNet; (**f**) The t-SNE visualized graphs of turning movement by camResNet.

From the figures, it can be seen that the camResNet model can classify all the working condition signals well and classify each kind of working condition signals well. In particular, the camResNet model can still classify the signals of acceleration conditions well when the acceleration conditions contain a large number of unstable Doppler shift signals. The classification maps show that the camResNet model can extract abstract and stable internal features under different conditions.

4.3.4. Recognition Results for Different Data of camResNet

Three different network models were used to compare the recognition results of underwater acoustic signals, which contained four working conditions. The models of DenseNet and SE_ResNet have a more extraordinary ability to recognize and were used for comparison with the camResNet model. The training method determines that the training and test data are the same—one of four working conditions. The recognition results were averaged by repeating the test five times, and the obtained experimental results are shown in Figure 9. The solid blue line is the recognition rate, which uses the data of straight motion working conditions as the training data and test data. The blue dotted line is the recognition rate, which uses the data of turn working conditions as the training data and test data. The solid red line is the recognition rate, which uses the data of deceleration working conditions as the training data and test data. The yellow dashed line is the recognition rate, which uses the data of acceleration working conditions as the training data and test data.

(1) The recognition rate of the camResNet model is higher than that of both the DenseNet model and the SE_ResNet model. The camResNet model can extract stable features that are effective for recognition.
(2) The recognition rate of the camResNet model under the straight motion condition is higher than under the other conditions, which indicates that the Doppler shift can affect the recognition of camResNet.
(3) There are different recognition rates with different working conditions containing different Doppler shifts. The maximum recognition rate of camResNet is 0,998; the minimum recognition rate is 0.994. The maximum recognition rate of DenseNet is 0.985, and the minimum recognition rate is 0.971. The decrease in recognition rate due to different Doppler shifts is smaller in the camResNet model than in the other models, which shows that the camResNet model has a better extraction of signals with Doppler shifts.

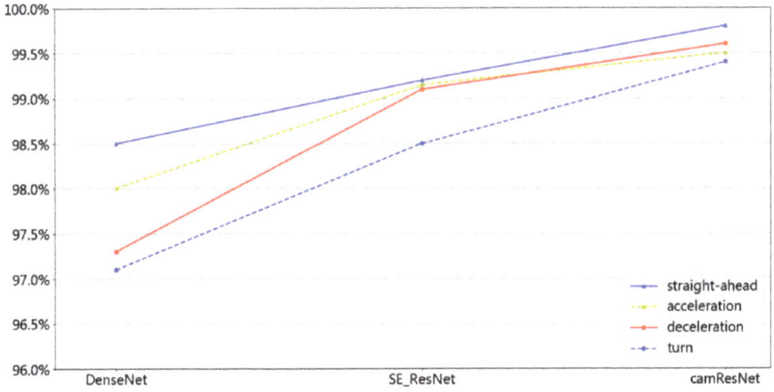

Figure 9. Recognition rate of cooperation targets.

The network is trained and tested using data under one working condition, which is easy to overfit by a deeper model of DenseNet. The SE_ResNet model uses self-coding to compress channel features but does not consider the sparse characteristics of underwater acoustic targets. The camResNet model builds two different channel attention mechanisms, which fully consider the sparsity of underwater acoustic signal and the continuity spectrum, and they have better recognition results than the other models.

The distributions of the training and test sets in the above experiments were identical. To further verify the recognition performance of the camResNet model, three network models were trained using four working conditions and tested under one working condition. The recognition results were averaged by repeating the test five times, and the obtained experimental results are shown in Figure 10. The solid blue line is the recognition rate, which uses the data of straight motion working conditions as the test data. The blue dotted line is the recognition rate, which uses the data of turn working conditions as the test data. The solid red line is the recognition rate, which uses the data of deceleration working conditions as the test data. The yellow dashed line is the recognition rate, which uses the data of acceleration working conditions as the test data.

(1) The maximum recognition rate of camResNet is 0,976; the minimum recognition rate is 0.965. The maximum recognition rate of DenseNet is 0.957, and the minimum recognition rate is 0.95. The recognition rate of the camResNet model is higher than that of the DenseNet model and the SE_ResNet model, and the performance is most evident under the deceleration condition.
(2) The recognition rates of the three network models vary smoothly under different working conditions, indicating that all three network models can extract stable signals from the initial signals and remove unstable frequency shifts. The camResNet model has the most robust ability from the recognition results.
(3) Compared with identical distributions of the training and test sets, the decrease in recognition rate due to different Doppler shifts becomes more prominent when the distributions of the training and test sets are not identical. This indicates that the recognition capabilities of the camResNet model with a Doppler shift are related to the distribution of training and test sets.

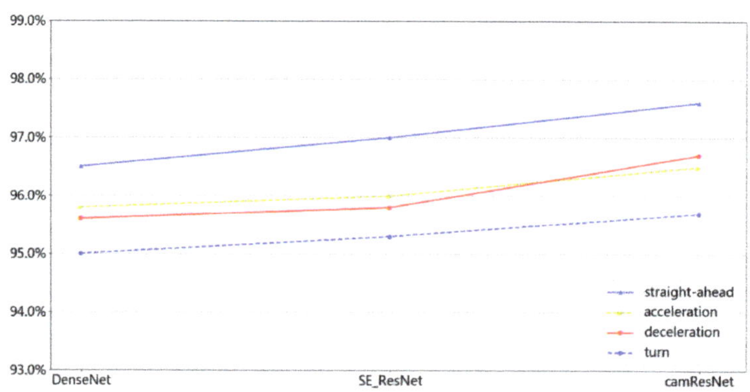

Figure 10. Recognition rate of non-cooperation targets.

SE_ResNet uses compressed information to obtain channel weights to obtain certain stable features, so the recognition ability under different working conditions is better than that of DenseNet. The stable signal of the Doppler shift represents multi-scale information, which causes extract information with one scale to lose helpful information. The camResNet model uses convolution operation to extract channel information from two aspects. The first part uses the convolution kernel superposition to expand the perceptual field and extract features of different scales. The second part extracts the feature from the local features of all information. Fusing the two features as the weights of channels can comprehensively extract the stable features under the Doppler frequency shift. Hence, the camResNet model has better recognition results for different working conditions data containing the Doppler frequency shift information.

5. Conclusions

The camResNet model adds a channel attention mechanism to the ResNet model based on the characteristics of underwater acoustic signals. This channel attention mechanism can enhance the stable spectral features and remove the unstable signals caused by the Doppler shifts. The experiments compare the recognition ability of six different deep-learning models under different Doppler shift frequencies. The results show that the recognition rate of the camResNet model is higher than that of the other network models. The camResNet model has a recognition rate of 98.2%, which is 1.1–15.8% higher than the other networks. The precision, recall rate, F1-score, and accuracy are used to demonstrate that the data used in the experiments are balanced between the classes and that the experimental results are valid. Test the effectiveness of the proposed method with the same distribution and different distributions for the training and test sets. The three network models with better recognition results are selected for testing. In the same training set and test set distribution, the recognition rate of camResNet varies from 0.003 to 0.023 for different working conditions. In contrast, the recognition rate of DenseNet varies from 0.015 to 0.019 for different distributions of the training set and test set. The results show that the proposed method is more suitable when the training and test sets are identically distributed. Further, using visualization methods to learn the features of the signal extracted by the camResNet model, the results show that the camResNet model can extract the stable multi-group harmonic signals and restore some weak high-frequency stable signals in the original signal.

The camResNet model can effectively extract the features of underwater acoustic signals with the Doppler shift. The following work will use the camResNet model to recognize the underwater acoustic signals with the Doppler shift for small samples, solving the problem of data-driven underwater acoustic signals in deep learning.

Author Contributions: Conceptualization, L.X.; methodology, L.X.; software, L.X.; validation, L.X.; formal analysis, L.X.; investigation, L.X.; resources, L.X.; data curation, L.X.; writing—original draft preparation, L.X.; writing—review and editing, X.Z.; visualization, A.J.; supervision, X.Z.; project administration, X.Z. All authors have read and agreed to the published version of the manuscript.

Funding: National Natural Science Foundation of China: 11774291.

Institutional Review Board Statement: Not applicable.

Informed Consent Statement: Not applicable.

Data Availability Statement: Not applicable.

Conflicts of Interest: The authors declare that they have no known competing financial interests or personal relationships that could appear to have influenced the work reported in this paper.

References

1. Li, Y.X.; Geng, B.; Jiao, S.B. Refined Composite Multi-Scale Reverse Weighted Permutation Entropy and Its Applications in Ship-Radiated Noise. *Entropy* **2021**, *23*, 476. [CrossRef] [PubMed]
2. Li, Y.X.; Geng, B.; Jiao, S.B. Dispersion Entropy-based Lempel-Ziv Complexity: A New Metric for Signal Analysis. *Chaos Solitons Fractals* **2022**, *161*, 112400. [CrossRef]
3. Li, Y.X.; Mu, L.; Gao, P. Particle Swarm Optimization Fractional Slope Entropy: A New Time Series Complexity Indicator for Bearing Fault Diagnosis. *Fractal Fract.* **2022**, *6*, 345. [CrossRef]
4. Li, Y.X.; Tang, B.; Yi, Y. A novel complexity-based mode feature representation for feature extraction of ship-radiated noise using VMD and slope entropy. *Appl. Acoust.* **2022**, *196*, 108899. [CrossRef]
5. Li, Y.X.; Wang, L.; Yang, X.H. A Novel Linear Spectrum Frequency Feature Extraction Technique for Warship Radio Noise Based on Complete Ensemble Empirical Mode Decomposition with Adaptive Noise, Duffing Chaotic Oscillator and Weighted-Permutation Entropy. *Entropy* **2019**, *21*, 507. [CrossRef] [PubMed]
6. Wang, L.; Wang, Q.; Zhao, L. Doppler-shift invariant feature extraction for underwater acoustic target classification. In Proceedings of the 2017 International Conference on Wireless Communications, Signal Processing and Networking (WiSPNET), Chennai, India, 22–24 May 2017; pp. 1209–1212.
7. Wang, Q.; Zeng, X.Y.; Wang, L. Passive Moving Target Classification Via Spectra Multiplication Method. *IEEE Signal Process. Lett.* **2017**, *24*, 451–455. [CrossRef]

8. Naderi, M.; Ha, D.V.; Nguyen, V.D.; Patzold, M. Modelling the Doppler power spectrum of non-stationary underwater acoustic channels based on Doppler measurements. In Proceedings of the OCEANS' 17, Aberdeen, UK, 19–22 June 2017; pp. 1–6.
9. Li, X.; Zhao, C.; Yu, J. Underwater Bearing-Only and Bearing-Doppler Target Tracking Based on Square Root Unscented Kalman Filter. *Entropy* **2019**, *21*, 740. [CrossRef] [PubMed]
10. Yang, H.H.; Xu, G.H.; Yi, S.Z. A New Cooperative Deep Learning Method for Underwater Acoustic Target Recognition. In Proceedings of the OCEANS 2019—Marseille, Marseille, France, 17–20 June 2019; pp. 1–4.
11. Hu, G.; Wang, K.; Liu, L. Underwater Acoustic Target Recognition Based on Depthwise Separable Convolution Neural Networks. *Sensors* **2021**, *21*, 1429. [CrossRef] [PubMed]
12. Wang, Q.; Zeng, X.Y. Deep learning methods and their applications in underwater targets recognition. *Tech. Acoust.* **2015**, *34*, 138–140.
13. Tian, S.; Chen, D.; Wang, H. Deep convolution stack for waveform in underwater acoustic target recognition. *Sci. Rep.* **2021**, *11*, 9614. [CrossRef]
14. Hong, F.; Liu, C.; Guo, L. Underwater Acoustic Target Recognition with a Residual Network and the Optimized Feature Extraction Method. *Appl. Sci.* **2021**, *11*, 1442. [CrossRef]
15. Xue, L.Z.; Zeng, X.Y. Underwater Acoustic Target Recognition Algorithm Based on Generative Adversarial Networks. *Acta Armamentarii* **2021**, *42*, 2444–2452.
16. Doan, V.-S.; Huynh-The, T. Underwater Acoustic Target Classification Based on Dense Convolutional Neural Network. *IEEE Geosci. Remote Sens. Lett.* **2020**, *99*, 1–5. [CrossRef]
17. Gao, Y.J.; Chen, Y.C.; Wang, F.Y. Recognition Method for Underwater Acoustic Target Based on DCGAN and DenseNet. In Proceedings of the 2020 IEEE 5th International Conference on Image, Vision and Computing (ICIVC), Beijing, China, 10–12 July 2020; pp. 15–21.
18. He, K.M.; Zhang, X.Y.; Ren, S.Q. Deep Residual Learning for Image Recognition. In Proceedings of the 2016 IEEE Conference on Computer Vision and Pattern Recognition (CVPR), Las Vegas, NV, USA, 26 June–1 July 2016; pp. 770–778.
19. He, K.M.; Zhang, X.Y.; Ren, S.Q. Identity Mappings in Deep Residual Networks. In Proceedings of the 2016 European Conference on Computer Vision, Amsterdam, The Netherlands, 11–14 October 2016; pp. 1–15.
20. Wu, Z.; Shen, C.; Hengel, A. Wider or Deeper: Revisiting the ResNet Model for Visual Recognition. *Pattern Recognit.* **2019**, *90*, 119–133. [CrossRef]
21. Liu, P.; Wang, G.; Qi, H. Underwater Image Enhancement With a Deep Residual Framework. *IEEE Access* **2019**, *7*, 94614–94629. [CrossRef]
22. Hu, J.; Shen, L.; Sun, G. Squeeze-and-Excitation Networks. In Proceedings of the IEEE Conference on Computer Vision and Pattern Recognition (CVPR), Salt Lake City, UT, USA, 18–22 June 2018; pp. 7132–7141.
23. Shen, S.; Yang, H.H.; Li, J.H. Auditory Inspired Convolutional Neural Networks for Ship Type Classification with Raw Hydrophone Data. *Entropy* **2018**, *20*, 990. [CrossRef] [PubMed]
24. Li, J.H.; Yang, H.H. The underwater acoustic target timbre perception and recognition based on the auditory inspired deep convolutional neural network. *Appl. Acoust.* **2021**, *182*, 108210. [CrossRef]
25. Arveson, P.T.; Vendittis, D.J. Radiated noise characteristics of a modern cargo ship. *J. Acoust. Soc. Am.* **2000**, *107*, 118–129. [CrossRef] [PubMed]
26. Jiang, J.; Wu, Z.; Lu, J. Interpretable features for underwater acoustic target recognition. *Measurement* **2020**, *173*, 108586. [CrossRef]
27. Hong, F.; Liu, C.W.; Guo, L.J. Underwater Acoustic Target Recognition with ResNet18 on ShipsEar Dataset. In Proceedings of the 2021 IEEE 4th International Conference on Electronics Technology (ICET), Chengdu, China, 7–10 May 2021; pp. 1240–1244.
28. Cheng, Y.S.; Li, Z.Z. *Underwater Acoustic Target Recognition*, 2nd ed.; Science Press: Beijing, China, 2018; pp. 45–48.

sensors

MDPI

Article

Study on a Detection Technique for Scholte Waves at the Seafloor

Minshuai Liang [1], Liang Wang [1,*], Gaokun Yu [1], Yun Ren [2] and Linhui Peng [1]

1 Department of Marine Technology, Ocean University of China, Qingdao 266100, China; lms2423@stu.ouc.edu.cn (M.L.); gkyu@ouc.edu.cn (G.Y.); penglh@ouc.edu.cn (L.P.)
2 State Key Laboratory of Acoustics, Institute of Acoustics, Chinese Academy of Sciences, Beijing 100190, China; renyun@mail.ioa.ac.cn
* Correspondence: wangliangher@ouc.edu.cn

Abstract: Scholte waves at the seafloor have significant potential for underwater detection and communication, so a study about detecting Scholte waves is very meaningful in practice. In this paper, the detection of Scholte waves at the seafloor is researched theoretically and experimentally. Acoustic models with the multilayer elastic bottom are established according to the ocean environment, and a tank experiment is designed and carried out to detect Scholte waves. Different from detecting Scholte waves in the seismic wavefield, a technique for detecting Scholte waves in the sound pressure field is proposed in this paper. The experimental results show that the proposed technique can detect Scholte waves effectively, and there are no problems such as seabed coupling and the effect of wave speeds. Furthermore, the results also show that this detection technique is still effective in conditions with a sediment layer. The existence of sediment layers changes the acoustic field conditions and affects the excitation of Scholte waves.

Keywords: Scholte wave detection; multilayer elastic bottom; acoustic pressure field; source depth; propagation distance

check for updates

Citation: Liang, M.; Wang, L.; Yu, G.; Ren, Y.; Peng, L. Study on a Detection Technique for Scholte Waves at the Seafloor. *Sensors* **2022**, *22*, 5344. https://doi.org/10.3390/s22145344

Academic Editors: Yuxing Li and Luca Fredianelli

Received: 13 June 2022
Accepted: 16 July 2022
Published: 18 July 2022

1. Introduction

In marine settings, the waves trapped near the fluid–solid interface are called Scholte waves [1]. Scholte waves are a kind of interface wave, and they are expected to have a longer traveling path or less transmission loss than waves in water [2,3]. These characteristics make Scholte waves a great prospect in underwater detection and communication, leading to research interests in ocean acoustics. Currently, Scholte waves have been applied in many areas, such as geoacoustic inversion [4–8], acoustic source localization [9,10], and imaging [11]. Therefore, how to detect Scholte waves effectively has significant meaning in ocean detection.

In existing methods, the seismic field is often applied to detect Scholte waves. Seismic measurement equipment such as Ocean Bottom Seismometers (OBSs) [12–14] and geophones [6,9] is used to record the seismic signal at the ocean bottom. Then, the velocity characteristics in the fields can be extracted from the recorded signal [2,8]. Finally, Scholte waves are identified according to the velocity characteristics.

In the measurement process, the seismic measurement equipment mainly records the seismic wave signal on the seafloor through the coupling between the equipment and seafloor, so their measurements are susceptible to the bottom type and seafloor topography [15,16]. Compared with seismic equipment, the hydrophone is immune to these issues, making it more suitable for measuring underwater acoustic signals.

In the stage of signal processing, the velocity characteristics of Scholte waves are used to identify Scholte waves as mentioned above. The velocities of Scholte waves have two features. One is slower speed, and their speeds are slower than other waves in the acoustic field. A later arriving signal with high energy can be observed in the time series [9,12,17,18]. Another is that the velocities of Scholte waves are dispersive in the waveguide [12]. Investigators have proposed many practical techniques to extract the dispersion curve of Scholte waves, the most extensive of which is the approach based on domain transformation [5,7,8,19–21]. The extracted dispersion curve can be used to identify Scholte waves at the seafloor. These velocity-based approaches can identify Scholte waves effectively when the Scholte wave velocity differs largely from the acoustic waves in water. However, they will be invalid when the velocities are close to each other.

This paper introduces a new technique to detect Scholte waves based on acoustic pressure field measurement and researches the detection of Scholte waves theoretically and experimentally. Acoustic models with the multilayer elastic bottom are established. The Scholte wave is identified according to the excitation and propagation characteristics. Scaled-down tank experiments for detecting Scholte waves are designed and carried out. Firstly, the detection principle is introduced. The characteristics of Scholte waves that vary with the source depth and propagation distance are researched. Then, the scaled-down experiment is introduced in detail. Acoustic field features in the water tank are analyzed. Measurement and analysis results of this experiment are presented. Furthermore, the research in conditions with sediment layers is introduced. Finally, the summary and conclusions are presented.

2. Detection Principle

2.1. Acoustic Model

Ocean acoustic studies generally assume the ocean bottom as the fluid bottom [22], but this assumption is not suitable for the research in this paper. The elastic bottom plays an indispensable role in the excitation and propagation of Scholte waves, so the ocean bottom is an elastic medium with shear waves in this research. The marine environment is simplified into a three-layer model consisting of seawater, basalt, and peridotite, as shown in Figure 1, based on Hamilton's studies [23] and the results of geological surveys [24].

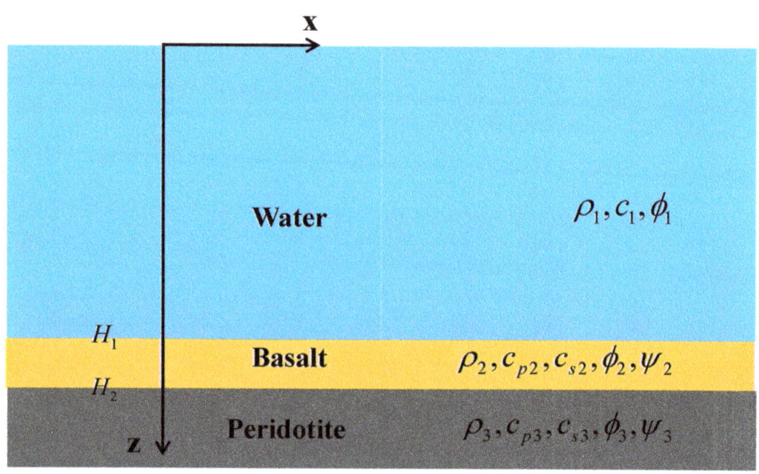

Figure 1. Schematic of the acoustic model for the marine environment. H_1 and H_2 are the interface depth for "water–basalt" and "basalt–peridotite".

The typical deep ocean environment is researched in this study. Referring to related research [25], the media parameters in the model are shown in Table 1. The KRAKENC program is used to analyze the acoustic field in this model. The phase velocities of the Scholte wave and normal modes at 10 Hz are calculated and listed in Table 2. The Scholte wave in the acoustic field is treated as the zeroth mode. These data show that the velocities of the Scholte wave and normal modes are extremely close, which means that the Scholte wave in the field cannot be identified using velocity features.

A new technique is proposed to detect Scholte waves according to the excitation and propagation characteristics.

Table 1. Parameters of the marine environment.

Media	Layer i	Depth	ρ (g/cm^3)	c_p (m/s)	c_s (m/s)
Seawater	1	3000	1	1500	-
Basalt	2	3100	2.7	5250	2500
Peridotite	3	-	3.28	6500	4000

Table 2. Phase velocities of the Scholte wave and normal modes at 10 Hz.

Order	Phase Velocity (m/s)
0	1489.741
1	1500.537
2	1502.139
3	1504.784
4	1508.457

2.2. Elastic Normal Modes

In order to interpret the detection principle, an elastic normal mode solution is presented for this range-independent model with the elastic bottom. Assume that all layers in the marine environment are isotropic media. Subscripts i = 1, 2, 3 are used to represent acoustic parameters in seawater, basalt, and peridotite, respectively. ρ_i is density; c_1 is seawater sound velocity; c_{p2} and c_{p3} are compressional wave velocities (P wave); c_{s2} and c_{s3} are shear wave velocities (S wave). H_1 and H_2 represent the interface depths for "water–basalt" and "basalt–peridotite".

The acoustic field is expressed by velocity potentials. ϕ_1 is the potential in water; ϕ_i and $\Psi_i\vec{e}_y$ $(i = 2, 3)$ are compressional and shear potentials in the elastic bottom, respectively. Potentials in the water and bottom satisfy the wave equations [26].

$$\frac{\partial^2 \phi_1}{\partial z^2} + \frac{\partial^2 \phi_1}{\partial x^2} = \frac{1}{c_1^2}\frac{\partial^2 \phi_1}{\partial t^2} \tag{1}$$

$$\frac{\partial^2 \phi_i}{\partial z^2} + \frac{\partial^2 \phi_i}{\partial x^2} = \frac{1}{c_{pi}^2}\frac{\partial^2 \phi_i}{\partial t^2} \tag{2}$$

$$\frac{\partial^2 \psi_i}{\partial z^2} + \frac{\partial^2 \psi_i}{\partial x^2} = \frac{1}{c_{si}^2}\frac{\partial^2 \psi_i}{\partial t^2} \tag{3}$$

A time–harmonic plane wave with the time dependence $e^{-i\omega t}$ is considered. Subsequently, all potentials can be represented as follows:

$$\phi_i = Z_i(z)e^{i(kx-\omega t)} \tag{4}$$

$$\psi_i = Z_{si}(z)e^{i(kx-\omega t)} \tag{5}$$

where ω=$2\pi f$ is the angular frequency and $k = \omega/c$ is the wavenumber. $Z_i(z)$ and $Z_{si}(z)$ are depth-dependent functions for compressional and shear waves; they satisfy the homogeneous depth-separated wave equations.

$$\left[\frac{d^2}{dz^2} + (k_i^2 - k^2)\right]Z_i(z) = 0 \tag{6}$$

$$\left[\frac{d^2}{dz^2} + (k_{si}^2 - k^2)\right]Z_{si}(z) = 0 \tag{7}$$

The general solutions for potentials can be obtained by the free surface condition and radiation condition at infinity.

$$\phi_1 = -2A\sinh(k\xi_1 z)e^{i(kx-\omega t)} \tag{8}$$

$$\phi_2 = \left(B_1 e^{-k\xi_2 z} + B_2 e^{k\xi_2 z}\right)e^{i(kx-\omega t)} \tag{9}$$

$$\psi_2 = \left(C_1 e^{-k\xi_{s2} z} + C_2 e^{k\xi_{s2} z}\right)e^{i(kx-\omega t)} \tag{10}$$

$$\phi_3 = \left(De^{-k\xi_3 z}\right)e^{i(kx-\omega t)} \tag{11}$$

$$\psi_3 = \left(Ee^{-k\xi_{s3} z}\right)e^{i(kx-\omega t)} \tag{12}$$

where $k_i^2 - k^2 \equiv -k^2\xi_i^2$, $k_{si}^2 - k^2 \equiv -k^2\xi_{si}^2$, and $k_1 = \omega/c_1$ is the wavenumber in water, $k_i = \omega/c_{pi}, i = 2,3$ are compressional wavenumbers, and $k_{si} = \omega/c_{si}, i = 2,3$ are shear wavenumbers in the elastic seabed. A, B_1, B_2, C_1, C_2, D, and E are undetermined coefficients for potentials.

Here, it is assumed that $\mathbf{u}$ is the displacement and $\mathbf{T}$ is the stress tensor. u_{ix} is the horizontal displacement; u_{iz} is the vertical displacement; T_{izz} is the normal stress; T_{izx} is the shear stress.

The displacements are determined from the velocity potentials as follows:

$$u_{ix} = \frac{1}{-i\omega}\left(\frac{\partial\phi_i}{\partial x} - \frac{\partial\psi_i}{\partial z}\right), u_{iz} = \frac{1}{-i\omega}\left(\frac{\partial\phi_i}{\partial z} + \frac{\partial\psi_i}{\partial x}\right) \tag{13}$$

According to the stress–strain constitutive relation, the following equations can be obtained:

$$T_{izz} = \rho_i c_{pi}^2\left(\frac{\partial u_{ix}}{\partial x} + \frac{\partial u_{iz}}{\partial z}\right) - 2\rho_i c_{si}^2\frac{\partial u_{ix}}{\partial x}, T_{izx} = \rho_i c_{si}^2\left(\frac{\partial u_{ix}}{\partial z} + \frac{\partial u_{iz}}{\partial x}\right) \tag{14}$$

The boundary condition at the interface between water and basalt ($z = H_1$) is satisfied by the continuity of the normal displacement, normal stress, and zero tangential stress.

$$T_{1zz} = T_{2zz}, u_{1z} = u_{2z}, T_{2xz} = 0 \tag{15}$$

The boundary condition at the interface between basalt and peridotite ($z = H_2$) is satisfied by the continuity of displacements and stresses.

$$u_{2z} = u_{3z}, u_{2x} = u_{3x}, T_{2zz} = T_{3zz}, T_{2xz} = T_{3xz} \tag{16}$$

Substituting the general solutions for potentials into the boundary conditions, Equations (15) and (16), a system of homogeneous linear equations can be obtained. Here, we present the equations in matrix form:

$$\begin{bmatrix} a_{11} & a_{12} & a_{13} & a_{14} & a_{15} & 0 & 0 \\ a_{21} & a_{22} & a_{23} & a_{24} & a_{25} & 0 & 0 \\ 0 & a_{32} & a_{33} & a_{34} & a_{35} & 0 & 0 \\ 0 & a_{42} & a_{43} & a_{44} & a_{45} & a_{46} & a_{47} \\ 0 & a_{52} & a_{53} & a_{54} & a_{55} & a_{56} & a_{57} \\ 0 & a_{62} & a_{63} & a_{64} & a_{65} & a_{66} & a_{67} \\ 0 & a_{72} & a_{73} & a_{74} & a_{75} & a_{76} & a_{77} \end{bmatrix} \begin{bmatrix} A \\ B_1 \\ B_2 \\ C_1 \\ C_2 \\ D \\ E \end{bmatrix} = \begin{bmatrix} 0 \\ 0 \\ 0 \\ 0 \\ 0 \\ 0 \\ 0 \end{bmatrix} \tag{17}$$

Elements in the first matrix represent the known coefficients in the equations of the boundary conditions. They are a function of k such that Equation (17) can be simplified as follows:

$$\mathbf{G}(k)\mathbf{M}=\mathrm{O} \tag{18}$$

O represents the zero vector. **M** represents the undetermined coefficients for the potentials, and the values cannot be zero. The non-zero solutions for the coefficients in vector **M** exist only when the determinant of matrix $\mathbf{G}(k)$ ($\det|\mathbf{G}(k)| = 0$) is zero. The values of k that make $\det|\mathbf{G}(k)| = 0$ are therefore the eigenvalues for this question. Once an eigenvalue has been found, the undetermined coefficients in **M** can be calculated by solving the linear equations. There have been many methods for figuring out the values of k, such as the Newton method [27] and bisection [25]. In order to accurately determine the wavenumber k, the KRAKENC program [28] is used. Substituting the wavenumber k into Equation (17), the undetermined coefficients in **M** can be calculated by solving the equations. Then, the normal stresses, T_{zz}, can be computed by the relations between the stresses and potentials. Acoustic pressure in water is just the negative of T_{zz}, according to the definition of acoustic pressure, $p = \frac{1}{i\omega}\rho_1 c_{p1}{}^2\nabla^2\phi_1$. Eventually, the pressure mode shape function for each k can be calculated. This processing of obtaining the pressure mode shape function will be used in subsequent analysis.

The pressure field for a single point source can be represented as a sum of the normal modes. The Scholte wave is treated as the zeroth mode. Then, the pressure can be written as [25]:

$$p(r,z) = \sum_{m=0}^{\infty} \Phi_m(r)\Psi_m(z) \tag{19}$$

where $\Psi_m(z)$ is a mode shape function, $\Phi_m(r)$ is a mode coefficient, and the subscript m is the order of modes. Here, the variable $\Phi_0(r)$ is used to represent the amplitude of the Scholte wave in the acoustic field. We assume that there is no continuous spectrum so that the modes form a complete set. The coefficient $\Phi_m(r)$ can be calculated by applying the operator (Equation (20)) to Equation (19), where $\rho(z)$ is the media density.

$$\int_0^{\infty} (\cdot)\frac{\Psi_m(z)}{\rho(z)}dz \tag{20}$$

Mode analyses based on the elastic normal mode solution are applied to the acoustic field in the model. The pressure mode shape functions $\Psi_m(z)$ at a source frequency of f = 10 Hz are illustrated in Figure 2, and the normal stress T_{zz} is shown in the basalt and peridotite layer. The mode shape of the Scholte wave shows that a part of the energy is distributed in the pressure field in seawater. It can be indicated that the excitation intensity of Scholte waves $\Phi_0(r)$ can be affected by source depths and propagation distances.

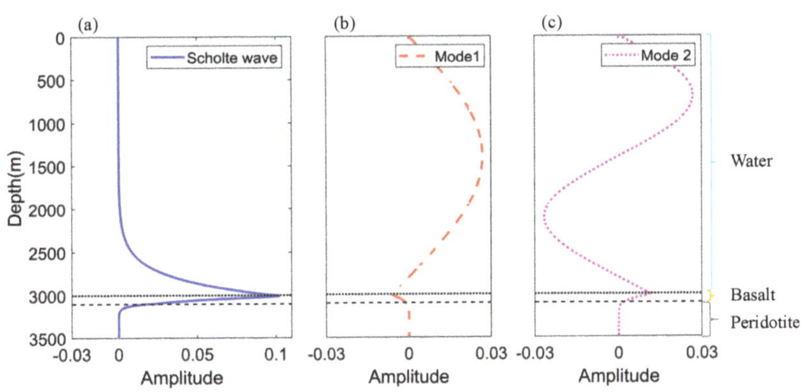

Figure 2. Pressure mode shape functions for a 10 Hz source in the ocean acoustic model. (**a**) Scholte wave. (**b**) Mode 1. (**c**) Mode 2.

The finite element method software, COMSOL Multiphysics [29], was applied to simulate the acoustic field in the constructed model. The environmental parameters in the simulation were kept in line with Table 1. The source frequency was f = 10 Hz, and a vertical receiving array was set to obtain pressure in water and normal stress in the bottoms at intervals of 10 m from 0 m to 3500 m. It should be noted that the upper limit of the integral in Equation (20) is infinity, but in this model, it can be found from the mode shape function in Figure 2 that the model distribution at a depth of 3500 m is approximate zero, so it is reasonable to truncate at this depth. Then, the Scholte wave amplitude in the pressure field can be obtained using the mode decomposition method. Figure 3a shows Scholte wave amplitudes, with different source depths, at a 3000-m horizontal distance from the source. The result indicates that the excitation intensity of the Scholte wave increases gradually with the increase of the source depth. It could also explain that Scholte waves can be excited efficiently when the source is at the seafloor or very close to the seafloor. Moreover, Scholte wave amplitudes can vary with range during propagation. When the source depth is SD = 2950 m, the Scholte amplitudes drop monotonically with propagation distance increasing, as shown in Figure 3b.

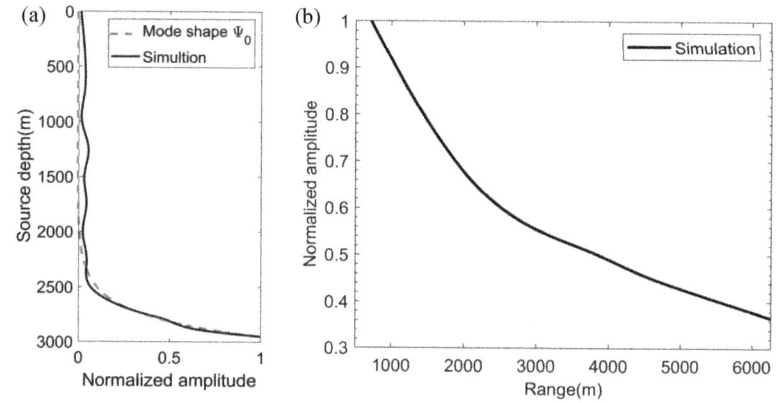

Figure 3. (**a**) Scholte wave amplitude versus source depth. The red dotted line is the mode shape for the Scholte wave and the black line is the simulation result. (**b**) Scholte wave amplitude versus propagation distance.

The simulation results indicate that the amplitude of the Scholte wave increases with an increase in source depths and decreases with an increase in propagation distance. These two characteristics can be applied to detect the Scholte wave, and a water tank experiment is designed according to this principle.

3. Tank Experiment

3.1. Acoustic Field Analysis for the Laboratory Environment

The acoustic model in Figure 1 is scaled down to laboratory size based on the similarity principle. Due to the close velocities, basalt and peridotite in the seabed are modeled by brass and iron slabs. Table 3 lists a scale model presented at a scale of 1:5000, where the depth and frequency were appropriately modified. Here, the source frequency is 50 kHz, and the water depth is 0.6 m.

Table 3. Parameters of the marine environment.

Media	**Layer i**	**Depth**	**ρ (g/cm^3)**	**c_p (m/s)**	**c_s (m/s)**
Water	1	0.6	1	1485	-
Brass	2	0.62	8.54	4640	2050
Iron	3	0.64	7.7	5850	3230

Figure 4 shows the pressure mode shape functions $\Psi_m(r)$ of the Scholte wave and normal modes for a 50 kHz source in the water tank. It is illustrated that the mode shape functions in two environments agree well by comparing the results in Figures 2 and 4. Moreover, it reveals that the parameters for the scale model experiment are correct.

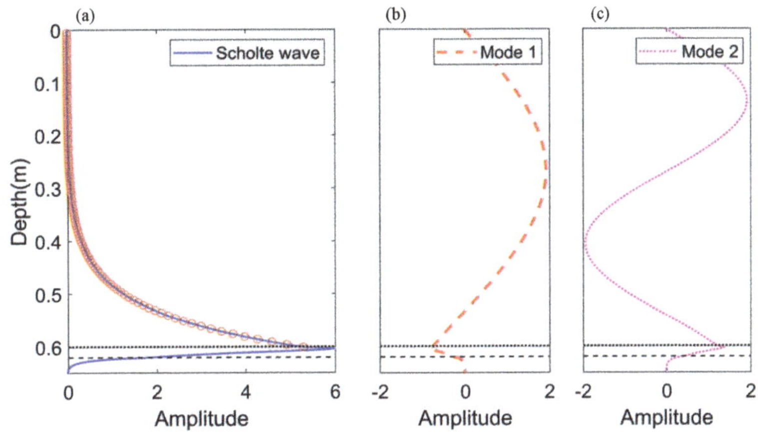

Figure 4. Pressure mode functions for a 50kHz source; the red circles represent the receiver positions. (**a**) Scholte wave. (**b**) Mode 1. (**c**) Mode 2.

3.2. Experiment Settings

The scale model experiment was performed in a water tank (L:3 m × W:2 m × H:1 m). Reflection waves from the tank's walls are absorbed by absorbing wedges. The tank is filled with water to a height of 0.6 m, where the sound speed in water is measured to be 1485 m/s. A spherical transducer, the source level of which is measured at approximately 140 dB (re 1 µPa.m/V) in the frequency range used in the experiment, is used as the source. An RHCA-7 hydrophone is used as the receiver. The sensitivities of the hydrophones are about −210 dB (re 1 V/µPa) in the frequency range from 20 to 100 kHz. The vertical array, including 119 receiver elements, is obtained by the synthetic aperture method. The source and receiver hydrophones are positioned in water using a robotic apparatus (with an accuracy of 0.01 mm), which allows for accurate positioning. The hydrophone measures depth from 0.5 to 59.5 cm; the depth interval is 0.5 cm. Figure 4 shows the receiver positions in the form of red circles, and the configuration of the experiment equipment is demonstrated in Figure 5.

3.3. Experimental Data Analysis

The source radiates ten cycles of sine waves at a frequency of f = 50 kHz in 1 s time intervals. Figure 6a shows the output signal of a power amplifier, which is used as a reference signal for a measurement with a vertical array of 119 hydrophones. Figure 6b demonstrates the received waveform of a hydrophone, which represents the temporal correlation between the received sound pressure by the hydrophone and the reference signal. By the temporal correlation, the virtual receiving array plays the same role as real receiving arrays of the same length. Therefore, the received waveforms by the virtual array are used to obtain the complex sound pressure required for identifying Scholte waves. To eliminate the effects of reverberation in the water tank, received waveforms are truncated by assuming that the pulse duration of the received signal from the source is almost the same for each hydrophone, and the truncated waveform is shown in Figure 6c.

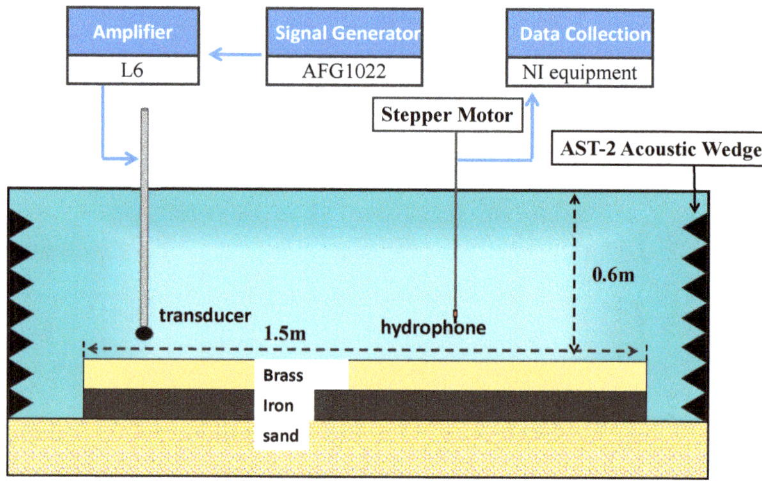

Figure 5. Diagram of the experimental system setup.

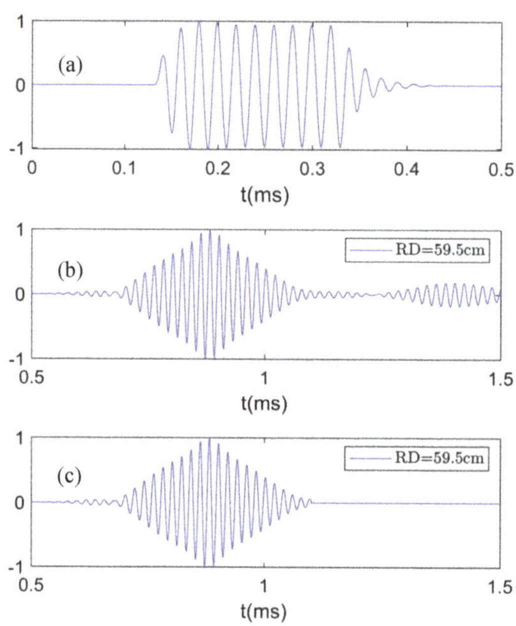

Figure 6. (**a**) Reference signal for a vertical array measurement; (**b**) the received waveform of a hydrophone with a source depth of SD = 55 cm (temporal correlation between the received sound pressure and the reference signal); (**c**) the truncated waveform adopted for identifying Scholte waves.

By the fast Fourier transform of truncated waveforms, we obtain the sound pressure $p(r, z)$ for the hydrophone at different positions. The sound pressure at 50 kHz is extracted to compose the matrix **P**, including 119 elements in the depth direction. The mode shape functions in the water tank are demonstrated in Figure 4. Finally, the amplitudes of the Scholte waves can be extracted using the mode decomposition.

Ten positions were set evenly for the source in the depth range from 0.1 to 0.55 m; the hydrophone array was at a horizontal distance of 0.6 m away from the source. The result in Figure 7a shows the normalized amplitudes of the Scholte wave and normal modes for a source depth of SD = 55 cm. It was demonstrated that the Scholte wave has the largest amplitude for this source depth, and the amplitudes of each mode in the experiment are in excellent agreement with the simulation result. This result shows that the water tank experiment is feasible and the processing for experimental data is correct. The Scholte wave amplitudes in the experiment field that vary with source depth are presented in Figure 7b. The results reveal that the tendency of the excitation amplitudes of Scholte waves in the experiment is consistent with the theoretical calculations.

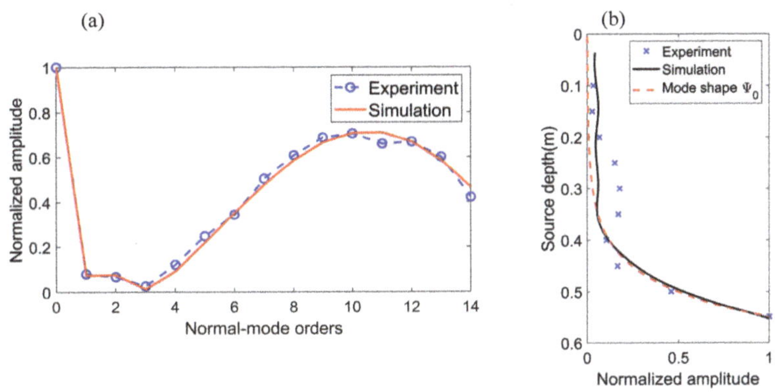

Figure 7. (**a**) Normalized amplitudes of the Scholte wave (0th order) and normal modes for the source depth, SD = 55 cm. The blue circle is the result of the water tank experiment and the red line is the simulation result. (**b**) Normalized amplitudes of the Scholte wave versus source depth. The blue cross is the result of the water tank experiment; the black line is the simulation result; the red dotted line is the mode shape function of the 0th mode.

Seven positions were set evenly for the receiving array within the horizontal range of 0.15–1.05 m away from the source, and the source was maintained at a depth of SD = 0.55 m. Scholte wave amplitudes in the experiment field that vary with the propagation distance are presented in Figure 8. It is shown that the Scholte wave amplitudes decrease as the propagation distance increases. The tendency of Scholte wave amplitudes in the experiment agrees well with the simulation result. These results show that the detection technique proposed in this paper can detect the Scholte wave at the seafloor successfully.

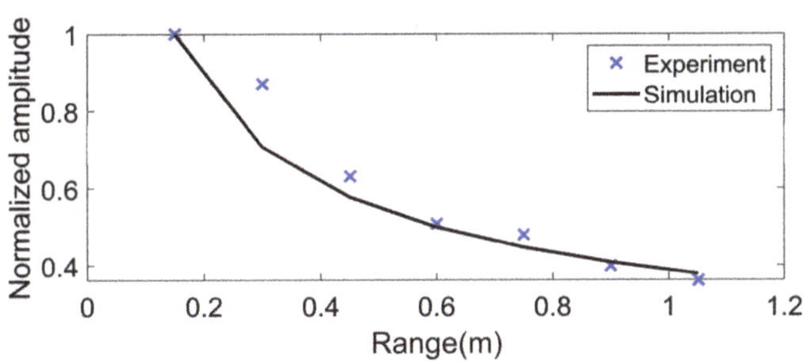

Figure 8. Scholte wave amplitude versus propagation distance. The blue cross is the result of the water tank experiment, and the black line is the simulation result.

4. Sediment Effect

4.1. Experiment Setting

In practice, there are many types of seafloor sediments. Here, silt was used as the sediment in the water tank experiment. Two cases, a sediment layer thickness of 1 mm and 12 mm, were considered in this research. The corresponding thickness is 5 m and 60 m in the ocean environment according to the scale of 1:5000. The sediment layer in the experiment was very thin, so it was treated as a liquid sediment layer. The model parameters for a 1 mm thickness are listed in Table 4. Here, the thickness of silt is 1 mm. The sediment layer in the water tank experiment and the result of thickness measurement are shown in Figure 9.

Table 4. Media parameters and sizes in the tank experiment.

Media	Layer i	Depth	ρ (g/cm^3)	c_p (m/s)	c_s (m/s)
Seawater	1	0.6	1	1500	-
Silt	2	0.601	1.2	1600	-
Basalt	3	0.621	8.54	4640	2050
Peridotite	4	0.641	7.7	5850	3230

Figure 9. (**a**) The silt sediment layer in the water tank experiment. (**b**) Silt sediment with a thickness of 1 mm.

The study above proved that the detection technique is valid for detecting Scholte waves at the seafloor. The same technique was applied to the condition with a sediment layer. Mode shapes in this acoustic model can be obtained by the elastic mode solution presented in the theoretical analysis. The process of experimental measurement is similar to the previous experiment, so no more details are provided here.

4.2. Experimental Results

Figure 10 shows the dependence of the mode amplitude for the Scholte wave and normal modes in the sound field on source depth. The hydrophone array was at a horizontal distance of 0.6 m away from the source. Here, the theory is the mode shape function $\Psi_m(r)$ for each mode. It can be seen from the figure that the results of the experiment and theory agree well with each other. This result indicates that treating the silt as fluid sediment is reasonable, and the Scholte wave can still be excited under this condition.

The experiment was repeated by changing the thickness of the sediment layer to 12 mm. Figure 11 shows the experimental results for the thickness of the sediment layer being 12 mm. It can be found that no Scholte wave is excited in the sound field under this condition, and the experimental results of normal waves in water are in good agreement with the theory.

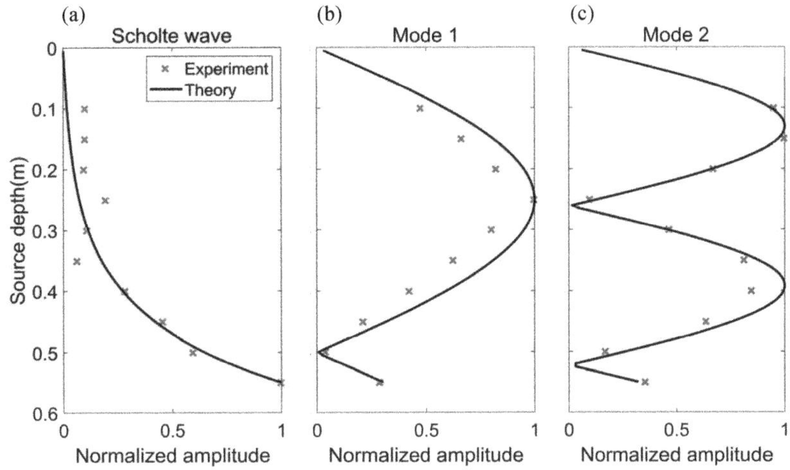

Figure 10. Mode amplitude versus source depth for the Scholte wave and normal modes (thickness of sediment is 1 mm, source frequency f = 50 kHz). The blue cross is the experimental result, and the black line is the theoretical result (mode shape function). (**a**) Scholte wave. (**b**) Mode 1. (**c**) Mode 2.

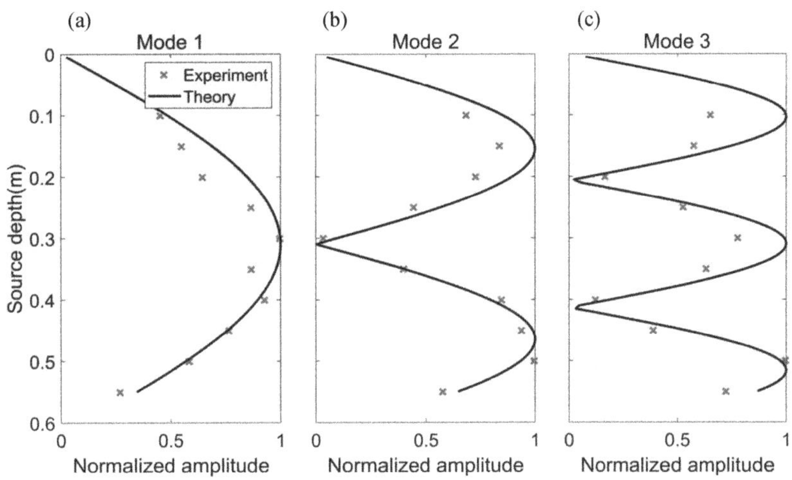

Figure 11. Mode amplitude versus source depth for the Scholte wave and normal modes (thickness of sediment is 12mm, source frequency f = 50 kHz). The blue cross is the experimental result, and the black line is the theoretical result (mode shape function). (**a**) Mode 1. (**b**) Mode 2. (**c**) Mode 3.

The presented results show that the sediment will affect the excitation of Scholte waves. When the sediment layer is thin (the thickness of sediment is 1 mm, and the acoustic wavelength is about 3 cm), the existence of the sediment layer does not affect the excitation of Scholte waves. When the thickness of the sediment layer increases, there is no Scholte wave mode in the acoustic field. Moreover, The results also show that the Scholte wave detection method proposed in this paper is feasible in the sedimentary layer environment.

5. Discussion

Unlike existing methods, this paper proposed a new detection technique based on the acoustic pressure field measurement for identifying Scholte waves according to the excitation and propagation characteristics of Scholte waves. The experimental results show that the detection technique can detect Scholte waves at the seafloor. There are certain discrepancies in the details between the experiments and calculations in this study. Here, the error analysis was performed. Due to the amplitudes of Scholte waves being extracted by mode decomposition in this research, the orthogonality between Scholte and normal modes waves is essential for the results' accuracy. Theoretically, the upper limit of the integral in Equation (20) is infinity, but only the range of the water depth can be measured in practice, so the modes cannot be strictly orthogonal. Figure 12 shows the orthogonal coefficients of the Scholte mode (0th mode) and each mode in the water column. The Scholte mode is not strictly orthogonal to higher-order normal modes. Therefore, this could be the cause of errors in the amplitudes of Scholte waves.

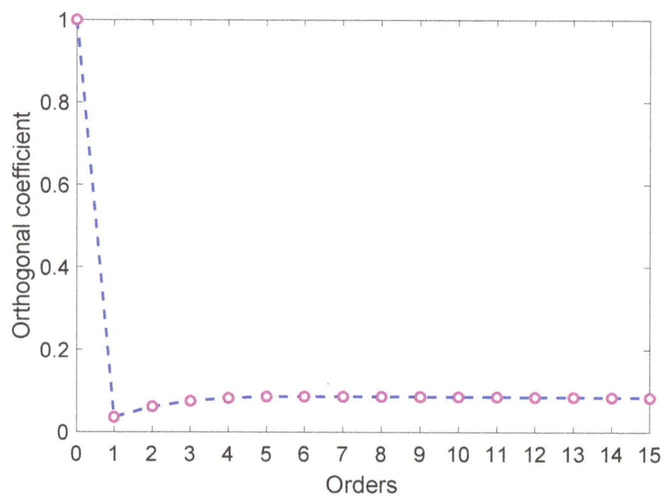

Figure 12. Orthogonal coefficients of the Scholte mode (0th mode) and each mode in water.

The acoustic field in this experiment would change when the sediment is added to the model as one layer of the multilayer medium. The normal modes were also changed due to the existence of the sediment layer. This problem still needs further theoretical and experimental research for different seafloor environmental conditions.

6. Conclusions

This paper introduces a new detection technique for Scholte waves at the seafloor. According to the actual ocean environment, a series of laboratory experiments with the scaled model of the elastic ocean bottom were designed and performed to detect Scholte waves at the seafloor. The study shows that Scholte wave amplitudes depending on different source depths and propagation distances are in good agreement with the theoretical results. These results indicate that the Scholte wave at the seafloor was detected successfully by the technique.

Furthermore, treating the silt layer in the laboratory experiment as fluid sediment is valid in this study. Some conclusions about the effect of sediment can be reached. When the sediment layer is thin, the existence of the sediment layer does not affect the excitation of Scholte waves. When the thickness of the sediment layer increases, there is no Scholte wave mode in the acoustic field. Moreover, The results also show that the Scholte wave detection method proposed in this paper is feasible in the sedimentary layer environment.

Author Contributions: Conceptualization, L.P. and G.Y.; methodology, M.L., G.Y. and L.W.; software, M.L. and L.W.; validation, M.L. and L.W.; formal analysis, M.L. and G.Y.; investigation, M.L., L.W. and L.P.; resources, G.Y., Y.R. and L.W.; writing—original draft preparation, M.L. and L.W.; writing—

review and editing, L.P., M.L. and L.W.; visualization, M.L., L.W. and Y.R.; supervision, L.P., G.Y. and Y.R.; project administration, L.P. and G.Y.; funding acquisition, L.P. and L.W. All authors have read and agreed to the published version of the manuscript.

Funding: This research was funded by the Nation Natural Science Foundation of China (Grant No. 11474258) and the State Key Laboratory of Acoustics, Chinese Academy of Sciences (Grant No. SKLA202206).

Institutional Review Board Statement: Not applicable.

Informed Consent Statement: Not applicable.

Data Availability Statement: Not applicable.

Conflicts of Interest: The authors declare no conflict of interest.

References

1. Scholte, J. The range of existence of Rayleigh and Stoneley waves. *Geophys. Suppl. Mon. Not. R. Soc.* **1947**, *5*, 120–126. [CrossRef]
2. Socco, L.V.; Foti, S.; Boiero, D. Surface-wave analysis for building near-surface velocity models—Established approaches and new perspectives. *Geophysics* **2010**, *75*, 75A83–75A102. [CrossRef]
3. Nguyen, X.N.; Dahm, T.; Grevemeyer, I. Inversion of Scholte wave dispersion and waveform modeling for shallow structure of the Ninetyeast Ridge. *J. Seismol.* **2009**, *13*, 543–559. [CrossRef]
4. Ritzwoller, M.H.; Levshin, A.L. Estimating shallow shear velocities with marine multicomponent seismic data. *Geophysics* **2002**, *67*, 1991–2004. [CrossRef]
5. Zywicki, D.J.; Rix, G.J. Mitigation of near-field effects for seismic surface wave velocity estimation with cylindrical beamformers. *J. Geotech. Geoenvironmental Eng.* **2005**, *131*, 970–977. [CrossRef]
6. Potty, G.R.; Miller, J.H. Measurement and modeling of Scholte wave dispersion in coastal waters. In Proceedings of the AIP Conference Proceedings, Ft. Worth, TX, USA, 5–10 August 2012; American Institute of Physics: NewYork, NY, USA, 2012; Volume 1495, pp. 500–507.
7. Dong, Y.; Piao, S.; Gong, L.; Zheng, G.; Iqbal, K.; Zhang, S.; Wang, X. Scholte wave dispersion modeling and subsequent application in seabed shear-wave velocity profile inversion. *J. Mar. Sci. Eng.* **2021**, *9*, 840. [CrossRef]
8. Godin, O.A.; Deal, T.J.; Dong, H. Physics-based characterization of soft marine sediments using vector sensors. *J. Acoust. Soc. Am.* **2021**, *149*, 49–61. [CrossRef] [PubMed]
9. TenCate, J.A.; Muir, T.G.; Caiti, A.; Kristensen, Å.; Manning, J.F.; Shooter, J.A.; Koch, R.A.; Michelozzi, E. Beamforming on seismic interface waves with an array of geophones on the shallow sea floor. *IEEE J. Ocean. Eng.* **1995**, *20*, 300–310. [CrossRef]
10. Ren, B.; Li, H. Characteristics of Scholte Wave for Target Detection. In *Advances in Wireless Communications and Applications;* Springer: Berlin/Heidelberg, Germany, 2021; pp. 115–121.
11. Williams, E.F.; Fernández-Ruiz, M.R.; Magalhaes, R.; Vanthillo, R.; Zhan, Z.; González-Herráez, M.; Martins, H.F. Scholte wave inversion and passive source imaging with ocean-bottom DAS. *Lead. Edge* **2021**, *40*, 576–583. [CrossRef]
12. Rauch, D. *Seismic Interface Waves in Coastal Waters: A Review, SACLANTCEN SR-42;* SACLANT ASW Research Centre: La Spezia, Italy, 1980.
13. Wang, X.; Xia, C.; Liu, X. A case study: Imaging OBS multiples of South China Sea. *Mar. Geophys. Res.* **2012**, *33*, 89–95. [CrossRef]
14. Du, S.; Cao, J.; Zhou, S.; Qi, Y.; Jiang, L.; Zhang, Y.; Qiao, C. Observation and inversion of very-low-frequency seismo-acoustic fields in the South China Sea. *J. Acoust. Soc. Am.* **2020**, *148*, 3992–4001. [CrossRef] [PubMed]
15. Zhu-Bo, L.I.; Pan, F.R. Present Situation and Prospect of Submarine Seismic Observation Technology. *N. China Earthq. Sci.* **2015**, *33*, 56–63.
16. Wang, S.; Qiu, X.; Zhao, M.; Li, P.; Liu, L.; Zhang, Y.; Xie, Z. Signal transfer and noise level of Ocean Bottom Seismometers. *Chin. J. Geophys.* **2019**, *62*, 3199–3207.
17. Cao, J.; Qi, Y.; Zhou, S.; Du, S.; Peng, Z.; Zhang, Y.; Qiao, C. Anomalous dispersion observed in signal arrivals at a deep-sea floor receiver. *JASA Express Lett.* **2021**, *1*, 076004. [CrossRef]
18. Dong, Y.; Piao, S.; Gong, L. Effect of three-dimensional seamount topography on very low frequency sound field in deep water. *J. Harbin Eng. Univ.* **2020**, *41*, 1464–1470.
19. Tang, K.; Cheng, G.; Liu, B. Seismo-acoustic signal extraction from shallow sea seabed seimic wave field based on FK method. *J. Phys. Conf. Ser.* **2022**, *2246*, 012023. [CrossRef]
20. Yuan, D.; Nazarian, S. Automated surface wave method: Inversion technique. *J. Geotech. Eng.* **1993**, *119*, 1112–1126. [CrossRef]
21. Meng, X.; Zhang, J.; Zhao, S.; Wang, X. Study on the Influence of Shallow Sea Sediments on the Frequency Dispersion of Scholte Wave in Ship Seismic Wave. In Proceedings of the ICMLCA 2021; 2nd International Conference on Machine Learning and Computer Application, VDE, Shenyang, China, 12–14 November 2021; pp. 1–4.
22. Schneiderwind, J.D.; Collis, J.M.; Simpson, H.J. Elastic Pekeris waveguide normal mode solution comparisons against laboratory data. *J. Acoust. Soc. Am.* **2012**, *132*, EL182–EL188. [CrossRef] [PubMed]
23. Hamilton, E.L. Geoacoustic modeling of the sea floor. *J. Acoust. Soc. Am.* **1980**, *68*, 1313–1340. [CrossRef]

24. Heacock, J.G. *The Earth's Crust: Its Nature and Physical Properties*; American Geophysical Union: Washington, DC, USA, 1977; Volume 20, pp. 16–24.
25. Jensen, F.B.; Kuperman, W.A.; Porter, M.B.; Schmidt, H.; Tolstoy, A. *Computational Ocean Acoustics*; Springer: Berlin/Heidelberg, Germany, 2011; Volume 794, pp. 375–380.
26. Ewing, W.M.; Jardetzky, W.S.; Press, F.; Beiser, A. *Elastic Waves in Layered Media*, 1st ed.; McGraw-Hill Book Company, Inc.: New York, NY, USA, 1957; pp. 156–189.
27. Hall, M.; Gordon, D.F.; White, D. Improved methods for determining eigenfunctions in multilayered normal-mode problems. *J. Acoust. Soc. Am.* **1983**, *73*, 153–162. [CrossRef]
28. Porter, M.B. *The KRAKEN Normal Mode Program*; Technical Report; Naval Research Lab: Washington, DC, USA, 1992.
29. Simon, B.; Isakson, M.; Ballard, M. Modeling acoustic wave propagation and reverberation in an ice covered environment using finite element analysis. In Proceedings of the Meetings on Acoustics 175ASA, Minneapolis, MN, USA, 7–11 May 2018; Acoustical Society of America: NewYork, NY, USA, 2018; Volume 33, p. 070002.

MDPI
St. Alban-Anlage 66
4052 Basel
Switzerland
www.mdpi.com

Sensors Editorial Office
E-mail: sensors@mdpi.com
www.mdpi.com/journal/sensors

www.ingramcontent.com/pod-product-compliance
Lightning Source LLC
Chambersburg PA
CBHW061508210926
51850CB00008B/1848
9783036591919